NUMERICALLY CONTROLLED MACHINE TOOLS

ELLIS HORWOOD SERIES IN AUTOMATED MANUFACTURING

Series Editor: COLIN BESANT, Professor of Computer-Aided Manufacture, Imperial College of Science and Technology, London University

Besant, C.B. & Lui, C.W.K.	Computer-aided Design and Manufacture, 3rd Edition
Ding, Q.L. & Davies, B.J.	Surface Engineering Geometry for Computer-aided Design and Manufacture
Gosman, B.E., Launder, A.D. & Reece, G.	Computer-aided Engineering: Heat Transfer and Fluid Flow
Gunasekera, J.S.	CAD/CAM of Dies
Vickers, G.W., Ly, M. & Oetter, R.G.	Numerically Controlled Machine Tools

ELLIS HORWOOD SERIES IN MECHANICAL ENGINEERING

Series Editor: J. M. ALEXANDER, formerly Stocker Visiting Professor of Engineering and Technology, Ohio University, Athens, USA, and Professor of Applied Mechanics, Imperial College of Science and Technology, University of London

The series has two objectives: of satisfying the requirements of postgraduate and mid-career engineers, and of providing clear and modern texts for more basic undergraduate topics. It is also the intention to include English translations of outstanding texts from other languages, introducing works of international merit. Ideas for enlarging the series are always welcomed.

Alexander, J.M.	Strength of Materials: Vol. 1: Fundamentals; Vol. 2: Applications
Alexander, J.M., Brewer, R.C. & Rowe, G.	Manufacturing Technology Volume 1: Engineering Materials
Alexander, J.M., Brewer, R.C. & Rowe, G.	Manufacturing Technology Volume 2: Engineering Processes
Atkins A.G. & Mai, Y.W.	Elastic and Plastic Fracture
Beards, C.F.	Vibration Analysis and Control System Dynamics
Beards, C.F.	Structural Vibration Analysis
Beards, C.F.	Noise Control
Beards, C.F.	Vibrations and Control Systems
Borkowski, J. and Szymanski, A.	Technology of Abrasives and Abrasive Tools
Borkowski, J. and Szymanski, A.	Uses of Abrasives and Abrasive Tools
Brook, R. and Howard, I.C.	Introductory Fracture Mechanics
Cameron, A.	Basic Lubrication Theory, 3rd Edition
Collar, A.R. & Simpson, A.	Matrices and Engineering Dynamics
Cookson, R.A. & El-Zafrany, A.	Finite Element Techniques for Engineering Analysis
Cookson, R.A. & El-Zafrany, A.	Techniques of the Boundary Element Method
Edmunds, H.G.	Mechanical Foundations of Engineering Science
Fenner, D.N.	Engineering Stress Analysis
Fenner, R.T.	Engineering Elasticity
Ford, Sir Hugh, FRS, & Alexander, J.M.	Advanced Mechanics of Materials, 2nd Edition
Gallagher, C.C. & Knight, W.A.	Group Technology Production Methods in Manufacture
Gohar, R.	Elastohydrodynamics
Haddad, S.D. & Watson, N.	Principles and Performance in Diesel Engineering
Haddad, S.D. & Watson, N.	Design and Applications in Diesel Engineering
Haddad, S.D.	Advanced Diesel Engineering and Operation
Hunt, S.E.	Nuclear Physics for Engineers and Scientists
Irons, B.M. & Ahmad, S.	Techniques of Finite Elements
Irons, B.M. & Shrive, N.G.	Finite Element Primer
Johnson, W. & Mellor, P.B.	Engineering Plasticity
Kleiber, M.	Incremental Finite Element Modelling in Non-linear Solid Mechanics
Kleiber, M. & Breitkopf, P.	Finite Element Methods in Structural Engineering: Turbo Pascal Programs for Microcomputers
Leech, D.J. & Turner, B.T.	Engineering Design for Profit
Leech, D.J. & Turner, B.T.	Project Management for Profit
Lewins, J.D.	Engineering Thermodynamics
Malkin, S.	Materials Grinding: Theory and Applications of Mechining with Abrasives
Maltbaek, J.C.	Dynamics in Engineering
McCloy, D. & Martin, H.R.	Control of Fluid Power: Analysis and Design, 2nd (Revised) Edition
Osyczka, A.	Multicriterion Optimisation in Engineering
Oxley, P.L.B.	The Mechanics of Machining
Piszcek, K. and Niziol, J.	Random Vibration of Mechanical Systems
Polanski, S.	Bulk Containers: Design and Engineering of Surfaces and Shapes
Prentis, J.M.	Dynamics of Mechanical Systems, 2nd Edition
Renton, J.D.	Applied Elasticity
Richards, T.H.	Energy Methods in Vibration Analysis
Ross, C.T.F.	Computational Methods in Structural and Continuum Mechanics
Ross, C.T.F.	Finite Element Programs for Axisymmetric Problems in Engineering
Ross, C.T.F.	Finite Element Methods in Structural Mechanics
Ross, C.T.F.	Applied Stress Analysis
Ross, C.T.F.	Advanced Applied Stress Analysis
Ross, C.T.F.	Finite Element Methods in Engineering Science
Roy, D. N.	Applied Fluid Mechanics
Roznowski, T.	Moving Heat Sources in Thermoelasticity
Sawczuk, A.	Mechanics and Plasticity of Structures
Sherwin, K.	Engineering Design for Performance
Stupnicki, J.	Stress Measurement by Photoelastic Coating
Szczepinski, W. & Szlagowski, J.	Plastic Design of Complex Shape Structured Elements
Thring, M.W.	Robots and Telechirs
Walshaw, A.C.	Mechanical Vibrations with Applications
Williams, J.G.	Fracture Mechanics of Polymers
Williams, J.G.	Stress Analysis of Polymers 2nd (Revised) Edition

NUMERICALLY CONTROLLED MACHINE TOOLS

G. W. VICKERS M.Sc., Ph.D.
M. LY B.Eng.
R. G. OETTER M.Sc.
all of Department of Mechanical Engineering, University of Victoria, Canada

ELLIS HORWOOD
NEW YORK   LONDON   TORONTO   SYDNEY   TOKYO   SINGAPORE

First published in 1990 by
ELLIS HORWOOD LIMITED
Market Cross House, Cooper Street,
Chichester, West Sussex, PO19 1EB, England

A division of
Simon & Schuster International Group
A Paramount Communications Company

© Ellis Horwood Limited, 1990

All rights reserved. No part of this publication may be
reproduced, stored in a retrieval system, or transmitted,
in any form, or by any means, electronic, mechanical,
photocopying, recording or otherwise, without the prior
permission, in writing, of the publisher

Printed in Great Britain by
Southampton Book Company.

British Library Cataloguing in Publication Data

Vickers, G. W.
Numerically controlled machine tools.
1. Machine tools. Numerical control
I. Title II. Ly, M. III. Oetter, R. G.
621.9023
ISBN 0-13-625526-4

Library of Congress Cataloging-in-Publication Data

Vickers, G. W. (Geoffrey W.), 1943–
Numerically controlled machine tools / G. W. Vickers,
M. Ly, R. G. Oetter.
p. cm.
ISBN 0-13-625526-4
1. Machine-tools — Numerical control. I. Ly, M. (Minh)
II. Oetter, R. G. (Rolf G.) III. Title.
TJ1189.V53   1990
621.9′023–dc20   90-44162
                                                 CIP

Table of Contents

Preface 9

Chapter 1 Introduction to Numerically Controlled Machines 11
    1.1   Machine Features 11
        1.1.1   Mechanical Drive Components 11
        1.1.2   Motor Drive and Feedback 14
        1.1.3   CNC Controller Unit 15
        1.1.4   Program Storage and Transmission 17
        1.1.5   Other Features 19
    1.2   Types of Programming 20

Chapter 2 Manual Part Programming 23
    2.1   Basic Machine Codes 25
        2.1.1   G-codes for CNC Milling Machines 25
        2.1.2   G-codes for CNC Lathes 41
        2.1.3   M-codes for CNC Machines 48
        2.1.4   F-, S- and T-codes for CNC Machines 51
        2.1.5   Examples of Basic Codes 52
    2.2   Advanced Machine Codes 56
        2.2.1   Canned Cycles 56
        2.2.2   Subprograms 61
        2.2.3   Custom Macros 63

Chapter 3 Computer Assisted Programming 71
    3.1   APT 71
        3.1.1   Geometry Statements 72
        3.1.2   Motion Statements 73
        3.1.3   Postprocessor and Auxiliary Statements ... 74
        3.1.4   Example of APT Programming 75
    3.2   Symbolic FAPT 81
        3.2.1   Execution of Symbolic FAPT 82
        3.2.2   Example of Symbolic FAPT 90
    3.3   Conversational Automatic Programming 92
        3.3.1   Execution of Conversational Programming . 93
        3.3.2   Example of Conversational Programming ... 94
        3.3.3   Example of Conversational Programming ... 97
    3.4   SmartCAM 101

Table of Contents

      3.4.1   Example of SmartCAM Programming 104

Chapter 4   **Integrated CAD/CAM Programming** 113
    4.1   Anvil-5000 114
        4.1.1   Part Definition 115
        4.1.2   Tool Path Definition 122
        4.1.3   Example of Anvil-5000 123

Chapter 5   **Direct Numerically Controlled Machining** 131
    5.1   DNC Remote Buffer Link 132
        5.1.1   Data Transmission 132
        5.1.2   Transmission Program. 136
        5.1.3   Transmission Problems................. 136
    5.2   Postprocessors 138
        5.2.1   Example of Program Conversion 141

Chapter 6   **General Curved Surface Machining** 143
    6.1   Data Input 144
    6.2   Surface Modelling 146
        6.2.1   Compiling and Using the Programs 150
        6.2.2   B-spline Lines 152
        6.2.3   B-spline Surfaces 153
        6.2.4   AutoCAD Surfaces 154
    6.3   Surface Orientation 157
    6.4   Surface Normal Calculation 159
    6.5   Cutter Location Calculation 162
        6.5.1   Cutter Offsets Program 166
        6.5.2   Cutter Location Program 167
    6.6   G-code Generation 167
    6.7   Tools for Surface Data Manipulation 170
        6.7.1   Inverting the Surface Normal Vector 170
        6.7.2   Transposing the Surface Matrix 170
        6.7.3   Reducing the Surface Resolution 171
        6.7.4   Adding Extensions to a Surface 172
    6.8   Comparison of Ball-mill and End-mill Cutters ... 173
    6.9   Machining Through Circular Arc Interpolation .. 177

Chapter 7   **Curved Surface Mahining: Case Studies** 185
    7.1   Marine Propellers 185
        7.1.1   Surface Definition 185
        7.1.2   Surface Machining 190
    7.2   Ship Hull Models 194
        7.2.1   Hull Definition 194
        7.2.2   Hull Machining 196
    7.3   Turbine Blades 199

Table of Contents

    7.4   Biomedical Applications 205
        7.4.1   Limb Shape Replication 205
        7.4.2   Shoe Lasts - Cylindrical Mill Turning 211
        7.4.3   Head Shapes - Laser Scanning Data 216
        7.4.4   Heart Valve Shapes - Surface Definition ... 221

Chapter 8 Integrated CAD/CAM For Shipyards: Case Study 223
    8.1   Data Input 226
    8.2   Line Fairing 227
    8.3   Surface Generation 228
        8.3.1   Compound Curve Surface Fitting 228
        8.3.2   Developable Surface Fitting 229
    8.4   Plate Expansion 230
    8.5   Internal Structure 231
    8.6   Automatic Cutout Insertion 231
    8.7   Data Export to CAD Programs 232
    8.8   G-code Generation 233

Chapter 9 Simulation Testing of CNC Part Programs 235
    9.1   Simulation Program 236
        9.1.1   Parser 237
        9.1.2   Error Processor 239
        9.1.3   Display Processor 240
        9.1.4   Enhancements 242

Chapter 10 Retrofitting Machine Tools 243
    10.1 Cabinet Door Manufacture 243
        10.1.1 Introduction 243
        10.1.2 Automated Woodshaper 244
        10.1.3 Automation Software 247
        10.1.4 Performance Evaluation 251

Appendix 1 C-Program: DNCLINK 252

Appendix 2 C-Program: POSTPRO 254

Appendix 3 C-Program: LIBRARY 263

Appendix 4 C-Program: FITLINE 273

Appendix 5 C-Program: FITSURF 287

Table of Contents

Appendix 6 C-Program: ROTATE 304

Appendix 7 C-Program: TRANSLATE 308

Appendix 8 C-Program: SURFNORM 312

Appendix 9 C-Program: OFFGEN 320

Appendix 10 C-Program: ADDOFFST 325

Appendix 11 C-Program: MAKEGCDC 330

Appendix 12 C-Program: INVNORM 348

Appendix 13 C-Program: TRANSPOS 351

Appendix 14 C-Program: RSKIP 355

Appendix 15 C-Program: RADDEND 362

Appendix 16 C-Program: SIMULATE 367

Bibliography 444

Index 448

Preface

Numerically controlled (CNC) machine tools form the basis of flexible manufacturing systems (FMS) and computer integrated manufacturing (CIM) systems. This book presents a comprehensive coverage of the operation and programming of CNC machine tools. Basic machine features and operation are discussed in Chapter 1 with manual, computer assisted, and integrated CAD/CAM programming in Chapters 2, 3 and 4. Direct numerically controlled machining is covered in Chapter 5, and general curved surface machining is covered in Chapter 6, together with listings of numerous computer programs. Case studies of curved surface machining applications and shipbuilding applications are given in Chapters 7 and 8. Simulation testing of CNC part programs is in Chapter 9 and an example of retrofitting manual machine tools is given in Chapter 10. The description of current CAD and CAM packages in Chapters 2, 3 and 4 is intended to indicate the type of procedures and decision processes that are required, rather than to emphasise product specific operations. Similarly the computer programs, given in Appendices 1 to 16, are all functional routines (which may be used directly for the specific applications) but are intended to illustrate an approach which may be readily amended, enhanced and combined for a variety of CNC machining applications.
 The emphasis on the practical aspects of programming and operation of CNC machine tools can be considered as complementary to other books which have concentrated on, for example, organizational and scheduling aspects of CNC machine tools, or on cutting tool technology (including chip formation, optimal cutting speeds and feeds, and plastic chip deformation). This book is intended for undergraduate and graduate student courses in manufacturing, at universities and colleges, as well for engineers involved in operation of CNC tools.
 The authors wish to express sincere thanks to Colin Bradley, Sanjev Bedi, Doug Dark, George Csanyi-Fritz, Doug Blake and Jim Duncan, and to other colleagues and students for their association with aspects of the research work that constitute sections of the book. Thanks to Suzanne Stevens and Shari Yore for assistance in writing, compiling and presenting the book. Also thanks to the various funding agencies that have supported much of the research work particularly

the Natural Sciences and Engineering Research Council of Canada and the Advanced System Institute of Canada.

Chapter 1
Introduction to Numerically Controlled Machines

The primary building blocks of flexible manufacturing and computer integrated manufacturing systems are numerically controlled (CNC) machine tools. These computer controlled and programmable machines are currently available in a wide variety of types and sizes. The most common are lathes and milling machines, although drilling, grinding, blanking, shearing and flame-cutting machines are also readily available. A typical CNC machining centre is shown in Figure 1.1 and a CNC turning centre is shown in Figure 1.2.

In many ways numerically controlled machines are similar to their manual counterparts. The primary relative generation motions of cutting tool and workpiece are identical, as are the machining techniques and approaches. The essential difference is that the manual positional movements are replaced by motor drives so that cutting tool positions are controlled by computers. This simple change has a dramatic effect on the capabilities of CNC machine tools.

1.1 MACHINE FEATURES

The main features of a typical CNC milling machine are shown in Figure 1.3. The machine consists of a rigid base with a moveable horizontal worktable, vertical tool spindle, automatic tool changer, and controller unit. It is designed and built to maintain a high accuracy of movement and to reduce static and dynamic deflection as well as vibration, resulting from high cutting loads and thermal distortion.

1.1.1 Mechanical Drive Components

The milling machine worktable is mounted on two sets of slideways which permit movement of the table, and hence the workpiece, in orthogonal horizontal directions. The vertical tool spindle is also mounted on a slideway to produce vertical movement and generate a third orthogonal motion. The conventional notation for the displacement of a cutting tool relative to a workpiece, in the X, Y and Z axes, is given in Figure 1.3. If a programmable rotary table is fitted to the machine, the rotation is referred to as the A axis. The corresponding notation for the X and Z axes in a CNC lathe is given in Figure 1.4. CNC machines are categorized by the number of axes of control milling machines normally have three, four or five axes while lathes normally have two axes.

Figure 1.1 CNC machining centre

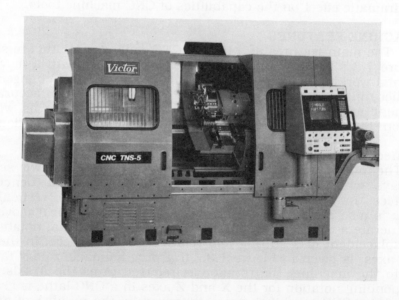

Figure 1.2 CNC turning centre

Sec. 1.1] **Machine Features** 13

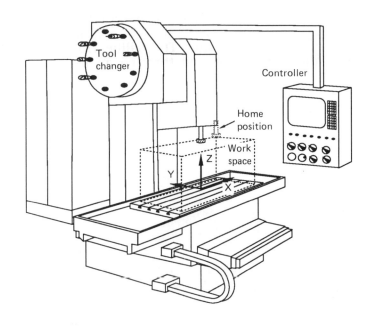

Figure 1.3 Main features of a CNC milling machine

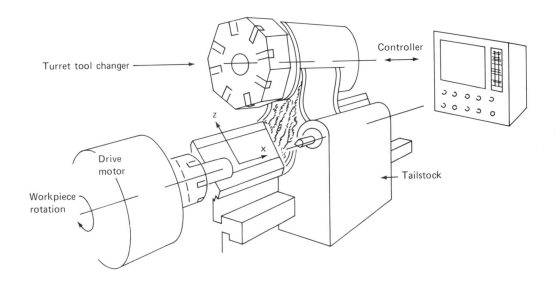

Figure 1.4 Main features of a CNC lathe

Table movement along a slideway is normally controlled by a leadscrew with a recirculating ballbearing nut, as shown in Figure 1.5. Both leadscrew and nut have a precision ground form into which ballbearings are allowed to run. The ballbearings progress along the thread form and are recirculated through an interior passage in the ball nut. This type of drive replaces the acme leadscrew and plain bronze nut and helps eliminate backlash.

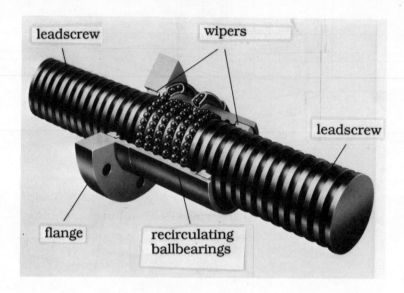

Figure 1.5 Leadscrew with recirculating ballbearing nut

1.1.2 Motor Drive and Feedback

Linear table motion is achieved by a controlled rotation of the leadscrew. Two types of motors are widely employed with CNC machines: DC servo motors and stepping motors.

Servo motors are the most widely used for CNC applications. They have high power and speed capabilities with low inertia and smooth operation. Stepping motors are relatively cheap, generate high torque at low speed, and require simpler electronic controls than servo motors.

Servo motors have a permanent magnet field housing and a wire wound, or printed circuit, armature. They are used in conjunction with rotary encoders which are mounted on the same shaft. Rotary encoders provide feedback to the DC servo motors on the absolute angular armature position. They consist of binary, or analogue, coded discs. The rotating discs can be optical or printed circuit type. In the case of optical encoders, the discs are divided into multiple track segments with transparent and opaque sections. Light sources and

Sec. 1.1] **Machine Features** 15

photocell receivers are placed on opposite sides of the rotating encoder disc. The pulses generated from the photocell receivers allow the absolute rotary position of the motor armature to be determined and thus control the actual rotational movement of the leadscrew.

Stepping motors operate on the transmission of pulses. The total number of pulses determines the amount of rotation, and hence distance travelled, while the frequency of pulses determines the speed of movement. Encoders can be used on stepping motors, but in general their absolute position is precisely proportional to the number of pulses applied to them.

A diagrammatic representation of a CNC machine table, leadscrew, recirculating ballbearing nut, DC servo motor and is given in Figure 1.6.

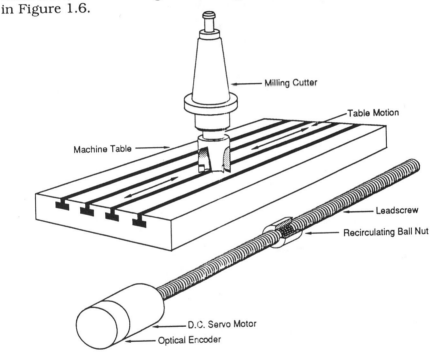

Figure 1.6 CNC machine table and drive components

1.1.3 CNC Controller Unit

Two photographs of typical CNC controller units are shown in Figure 1.7. They consist of an alphanumeric keyboard, dial control knobs, over-ride switches, and a CRT screen. Controller units contain all the electronics, control hardware, and computers required to read and interpret a program of instructions and to convert these instructions into machine tool movements and functions. They also enable programs to be stored, retrieved, and executed.

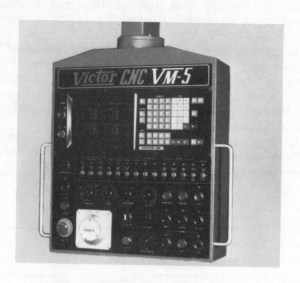

Figure 1.7(a) Fanuc Series 6-MB controller unit

Figure 1.7(b) Fanuc Series 15-M controller unit

Controllers operate in the series of input modes as follows:
- manual - where movement of the dial control knobs moves the machine table in real time. This is used primarily for setting up the machine and workpiece.
- manual data input (MDI) - where machine code data (G-code) is entered into the CNC control unit via the console keyboard. Although complete part programs may be entered at the machine, MDI is most commonly used for editing part programs already resident in the controller's memory.
- tape - where paper tape programs entered at the tape reader are read and stored in the controller's memory.
- memory - where programs stored in memory can be accessed for running or editing.
- edit - where programs can be interactively altered and downloaded to punch paper tape for backup storage.

An additional and increasingly common feature is the MDI conversational programming capability. Conversational programming is really a computer assisted method of programming and is discussed in detail in Chapter 3. Programming is carried out at the CNC controller console with the operator responding to a series of questions. In this way, the part material and cutting tool characteristics can be identified, a part shape and cutter path defined, and a coded program generated. The operator may display a graphical simulation of the part and cutter path with associated programming code on the CRT screen.

1.1.4 Program Storage and Transmission

Programs for producing parts are arranged in the form of coded blocks of information. This code is interpreted by the machine tool controller in order to control the table, spindle movements and all auxiliary functions. Standard machine codes, called G-codes, and the development of manual part programs are given in Chapter 2.

Part programs are usually fed into the machine in the form of punched paper tape, magnetic tape or through a standard RS-232 serial communication link, and are stored in the controller memory. The most common transmission medium for CNC machine tools is punched paper tape. An optical tape reader mounted on the machine is used to sense punched holes in the paper tape and transmit the data to the machine controller. As shown in Figure 1.8, the paper tape has eight tracks of coded holes and an additional feed sprocket hole. It is usually 25 mm wide and has 10 characters per 25 mm length.

The alphanumeric code characters for paper tape are expressed in binary form in the ISO (International Organization for Standardization) format, based on the ASCII (American Standard Code for Information Interchange) format. As shown in Figure 1.9, the code is a 7-bit binary code with an eighth parity bit which is used as a means of checking the accuracy of transmission. In the binary system there are two numbers: 0 and 1. Successive digits are represented by the number 2 raised to successive powers. The values of the powers of 2 are $2^0 = 1$, $2^1 = 2$, $2^2 = 4$, $2^3 = 8$ etc., and so the binary equivalent of the decimal number 5 is 0101 as follows:

$$1 \times 2^0 + 0 \times 2^1 + 1 \times 2^2 + 0 \times 2^3 = 1 \times 1 + 0 \times 2 + 1 \times 4 + 0 \times 8$$
or   decimal number 5 ~ binary number 0101

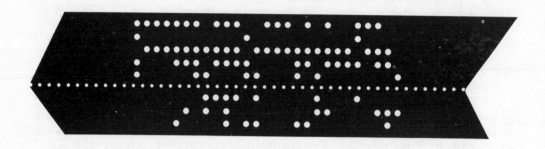

Figure 1.8 Paper tape hole configuration

Character	8	7	6	5	4	3	2	1	Character	8	7	6	5	4	3	2	1
0			O	O	O	°			I	O	O		O	°			O
1		O		O	O	°			O	J	O	O		O	°		O	
2		O		O	O	°		O		K			O	O	°		O	O
3				O	O	°		O	O	L	O	O		O	°	O		
4	O			O	O	°	O			M			O	O	°	O		O
5				O	O	°	O		O	N			O	O	°	O	O	
6				O	O	°	O	O		O	O	O		O	°	O	O	O
7	O			O	O	°	O	O	O	P			O		O	°		
8	O		O	O	O	°				Q	O	O	O		O	°			O
9			O	O	O	°			O	R	O	O	O			°		O	
A		O				°			O	S			O		O	°		O	O
B		O				°		O		T	O	O	O			°	O		
C	O	O				°		O	O	U			O		O	°	O		O
D		O				°	O			V			O		O	°	O	O	
E	O	O				°	O		O	W	O	O	O			°	O	O	O
F	O	O				°	O	O		X	O	O		O	O	°			
G		O				°	O	O	O	Y			O	O	O	°			O
H	O	O			O	°				Z			O	O	O	°		O	

Figure 1.9 ISO alphanumeric code characters

Sec. 1.1]   **Machine Features**   19

With the transfer of part programs to the machine tool through an RS-232 serial communication link, coded data is transferred in ASCII code with each '0' bit being represented by a +3 volt signal and each '1' bit being represented by a zero volt signal. The transfer speed of data, called the baud rate and measured in bits per second, between the transmitting computer and the CNC machine has to be set at a compatible level.

This method is the normal means of transferring data between electronic devices and is becoming increasingly common with CNC machines. It also has the potential for handling and editing large programs easily and for running the machine in real time, as discussed in Chapter 5.

1.1.5 Other Features

Automatic tool changers are frequently fitted on machining centres to enable a variety of tools to be selected during machining. An automatic tool changer for storing and dispensing twenty tools is shown in Figure 1.10.

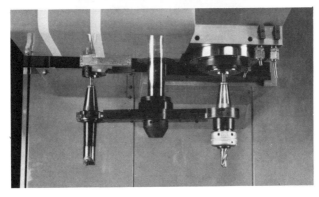

Figure 1.10 Automatic tool changer

20   **Introduction to Numerically Controlled Machines**   [Ch.1]

Removable worktables are also used to reduce downtime on the machine. The tables operate on an automatic feed mechanism, which allows them to be routed to a number of machine tools. Component parts are located on the tables away from the machine tools and are inserted at the appropriate machine for manufacture. Devices for the automatic removal of swarf are also common.

Parts can be inspected, while located on a CNC machine table, with the use of inspection probes. Probes are mounted in the machine tool spindle from the automatic tool changer and used to inspect any feature of the machined part. Sensors in the probe detect when contact has been made with the surface of the workpiece, and these signals are transmitted to the machine controller. An inspection probe is shown in Figure 1.11.

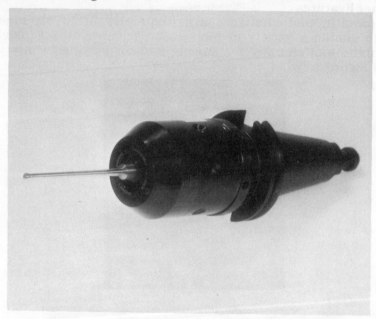

Figure 1.11   CNC inspection probe

1.2 TYPES OF PROGRAMMING

From a programmer's point of view, the control aspects of the CNC machine tool are all taken care of by the machine tool manufacture. Essentially, part programming consists of specifying geometric point-to-point moves of the tool in relation to the workpiece. All interpolated motion, either linear or circular, between the defined end points is provided by the machine controller. Velocity of motion is determined by the specified program feedrate, or by the fast traverse mode; and the acceleration profile of all motion is preset within the machine controller.

There are a number of ways of programming a CNC machine tool. The most basic way is to program manually using standard G-codes. While this is the slowest and most laborious method, it is still commonly used in job shops and with standalone machines. The manual programming approach is given in Chapter 2 and is useful for a full understanding of machine tool operations and interpretation of machine codes which are eventually used by all other methods.

The first, and until recently the most comprehensive, computer assisted method of programming was APT (Automatically Programmed Tools). It consists of English-like geometry statements to describe the part shape in terms of points, lines and circles etc., and motion statements to describe the cutter path. A machine tool code program is thus developed automatically by computer processing of the APT program. However, the APT approach is non-graphical and does not use the more convenient and standard CAD means of defining parts and tool paths.

Interactive computer aided graphic programs are being increasingly used to define parts and cutter paths. They are much faster and more reliable than the manual methods. One approach which is available with CNC machine tools is the interactive graphics or conversational programming method. In this method the programmer responds to questions and instructions at the control console in order to define the part and the tool path. The conversational method produces G-code programs directly in the controller memory. The G-code program may be viewed, edited and transmitted in exactly the same way as a manual G-code program. Also, the programming can be accomplished while the machine is running on another job (i.e., foreground\background mode), thus avoiding unnecessary machine downtime. One difficulty with this approach is that, in general, CNC controllers tend to have relatively small, special purpose, computer memory, which is adequate for storing only a few medium length programs. In fact modern machines have a storage memory equivalent to approximately 100 m of paper tape. Extended memory for CNC controllers is expensive, particularly when compared with the cost of memory for PC computers.

There are other computer assisted CAM programs which run on personal computers. They use the part definition output from a standard CAD package, such as AutoCAD, and with interactive instructions produce a graphically displayed cutter path and a file defining the cutter motions, called a cutter location file (CLFile). An additional program, called a postprocessor, is then used to convert the CLFile to G-code instructions for controlling the machine. All of the above computer assisted programming methods are discussed in Chapter 3. The development of postprocessor programs and graphical simulation programs is given in Chapter 5. It is interesting to note that the trend in all CAD/CAM applications is away from packaged programs and special purpose computers towards the use of standard computers. With this trend it is likely that standard computers will be used increasingly for part and machine path definition and manipulation, program storage and other operations, while the machine controller will be restricted to controlling the hardware, i.e., the intelligence may move from the controller to external computers.

An integrated approach to part definition (CAD) and machining definition (CAM) together with information management, data base support, and engineering analysis is available in a number of comprehensive although more expensive packages. In these programs the part can be defined in wire frame or solids modelling mode; it can be analysed and modified through finite element programs and the machining paths, with graphical interaction, produced. An example of an integrated CAD/CAM program, called Anvil 5000, is given in Chapter 4.

The means for developing direct numerical link (DNC) programs, to enable programs to be downloaded to machine controllers or to be fed directly to a buffer behind the tape reader, is given in Chapter 5. In the latter approach, the data is transmitted to the controller as though it were reading paper tape, although it is actually taking data from a computer and thus the machine is effectively run in real time (or a block of code at a time) from a remote computer. These programs can be used to control a number of machines through a central computer and are the basic programming techniques for flexible manufacturing cells and integrated manufacturing systems.

Definition and machining of curved surfaces is discussed in Chapter 6. The positioning of a variety of shaped milling cutters in relation to arbitrarily curved surface is considered and equations derived. The description and listing of a series of C-based computer programs are given for all aspects of surface definition and machining. Case studies of curved surface machining applications (from marine propellers to heart valves shapes) are given in Chapter 7 and integrated lofting, fairing, and manufacturing for small shipyards (based on many of the programs in Chapter 6) is given in Chapter 8.

Development of a computer simulation program for checking the accuracy of all statements in G-code files and displaying the output on a graphics screen is given in Chapter 9. In Chapter 10 the means of retrofitting manual machine tools with motors and controllers in order to obtain programmable CNC machine tools is discussed.

Chapter 2
Manual Part Programming

Machine tool controllers are able to interpret a series of simple standard command codes in order to control the table and spindle movements and all auxiliary functions. The prime code is the G-code, which can be used, for example, to define fast traverse motion (G00), linear (G01) and circular (G02 and G03) tool motion under controlled feed rates, millimetre and inch motion (G20 and G21), tool length compensation (G43), and preprogrammed function calls (G65). Other codes include feedrate F-codes, cutting speed S-codes, tool selection T-codes and auxiliary functions, such as spindle start and stop, M-codes. These machine codes, which are outlined below, are collectively referred to as G-codes.

The program for producing a part is arranged in the form of coded blocks or lines of information. Each block contains a group of commands sufficient for one individual machining operation. A typical block is given as follows:

N200 G01 X120 Y100 Z80 F150;

and is interpreted that in block 200 the tool should move from its present position in a straight line to X = 120 mm, Y = 100 mm and Z = 80 mm at a feedrate of 150 mm/min.

Some typical beginning and end sequences for CNC part programs are given in Table 2.1. Each program should have an identifying number, in this case O006. It is good practise to start the program with a return home (G91 G28) instruction so that all subsequent programming is from a known datum. This home datum point is determined by the machine tool manufacturer and is normally at Xmax, Ymax and Zmax. For safety the Z axis is usually returned to the home position first. The G92 X250 Y125 Z75 instruction establishes the position of the workpiece in relation to the home position. The tool defined in storage location T18 is inserted in the spindle with the M06 instruction, and the tool in storage location T19 is positioned on standby. Absolute coordinates, millimetre measurements and tool length compensation are selected with G90, G21 and G43, respectively. The cutter is moved with rapid traverse to the position Z = 100 mm and then to the position X = 0 mm, Y = 0 mm (or to any other starting position). A spindle speed of 2000 rpm is selected with S2000 and the spindle turned on with M03.

Table 2.1
G-code Program - Beginning and End Sequences

O006;	program number 0006
N10 G91 G28 Z0;	return home Z
N20 G91 G28 X0 Y0;	return home X and Y
N30 G92 X250 Y125 Z75;	establish workpiece position
N40 T18;	call for tool 18 in the automatic tool changer
N50 T19 M06;	insert tool 18 in spindle, tool 19 on standby
N60 G90 G21 G43 G00 Z100 H18;	absolute coords, mm, tool length compensation rapid traverse to Z100
N70 G00 X0 Y0;	rapid traverse to X0 and Y0
N80 S2000 M03;	set spindle speed and turn spindle on
N90	
N100	
N......	program body
N190	
N200	
N210 G91 G28 Z0;	return home Z
N220 G91 G28 X0 Y0;	return home X and Y
N230 M05;	turn spindle off
N240 M30;	end of program and rewind
N250 %	end of tape

The basic G-codes, with examples, are given in Section 2.1. The advanced G-codes, with examples, are given in Section 2.2.

2.1 BASIC MACHINE CODES

2.1.1 Basic G-Codes for CNC Milling Machine

G00 Positioning (Rapid traverse)
    G00 X__ Y__ Z__ ;

The tool is moved in rapid traverse mode to the point programmed by the address X, Y and Z. This is used for non-cutting motion, as the move is achieved at full traverse rate. The tool path is not necessarily a straight line. The move can be programmed in either absolute (G90) or incremental (G91) mode.

Block format (example)

N60 G90 G00 X8.0 Y7.0 Z10.0 ;

N60 Sequence number
G90 Absolute dimensions
G00 Linear interpolation in rapid traverse
X   X-coordinate of end position
Y   Y-coordinate of end position
Z   Z-coordinate of end position

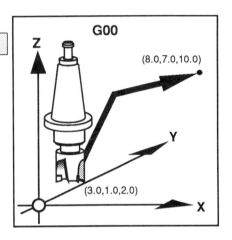

G01 Linear Interpolation (Feed)
    G01 X__ Y__ Z__ F__ ;

The tool is moved in a straight line at a given or predefined feedrate. The speed of each axis is automatically controlled so that the tool moves in a straight line. This is used for linear cutting motion. The move can be programmed in either absolute (G90) or incremental (G91) mode.

Block format (example)

N70 G91 G01 X5.0 Y6.0 Z8.0 F100 ;

N70 Sequence number
G91 Incremental dimensions
G01 Linear interpolation
X   Distance from start to end position along X-axis
Y   Distance from start to end position along Y-axis
Z   Distance from start to end position along Z-axis
F   Feedrate

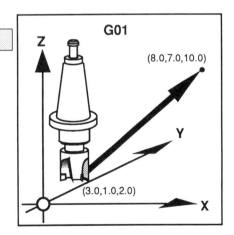

G02 Circular Interpolation (Clockwise)

G02 X__ Y__ R__ F__ ;   Radius and end position
G02 X__ Y__ I__ J__ F__ ;   Arc centre and end position

The tool is moved in a clockwise circular arc by specifying either the radius and the end position or the arc centre and the end position. This is used for circular cutting at a specified feedrate.

Block format (radius and end position example)

```
N80 G90 G02 X7.6 Y8.2 R3.5 F100 ;
```

N80 Sequence number
G90 Absolute dimensions
G02 Circular interpolation, clockwise
X   X-coordinate of end position
Y   Y-coordinate of end position
R   Radius of arc, arc less than 180 degrees (see below)
F   Feedrate

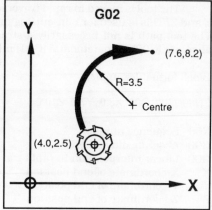

Block format (arc centre, end position example)

```
N90 G90 G02 X7.6 Y8.2 I2.4 J2.3 F100. ;
```

N90 Sequence number
G90 Absolute dimensions
G02 Circular interpolation, clockwise
X   X-coordinate of end position
Y   Y-coordinate of end position
I   Distance along X-axis from start to arc centre
J   Distance along Y-axis from start to arc centre
F   Feedrate

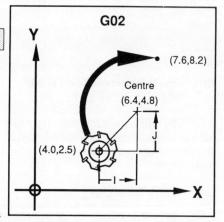

When using the radius and end position option, the radius is assigned either a positive or a negative value. An arc equal to or less than 180 degrees has a positive radius (1) and an arc greater than 180 degrees has a negative value (2).

Block format (radius and end position examples)

```
N100 G02 X__ Y__ R__ F__ ;
```

Radius and end position with arc less than 180 degrees.

```
N110 G02 X__ Y__ R-__ F__ ;
```

Radius and end position with arc greater than 180 degrees.

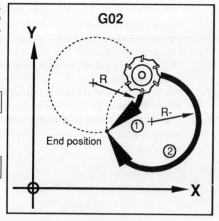

Sec. 2.1]   **Basic Machine Codes**   27

G03 Circular Interpolation (Counter-clockwise)
   G03 X__ Y__ R__ F__ ;   Radius and end position
   G03 X__ Y__ I__ J__ F__ ;   Arc centre and end position

The tool is moved in a counter-clockwise circular arc by specifying either the radius and the end position or the arc centre and the end position. This is used for circular cutting at a specified feedrate.

Block format (radius and end position example)

   N120 G90 G03 X3.3 Y7.8 R3.5 F100. ;

N120 Sequence number
G90   Absolute dimensions
G03   Circular interpolation, counter-clockwise
X   X-coordinate of end position
Y   Y-coordinate of end position
R   Radius of arc, arc less than 180 degrees

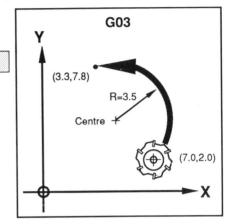

Block format (radius and end position example)

   N130 G91 G03 X-5.7 Y1.5 R-3.5 F100. ;

N130 Sequence number
G91   Incremental dimensions
G03   Circular interpolation, counter-clockwise
X   Distance along X-axis from start to end position
Y   Distance along Y-axis from start to end position
R-   Radius of arc, arc greater than 180 degrees
F   Feedrate

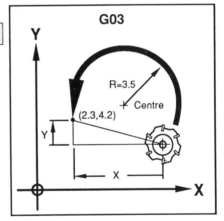

Block format (arc centre, end position example)

   N140 G90 G03 X1.8 Y4.1 I-2.4 J1.5 F100. ;

N140 Sequence number
G90   Absolute dimensions
G03   Circular interpolation, counter-clockwise
X   X-coordinate of end position
Y   Y-coordinate of end position
I   Distance along X-axis from start to arc centre
J   Distance along Y-axis from start to arc centre
F   Feedrate

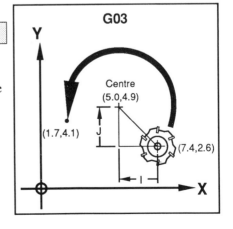

G04 Dwell (Time delay)
    G04 X__ ; (seconds)
    G04 P__ ; (milliseconds)

The tool is held stationary for a period of time, determined by address X in seconds or P in milliseconds. This can be used for drilling and boring cycles to ensure true precision of hole depth.

Block format (second example)

```
N150 G04 X1.5 ;
```

N150 Sequence number
G04   Dwell
X   Time delay in seconds

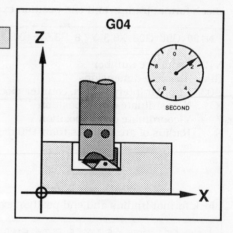

The following example is equivalent to the one above.

Block format (millisecond example)

```
N160 G04 P1500 ;
```

N160 Sequence number
G04   Dwell

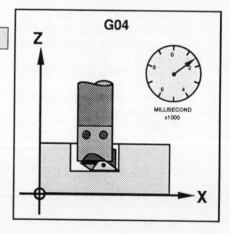

G17, G18, G19 Plane Selection
    G17 :   X-Y plane
    G18 :   X-Z plane
    G19 :   Y-Z plane

The tool moves in one of these planes. This is used for determining the plane in which circular interpolation or cutter compensation occur. The default is G17 in the X-Y plane.

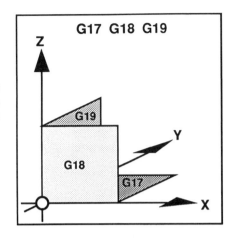

Block format (G02 example)

    N170 G17 G02 X__ Y__ R__ F__ ;
    N180 G17 G02 X__ Y__ I__ J__ F__ ;

    N190 G18 G02 X__ Z__ R__ F__ ;
    N200 G18 G02 X__ Z__ I__ K__ F__ ;

    N210 G19 G02 Y__ Z__ R__ F__ ;
    N220 G19 G02 Y__ Z__ J__ K__ F__ ;

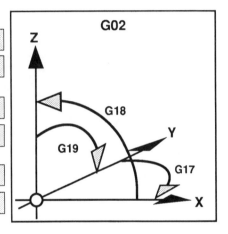

Block format (G03 example)

    N230 G17 G03 X__ Y__ R__ F__ ;
    N240 G17 G03 X__ Y__ I__ J__ F__ ;

    N250 G18 G03 X__ Z__ R__ F__ ;
    N260 G18 G03 X__ Z__ I__ K__ F__ ;

    N270 G19 G03 Y__ Z__ R__ F__ ;
    N280 G19 G03 Y__ Z__ J__ K__ F__ ;

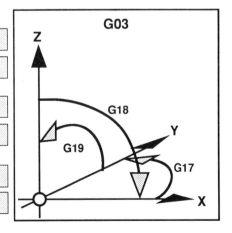

G20, G21 Inch/Millimetre Units
   G20 ;   Inch unit
   G21 ;   Millimetre unit

The units are specified at the beginning of the program before setting the work coordinate system (G92) and must be the only instruction in the block. Once G20 or G21 is specified, it should not be changed during the program.

Block format (inch example)

```
N30 G20 ;
```

```
N40 G92 X4.0 Y5.0 Z4.0 ;
```

N30   Sequence number
G20   Inch units

N40   Sequence number
G92   Setting the work coordinate system

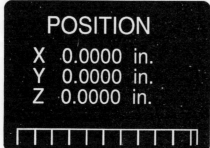

Block format (millimetre example)

```
N30 G21 ;
```

```
N40 G92 X100. Y150. Z100. ;
```

N30   Sequence number
G21   Millimetre units

N40   Sequence number
G92   Setting the work coordinate system

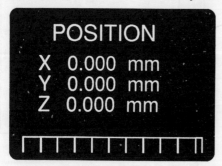

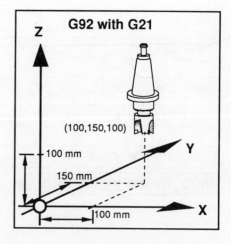

G22, G23 Stored Stroke Limit

    G22 X_ Y_ Z_ I_ J_ K_ ; Stored stroke limit ON
    G23 ;   Stored stroke limit OFF

The tool movements are restricted to the inside or the outside of a specified rectangular volume. The volume is determined by addresses X, Y, Z, I, J and K. It is used as a safety precaution to guard the tool from any obstruction on the machine table. Parameter RWL in the CNC controller memory selects either the inside or the outside as the restricted volume. An RWL parameter value of 0 selects the inside and an RWL parameter value of 1 selects the outside.

Block format (RWL parameter 0 example)

    N470 G22 X100. Y70. Z0. I250. J210. K100. ;

N470   Sequence number
G22   Stored stroke limit ON
X   X-coordinate point of boundary
Y   Y-coordinate point of boundary
Z   Z-coordinate point of boundary
I   X-coordinate point of boundary
J   Y-coordinate point of boundary
K   Z-coordinate point of boundary

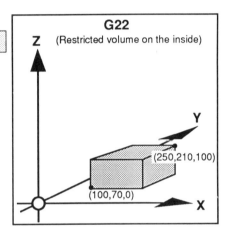

Block format (RWL parameter 1 example)

    N480 G22 X100. Y210. Z100. I250. J70. K0. ;

N480   Sequence number
G22   Stored stroke limit ON
X   X-coordinate point of boundary
Y   Y-coordinate point of boundary
Z   Z-coordinate point of boundary
I   X-coordinate point of boundary
J   Y-coordinate point of boundary
K   Z-coordinate point of boundary

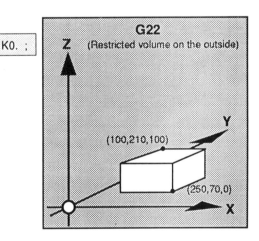

Block format (cancel stroke limit example)

    N490 G23 G00 X50. ;

N490   Sequence number
G23   Stored stroke limit OFF
G00   Positioning
X   X-coordinate of end position

G28 Return to Home Position
G28 X__ Y__ Z__ ;

The tool is moved to home position via an intermediate point defined by address X, Y and Z. The intermediate point is determined by G90 or G91 function code. In combination with G91, the G28 command is used at the beginning of each program to return the tool to the home position or to make a tool change during the program. It is advisable to return the Z-axis first in order to clear the workpiece and clamps, then return the X and Y axes. All movement is at rapid traverse.

Block format (beginning of program example)

```
N10 G91 G28 Z0.0 ;
```

```
N20 G91 G28 X0.0 Y0.0 ;
```

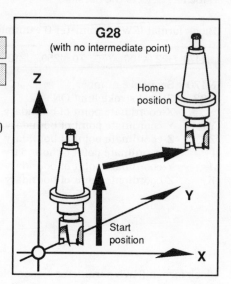

N10 Sequence number
G91 Incremental dimensions
G28 Return to reference point (Z home position)
Z Incremental Z-distance to intermediate point

N20 Sequence number
G91 Incremental dimensions
G28 Return to reference point (X and Y home position)
X Incremental X-distance to intermediate point
Y Incremental Y-distance to intermediate point

When performing a tool change, it is only necessary to return the Z axis to the home position.

Block format (tool change example)

```
N400 G90 G28 Z100. ;
```

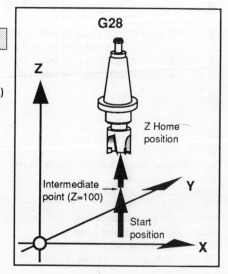

N400 Sequence number
G90 Absolute dimensions
G28 Return to reference point (Z home position)
Z Z-coordinate point of intermediate point

G29 Return from Home Position
G29 X__ Y__ Z__ ;

The tool is moved from the home position to an intermediate point and then to the point defined by addresses X, Y and Z. It is used after a tool change and the intermediate point is the one defined earlier by the G28 move. When using incremental programming, an incremental distance from the intermediate point to the end point is given by addresses X, Y and Z.

Block format (absolute example)

N410 G90 G28 X60. Y100. ;

N420 G29 X120. Y40. ;

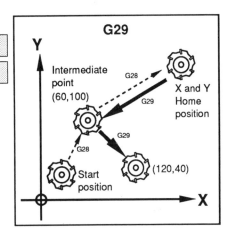

N410 Sequence number
G90 Absolute dimensions
G28 Return to reference point
X   X-coordinate of intermediate point
Y   Y-coordinate of intermediate point

N420 Sequence number
G29 Return from reference point
X   X-coordinate of end point
Y   Y-coordinate of end point

Block format (incremental example)

N430 G90 G28 X80. Y110. ;

N440 G91 G29 X70. Y-60. ;

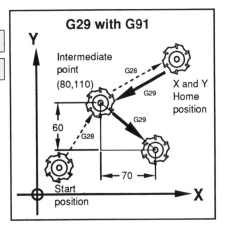

N430 Sequence number
G90 Absolute dimensions
G28 Return to reference point
X   X-coordinate of intermediate point
Y   Y-coordinate of intermediate point

N440 Sequence number
G91 Incremental dimensions
G29 Return from reference point
X   X-distance from intermediate point to end point
Y   Y-distance from intermediate point to end point

G40, G41, G42 Tool Radius Compensation

G40 ;	Tool radius compensation cancel
G41 D__ ;	Tool radius compensation left
G42 D__ ;	Tool radius compensation right

The tool is moved to the left (G41) or to the right (G42) of the program path by the amount of the cutter radius. The offset amount is registered with a D address.

Block format (tool radius compensation left example)

```
N270 G90 G17 G01 G41 D10 X30. Y30. F100. ;
```
```
N280 G01 Y55. ;
```

N270 Sequence number
G90 Absolute dimensions
G17 X-Y plane selection
G01 Linear interpolation
G41 Tool radius compensation left
D Offset amount
X X-coordinate of end point
Y Y-coordinate of end point
F Feedrate

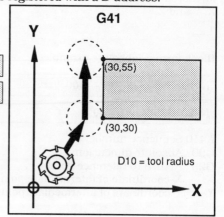

Block format (tool radius compensation right example)

```
N290 G90 G17 G01 G42 D11 X30 Y30 F100. ;
```
```
N300 G01 X60. ;
```

N270 Sequence number
G90 Absolute dimensions
G17 X-Y plane selection
G01 Linear interpolation
G42 Tool radius compensation right
D Offset amount
X X-coordinate of end point
Y Y-coordinate of end point
F Feedrate

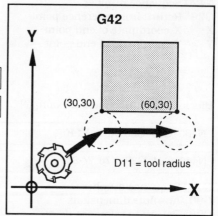

Block format (tool radius compensation cancel example)

```
N310 G00 G40 X10. Y10. ;
```

N310 Sequence number
G00 Positioning at rapid traverse
G40 Tool radius compensation cancel
X X-coordinate of end point
Y Y-coordinate of end point

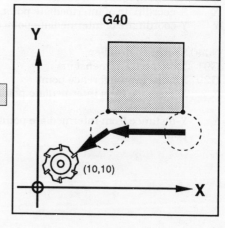

Sec. 2.1] **Basic Machine Codes** 35

Examples of **G41, G42 Tool Radius Compensation**

Block format (outside corner at an obtuse angle)

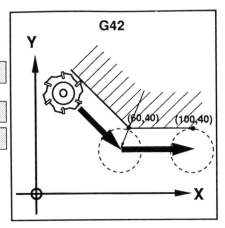

```
N600 G42 D16 ;
    .
    .
N650 G01 X60. Y40. F100. ;

N660 X100. Y40. ;
```

N600 Sequence number
G42 Cutter compensation right
D   Offset address 16
G01 Linear interpolation
X, Y Coordinates of end point

Block format (inside corner at an acute angle)

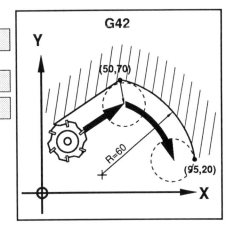

```
N700 G42 D17 ;
    .
    .
N750 G01 X50. Y70. F100. ;

N760 G02 X95. Y20. R60. ;
```

N700 Sequence number
G42 Cutter compensation right
D   Offset address 17
X, Y Coordinates of end point
G02 Circular interpolation CW
X, Y Coordinates of end point
R   Radius of arc

Block format (Outside corner at an acute angle)

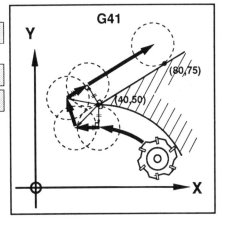

```
N800 G41 D18 ;
    .
    .
N850 G03 X40. Y50. R60. F100. ;

N860 G01 X80. Y75. ;
```

N800 Sequence number
G41 Cutter compensation left
D   Offset address 18
G03 Circular interpolation CCW
X, Y Coordinates of end point
R   Radius of arc
X, Y Coordinates of end point

G43, G44, G49 Tool Length Compensation

G43 H__ ; or G43 Z__ H__ ;
Tool length compensation positive direction
G44 H__ ; or G44 Z__ H__ ;
Tool length compensation negative direction

The tool is moved above or below some standard tool position in order to compensate for cutter length. The offset direction is set by either G43 or G44 and the amount is registered with an H address in the CNC controller memory. The value specified by address Z is either added to or subtracted from the offset amount. The offset amount registered to H00 is always 0. Tool length compensation is used when two or more tools of different length are used in a program. This makes it possible to apply length compensation without changing the program coordinates.

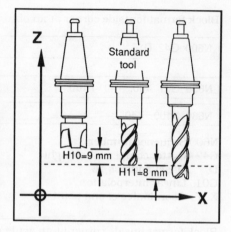

Block format (G43 example) H11=8 mm

```
N900 G43 H11 ;
```

N900 Sequence number
G43 Tool length compensation, positive direction
H   Offset address 11

Block format (G44 and Z example) H10 = 9mm

```
N910 G44 H10 ;
```
or
```
N910 G44 Z-9.0 H00 ;
```

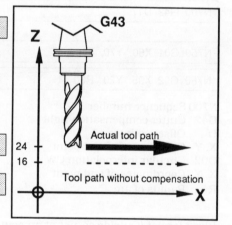

N910 Sequence number
G44 Tool length compensation, negative direction
H   Offset address 10

Z   Compensation amount
H00 Offset amount is zero

Block format (tool length compensation cancel example)

```
N920 G49 ;
```
or
```
N920 H00 ;
```

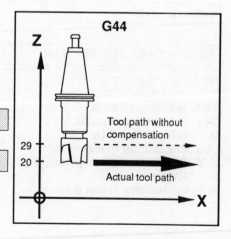

G49 Tool length compensation cancel
H00 Tool length compensation set to zero

G50, G51 Scaling

    G50 ;   (Scaling OFF)
    G51 I__ J__ K__ P__ ;   (Scaling ON)

The tool is moved by a scaling factor P about a scaling point defined by address I, J and K.

Block format (scaling example)

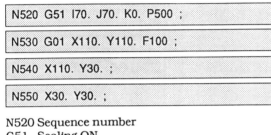

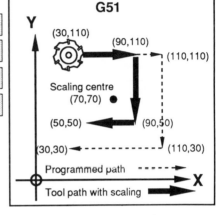

N520 Sequence number
G51 Scaling ON
I   X-coordinate of scaling centre
J   Y-coordinate of scaling centre
K   Z-coordinate of scaling centre
P   Scaling factor

Block format (scaling cancel example)

N560 G50 ;

N540 Sequence number
G50 Scaling OFF

G54, G55, G56, G57, G58, G59   Work Coordinate System

G54 ;   Work coordinate system #1
G55 ;   Work coordinate system #2
G56 ;   Work coordinate system #3
G57 ;   Work coordinate system #4
G58 ;   Work coordinate system #5
G59 ;   Work coordinate system #6

The tool is moved to a coordinate system by setting the distance in each axis from the current tool position to zero point of the new coordinate system. All moves are made in incremental mode. G54 is selected when the power is turned on.

Block format (example)

    N460 G55 G00 X30. Y20. ;

    N470 G01 X5. Y20. F100. ;

    N480 G01 X20. Y0. ;

    N490 G56 G00 X20. Y25. ;

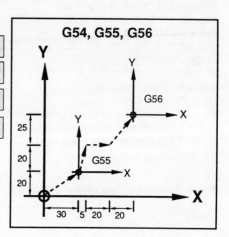

N460 Sequence number
G55 Work coordinate system #2
G00 Positioning
X   X-distance to zero point of coordinate system #2
Y   Y-distance to zero point of coordinate system #2

N490 Sequence number
G56 Work coordinate system #3
G00 Positioning
X   X-distance to zero point of coordinate system #2
Y   Y-distance to zero point of coordinate system #2

Sec. 2.1]   **Basic Machine Codes**   39

G90 Absolute Dimensions
G90 ;

All points and tool positions are taken from an origin which is set by the G92 command. When moving the tool from one point to another, the coordinate is entered using address X, Y and Z.

Block format (example)

```
N50 G90 G01 X25. Y35. Z30. ;
```

N50 Sequence number
G90 Absolute dimensions
X   X-coordinate of end point
Y   Y-coordinate of end point
Z   Z-coordinate of end point

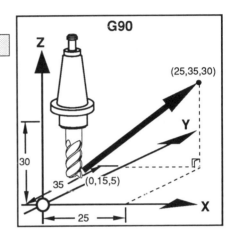

G91 Incremental Dimensions
G91 ;

All points and tool positions are measured from the current point. When moving the tool from one point to another, the distance between the points is entered using address X, Y and Z. Plus or minus values are dependent on the new point in relation to the current point.

Block format (example)

```
N60 G91 G01 X25. Y20. Z25. F100 ;
```

N60 Sequence number
G91 Incremental dimensions
G01 Linear interpolation
X   X-distance from current point to end point
Y   Y-distance from current point to end point
Z   Z-distance form current point to end point

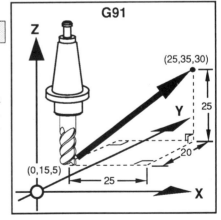

G92 Programming of Absolute Zero Point
  G92 X__ Y__ Z__ ;

This command sets the origin of the coordinate system a specified distance from the home position or current tool position. Address X, Y and Z defines the location of the origin in relation to the current tool position. It is used at the start of the program to establish the coordinate system in relation to the CNC machine home position or can be used to establish the coordinate system in relation to some arbitrary position.

Block format (home position example)

| N40 G91 G28 X0. Y0. Z0. ; |

| N80 G92 X350. Y350. Z250. ; |

N80   Sequence number
G92   Programming of absolute zero point
X   X-distance from tool position to zero point
Y   Y-distance from tool position to zero point
Z   Z-distance from tool position to zero point

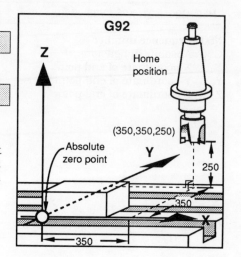

Block format (arbitrary position example)

| N100 G92 X200. Y200. Z100. ; |

N100 Sequence number
G92   Programming of absolute zero point
X   X-distance from tool position to zero point
Y   Y-distance from tool position to zero point
Z   Z-distance from tool position to zero point

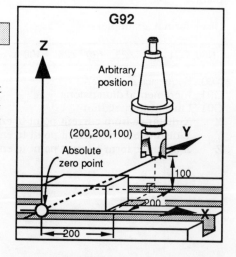

2.1.2 Basic G Codes for CNC Lathe Machine

G00 Positioning (Rapid traverse)
  G00 X__ Z__ ;

The tool is moved in rapid traverse mode to the point programmed by address X and Z. Address X is a diameter value. This is used for non-cutting motion, as the move is achieved at full traverse rate. The tool path is not necessarily a straight line.

Block format (example)

N110 G90 G00 X60. Z10. ;

N110 Sequence number
G90 Absolute dimensions
G00 Positioning
X   X-diameter of end point
Z   Z-coordinate of end point

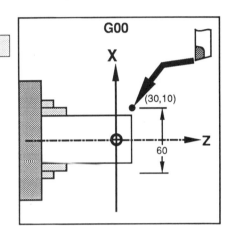

G01 Linear Interpolation (Feed)
  G01 X__ Z__ F__ ;

The tool is moved in a straight line at a given or predefined feedrate. The speed of each axis is automatically controlled so that the tool moves in a straight line. This is used for linear cutting motion.

Block format (example)

N120 G90 G01 X80. Z-100. F100. ;

N120 Sequence number
G90 Absolute dimensions
G01 Linear interpolation
X   X-diameter of end point
Z   Z-coordinate of end point
F   Feedrate

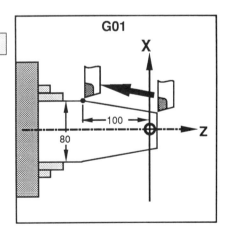

G03 Circular Interpolation (Counter-clockwise)
   G03 X__ Z__ R__ F__ ;   Radius and end position
   G03 X__ Z__ I__ K__ F__ ;   Arc centre and end position

The tool is moved in a counter-clockwise circular arc by specifying either the radius and end point or the arc centre and end point. This is used for circular cutting at a specified feedrate.

Block format (radius and end position example)

```
N90 G91 G03 X100. Z-50. R60. F100. ;
```

N90   Sequence number
G91   Incremental dimensions
G03   Circular interpolation CCW
X   Incremental X-diameter from start to end position
Z   Z-distance from start to end position
R   Radius of arc
F   Feedrate

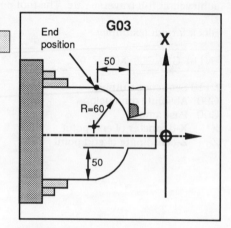

Block format (arc centre, end position example)

```
N100 G91 G03 X100. Z-40. I25. K80. F100. ;
```

N100 Sequence number
G91   Incremental dimensions
G03   Circular interpolation CCW
X   Incremental X-diameter from start to end position
Z   Z-distance from start to end position
I   X-distance from start to arc centre
K   Z-distance from start to arc centre
F   Feedrate

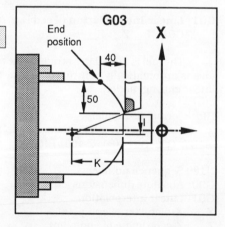

G03 Circular Interpolation (Counter-clockwise)
  G03 X__ Z__ R__ F__ ;   Radius and end position
  G03 X__ Z__ I__ K__ F__ ; Arc centre and end position

The tool is moved in a counter-clockwise circular arc by specifying either the radius and end point or the arc centre and end point. This is used for circular cutting at a specified feedrate.

Block format (radius and end position example)

N90 G91 G03 X100. Z-50. R60. F100. ;

N90 Sequence number
G91 Incremental dimensions
G03 Circular interpolation CCW
X   Incremental X-diameter from start to end position
Z   Z-distance from start to end position
R   Radius of arc
F   Feedrate

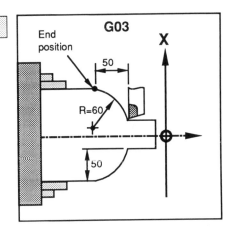

Block format (arc centre, end position example)

N100 G91 G03 X100. Z-40. I25. K80. F100. ;

N100 Sequence number
G91 Incremental dimensions
G03 Circular interpolation CCW
X   Incremental X-diameter from start to end position
Z   Z-distance from start to end position
I   X-distance from start to arc centre
K   Z-distance from start to arc centre
F   Feedrate

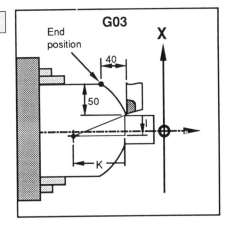

G40, G41, G42 Tool Nose Radius Compensation

G40 X__ Z__ I__ K__ ;
Tool nose radius compensation cancel
G41 ;
Tool nose radius compensation right
G42 ;
Tool nose radius compensation left

The tool is moved to the right (G41) or to the left (G42) of the programmed path by the amount set in the tool offset table. Although the two cutting edges of a turning tool appear to come to a point, the point of the tool is a radius, called the tool nose radius.

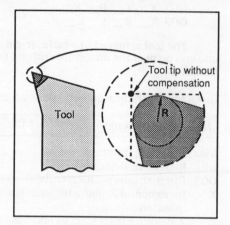

Block format (G41 example)

N190 G41 G01 X40. Z-50. ;

N200 G01 X60. Z-30. ;

N210 G01 X60. Z__ ;

N190 Sequence number
G41 Tool nose radius compensation right
G01 Linear interpolation
X   X-diameter of end point
Z   Z-coordinate of end point

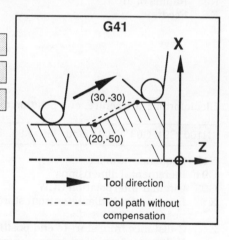

Block format (G42 example)

N220 G42 G01 X60. Z-50. ;

N230 G01 X30. Z-30. ;

N240 G01 X30. Z__ ;

N220 Sequence number
G42 Tool nose radius compensation left
G01 Linear interpolation
X   X-diameter of end point
Z   Z-coordinate of end point

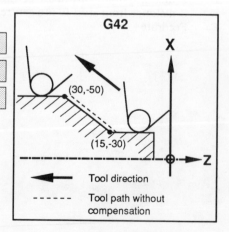

Sec. 2.1]   **Basic Machine Codes**   45

G74 End Face Peck Drilling Cycle
   G74 Z__ K__ F__ ;   Peck drilling
   G74 X__ Z__ I__ K__ F__ D__ ; End facing

The tool moves in a forward motion then backward motion along the Z-axis. This is used for peck drilling. When address X, I and D are included with G74, end facing can be performed.

Block format (peck drilling example)

| N150 G90 G74 Z-50. K5. F50. ; |

N150 Sequence number
G90 Absolute dimensions
G74 Peck drilling
Z   Z-coordinate of hole bottom
K   Depth of cut in Z direction
F   Feedrate

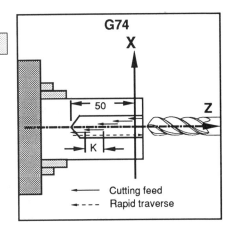

Block format (end facing example)

| N160 G90 G74 X0. Z-8. I3. K5. F100. D1. ; |

N160 Sequence number
G90 Absolute dimensions
G74 End facing
X   X-coordinate for depth of face
Z   Facing amount in Z direction
I   Depth of each face cut (incremental X amount)
K   Depth of each cut in Z direction
F   Feedrate
D   Relief amount at end of each cut

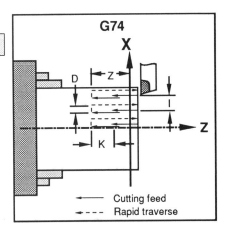

G75 Grooving in X Axis
G75 X__ Z__ I__ K__ F__ D__ ;

The tool moves in a downward then an upward motion along the X-axis. This is used for grooving.

Block format (example)

N170 G90 G75 X20. Z8. I3. K-0.5 F100. D0. ;

N170 Sequence number
G90 Absolute dimensions
G75 Grooving in X-axis
X X-coordinate for groove depth
Z Z-coordinate for groove length
I Depth of each peck cut (in X direction)
K Movement amount in Z direction
F Feedrate
D Relief amount at bottom of each groove cut

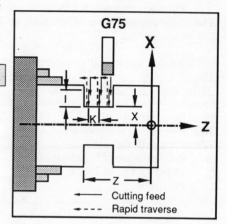

G78 Thread Cutting Cycle
G78 X__ Z__ F__ ;
Straight thread cutting
G78 X__ Z__ I__ F__ ;
Taper thread cutting

The tool performs straight thread cutting or taper cutting depending whether address I is specified. Address X is a diameter value and F is a feedrate value to specify the lead. At the end point the thread is chamfered at a 45 degree angle.

Block format (straight thread cutting example)

N180 G90 G78 X50. Z-30. F1.5 ;

N180 Sequence number
G90 Absolute dimensions
G78 Thread cutting cycle
X X-diameter value for depth of thread
Z Z-coordinate of end point
F Feedrate determines lead

Block format (taper thread cutting example)

N190 G90 G78 X50. Z-30. I10. F1.5 ;

N180 Sequence number
G90 Absolute dimensions
G78 Thread cutting cycle
X X-diameter value for depth of thread, at end point
Z Z-coordinate of end point
I Taper amount in X direction
F Feedrate determines lead

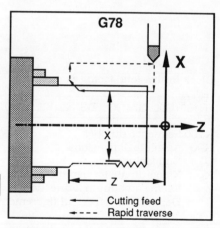

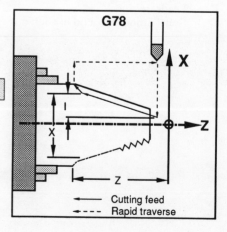

G96, G97 Constant Surface Speed Control

G96 ;   Constant surface speed control
G97 ;   Constant surface speed control cancel

The spindle speed either remains constant or varies with the radius of the workpiece. The spindle speed defined by address S is given with or before a G96 or G97.

Block format (G96 example)

N60 G96 S0300 ;

N60 Sequence number
G96 Constant surface speed control
S   Surface speed (m/min)

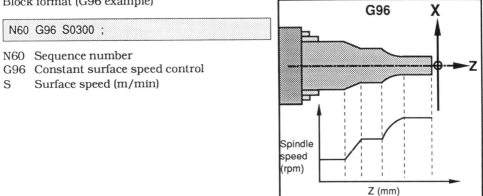

Block format (G97 example)

N70 G97 S1000 ;

N70 Sequence number
G97 Constant surface speed control cancel
S   Spindle speed (rpm)

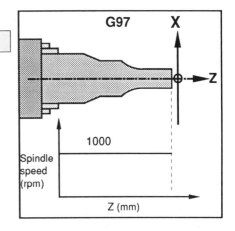

2.1.3 M-Codes for CNC Machines

M-codes are used to control various functions on CNC machines. The following M-codes are the same for both the lathe and milling machines, unless otherwise stated.

M00 Program Stop
M00 ;

Machining is stopped after a block containing M00. This can be used to make an inspection before the machining is finished.

M01 Optional Stop
M01 ;

Machining is stopped after a block containing M01 only if the Optional Stop switch on the CNC control panel is turned on.

M02 End of Tape, Restart
M02 ;

The end of an NC tape is indicated by M02. This is used when a program requires two or more tapes. After reading an M02 the tape reader waits for the next tape to be inserted.

M03 Spindle Start Clockwise
M03 ;

The spindle rotation is started in a clockwise direction. The spindle speed specified with an S address is given before or with M03.

Block format (example)

    N50 S0300 M03 ;

N50   Sequence number
S   Spindle speed (m/min.)
M03   Spindle start CW

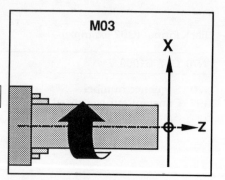

M04 Spindle Start Counterclockwise
M04 ;

The spindle rotation is started in a counter-clockwise direction. The spindle speed specified with an S address is given before or with M04.

Block format (example)

    N60 S0200 M04 ;

N60   Sequence number
S   Spindle speed (m/min.)
M03   Spindle start CCW

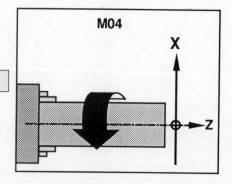

Sec. 2.1] **Basic Machine Codes** 49

M05 Spindle Stop
M05 ;

The spindle rotation is stopped. The spindle stops after a move command, if the move and the spindle stop commands are both specified in the same block.

Block format (example)

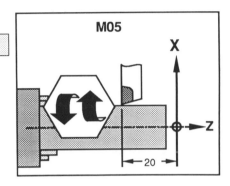

N80 G01 Z-20. M05 ;

N80 Sequence number
G01 Linear interpolation
Z   Z-coordinate of end point
M05 Spindle stops at Z=-20

M06 Tool Change
M06 T___ ;

The current tool in use is replaced by the tool with the tool identification number given in the T address. To perform a tool change on the milling machine, the Z-axis must be returned home first. The tool is then inserted in the spindle arm. A tool change on the lathe machine is performed by a rotation of the turret. The turret should be far enough away from the workpiece to avoid a collision.

Block format (example)

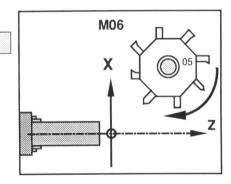

N90 M06 T0505 ;

N90   Sequence number
M06   Tool change
T0505 Tool number 05 with offset amount
       stored with 05

M07 Oil Mist On
M07 ;

The tool and workpiece are lubricated and cooled by an oil to improve the finish and reduce tool wear and breakage.

Block format (example)

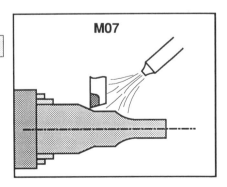

N100 M07 ;

N100 Sequence number
M07 Oil mist on

M09 Coolant, Oil Unit Off
M09 ;

The coolant is turned off when cutting is not being performed.

M30 End of Program and Rewind
M30 ;

The end of a program is indicated by M30. This rewinds the NC tape and the program in the CNC controller memory to its start after the program is completed.

M41 Low Range (for the lathe only)
M41 ;

The spindle rotates in the low gear. Its use depends on the material of the workpiece and tool being used, and the required quality of the finished product.

M41 High Range (for the lathe only)
M42 ;

The spindle rotates in the high gear. Its use depends on the material of the workpiece and tool being used, and the requirement for a high quality finish.

M98 Subprogram Call
M98 P____ L__ ;

The program flow is transferred from the main program to the subprogram. Address P specifies the subprogram being called and address L indicates the number of times the subprogram is executed. See the subprogram section for more details.

M99 End of Subprogram
M99 ;

The program flow is transferred from the subprogram back to the main program. This is used at the end of a subprogram. See the subprogram section for more details.

Sec. 2.1] **Basic Machine Codes** 51

2.1.4 F-, S- and T-codes for CNC Machines

F   **Feedrate Function**
   F___ ;

The tool is moved at a speed specified by address F. The feedrate units are inch/min when programmed with G20 and mm/min when programmed with G21.

Block format (example)

```
N70 G21 G01 X10. F100. ;
```

N70   Sequence number
G21   Metric units
G01   Linear interpolation
X   X coordinate of end point
F   Feedrate (mm/min)

S   **Spindle Speed Function**
   S___ ;

The spindle speed is specified by address S and is given in rpm.

Block format (example)

```
N40 S1000 ;
```

N40   Sequence number
S   Spindle speed (rpm)

T   **Tool Function**
   T___ ;

The tool is called by its tool number and its tool offset number. Both numbers are given in the T address and are registered in the tool table in the CNC controller memory.

Block format (example)

```
N30 T0104 ;
```

N30   Sequence number
T   Tool number (01), tool offset number (04)

2.1.5 Examples of Basic Codes

Milling Example

The operations to machine the component shown in the figure below are as follows:

1) Initialization
2) Facing
3) Deep side cutting
4) Track pocketing
5) Side cutting
6) Hole pattern
7) Program end

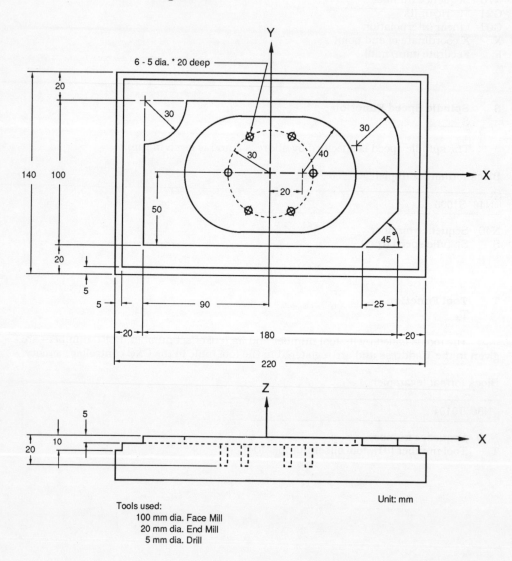

Tools used:
  100 mm dia. Face Mill
  20 mm dia. End Mill
  5 mm dia. Drill

Unit: mm

Sec. 2.1] **Basic Machine Codes** 53

Listing of G code

1) Initialization

O23;
N10 G91 G28 Z0.0;
N20 G28 X0.0 Y0.0;
N30 G21;
N40 G92 X500.0 Y250.0 Z400.0;
N50 T14;
N60 M06;
N70 S1400 M03;

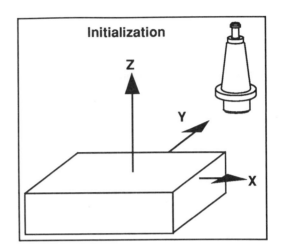

2) Facing

N80 G90 G43 G00 Z100.0 H14;
N90 G00 X-170.0 Y-70.0;
N100 G01 Z0.0 F500.0;
N110 G01 X170.0 F1000.0;
N120 Y-20.0;
N130 X-170.0;
N140 Y80.0;
N150 X170.0;
N160 Y80.0;
N170 X-170.0;
N180 G91 G28 Z0.0;
N190 G28 X0.0 Y0.0;
N200 M05;

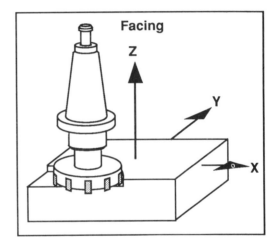

3) Deep side cutting

N210 T15;
N220 M06;
N230 G49;
N240 S1400 M03;
N250 G90 G43 G00 Z100.0 H15;
N260 G00 X-130.0 Y-70.0;
N270 G01 Z-10.0 F500.0;
N280 G42 G01 X-110.0 Y-70.0 D25
     F1000.0;
N290 G01 X110.0 F1000.0;
N300 Y70.0;
N310 X-110.0;
N320 Y-70.0;
N330 G01 G40 Y-100.0;
N340 Z10.0;

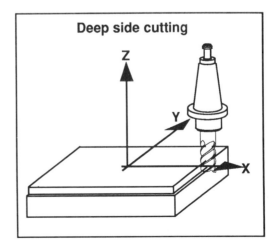

4) Track pocketing

N350 G00 X0.0 Y0.0;
N360 G01 Z-5.0 F500.0;
N370 G42 G01 Y-40.0 D25 F1000.0;
N380 X20.0;
N390 G03 X20.0 Y40.0 R40.0;
N400 G01 X-20.0;
N410 G03 X-20.0 Y-40.0 R40.0;
N420 G01 X10.0;
N430 G01 G40 X20.0 Y0.0;
N440 G01 Z10.0;

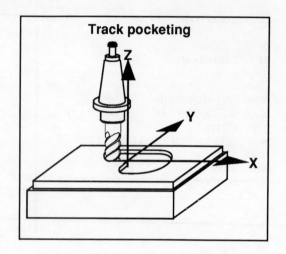

5) Side cutting

N450 G00 X-140.0 Y-70.0;
N460 G01 Z-5.0 F1000.0;
N470 G42 G01 X-90.0 Y-50.0 D25
      F1000.0;
N480 X65.0;
N490 X90.0 Y-25.0;
N500 Y20.0;
N510 G03 X60.0 Y50.0 R30.0;
N520 G01 X-60.0;
N530 G02 X-90.0 Y20.0 R30.0;
N540 G01 Y-50.0;
N550 G01 G40 X-110.0 Y-90.0;
N560 G01 Z10.0;
N570 G91 G28 Z0.0;
N580 G20 X0.0 Y0.0;

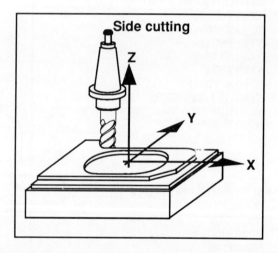

6) Hole pattern

N590 T16;
N600 M06;
N610 S1400 M03;
N620 G49;
N630 G90 G43 G00 Z100.0 H16;
N640 G01 X20.0 Y0.0 F1000.0;
N650 G01 Z-20.0 F500.0;
N660 G04 X1.0;
N670 G01 Z0.0;
N680 G00 Y17.32 X10.0;
N690 G01 Z-20.0 F500.0;
N700 G04 X1.0;
N710 G01 Z0.0;
N720 G01 X-10.0;
N730 G01 Z-20.0 F500.0;
N740 G04 X1.0;
N750 G01 Z0.0;
N760 G01 X-20.0 Y0.0;
N770 G01 Z-20.0 F500.0;
N780 G04 X1.0;
N790 G01 Z0.0;
N800 X-10.0 Y-17.32;
N810 G01 Z-20.0 F500.0;
N820 G04 X1.0;
N830 G01 Z0.0;
N840 G01 X10.0;
N850 G01 Z-20.0 F500.0;
N860 G04 X1.0;
N870 G01 Z0.0;

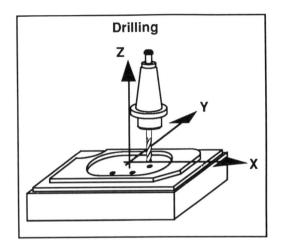

7) Program end

N880 G91 G28 Z0.0;
N890 G91 G28 X0.0 Y0.0;
N900 M05;
N910 G40;
N920 G49;
N930 M30;
%

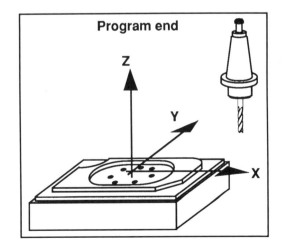

2.2 ADVANCED MACHINE CODES

2.2.1 Canned Cycles

A canned cycle is a fixed sequence of operations that can be used with a single G-code. It is used to reduce programming time on repetitive and commonly used machine operations. G73 to G89 are pre-programmed in the CNC controller memory. G80 is the canned cycle cancel function. All canned cycles are modal functions.

Canned Cycle Functions		
G-code	Operation at hole bottom	Function
G73	-	High speed peck drilling
G74	CW	Counter tapping cycle
G76	Stop	Fine boring cycle
G80	-	Canned cycle cancel
G81	-	Drilling cycle, spot boring
G82	Dwell	Drilling cycle, counter boring
G83	-	Peck drilling cycle
G84	CCW	Tapping cycle
G85	-	Boring cycle
G86	Stop	Boring cycle
G87	Stop	Boring cycle, back boring
G88	Dwell, stop	Boring cycle
G89	Dwell	Boring cycle

Canned cycles generally include the following six operations:

① Positioning on X and Y axes
② Rapid traverse to point R
③ Drilling, boring or tapping
④ Operation at the hole bottom
⑤ Return to R point level
⑥ Return to initial point level

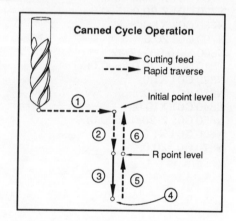

Canned cycles can be programmed with either G90 or G91. In G90 programming, point R and point Z are coordinate values on the Z-axis. In G91 programming, point R is an incremental distance measured from the current Z position, and point Z is an incremental distance measured from point R.

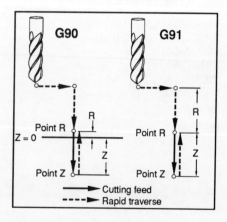

Sec. 2.2]   **Advanced Machine Codes**   57

G80 Canned Cycle Cancel
G80 ;

All canned cycles are modal and are cancelled with G80.

G81 Drilling Cycle, Spot Boring
G81 X_ Y_ Z_ R_ P_ F_ L_ ;

The tool is moved in rapid traverse to the X-Y position and then down to the R point. Drilling occurs between point R and Z at the specified feedrate. The tool is returned at rapid traverse to the initial point or the R point depending upon G98 or G99.

Block format (example)

N270 G90 G98 G81 X20. Y10. Z-5.0 R3.0 F100. ;

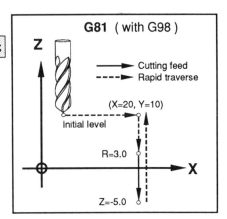

N270 Sequence number
G90 Absolute dimensions
G98 Initial point return level
G81 Drilling cycle , counter boring
X X-coordinate of hole position
Y Y-coordinate of hole position
Z Z-coordinate of hole bottom
R Z-coordinate of point R
F Feedrate

G82 Drilling Cycle, Counter Boring
G82 X_ Y_ Z_ R_ P_ F_ L_ ;

The tool is moved exactly the same as G81 except at the bottom of the hole a dwell, defined by address P, is performed.

Block format (example)

N270 G90 ;

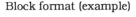
N280 G99 G82 X16 Y20. Z-5.0 R3.0 P1000 F100 ;

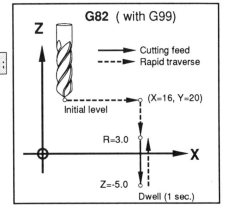

N280 Sequence number
G99 R point level return
G82 Drilling cycle, counter boring
X X-coordinate of hole position
Y Y-coordinate of hole position
Z Z-coordinate of hole bottom
R Z-coordinate of point R
P Dwell (milliseconds)
F Feedrate

G83 Peck Drilling Cycle
G83 X_ Y_ Z_ Q_ R_ F_ L_ ;

The tool is moved in rapid traverse to the X-Y position and then down to the R point. Drilling occurs between point R and Z in steps defined by address Q. This pecking amount is an incremental value. The tool changes from rapid traverse to the cutting feedrate a distance 'd' above the the last peck. The tool is returned at rapid traverse to the initial point or to the R point depending upon G98 or G99.

Block format (example)

N280 G90 ;

N290 G98 G83 X12. Y25. Z-5.0 Q2.0 R1.0 F100 ;

N290	Sequence number
G98	Initial point return level
G83	Peck drilling cycle
X	X-coordinate of hole position
Y	Y-coordinate of hole position
Z	Z-coordinate of hole bottom
Q	Peck amount
R	Z-coordinate of point R
F	Feedrate

There is no movement in the X or Y direction once the drilling has started

G84 Tapping Cycle
G84 X_ Y_ Z_ R_ F_ L_ ;

The tool is moved in rapid traverse to the X-Y position and then down to the R point. Tapping between point R and Z occurs in a clockwise direction, and between point Z and R occurs in a counter-clockwise direction.

Block format (example)

N300 G91 ;

N310 G98 G84 X10. Y10. Z-10. R-8. F100. L2 ;

N310	Sequence number
G98	Initial point return level
G84	Tapping cycle
X	X-distance to hole position
Y	Y-distance to hole position
Z	Z-distance from R point to hole bottom
R	Z-distance from initial level to R point
F	Feedrate
L	Repeat cycle twice

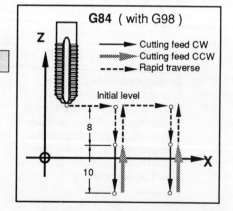

G85 Boring Cycle
G85 X__ Y__ Z__ R__ F__ ;

The tool is moved exactly the same as G84 except the spindle does not reverse direction at the bottom of the hole.

Block format (example)

N310 G91 ;

N320 G99 G85 X8. Y12. Z-10. R-8.F100. L2 ;

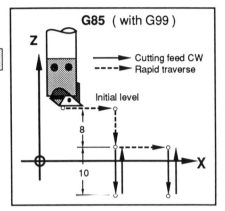

N320   Sequence number
G99   R point level return
G85   Boring Cycle
X   X-distance to hole position
Y   Y-distance to hole position
Z   Z-distance from R point to hole bottom
R   Z-distance from initial level to R point
F   Feedrate
L   Repeat cycle twice

Example of Canned Cycle

The following example is a sequence of holes all drilled to the same depth. The holes are spaced evenly apart.

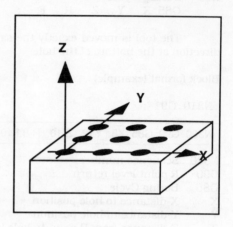

Block format (example for one hole)

```
N80  G90 .......
         .
         .
N150 G99 G81 X0.0 Y0.0 Z-5.0 R5.0 F100. ;
         .
         .
```

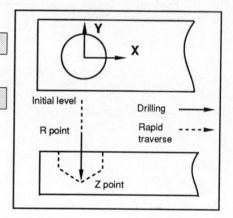

Block format (example for all holes)

```
N150 G99 G81 X0.0 Y0.0 Z-5.0 R5.0 F100. ;
N160 X67.5 ;
N170 X135. ;
N180 Y67.5 ;
N190 X67.5 ;
N200 X0.0 ;
N210 Y135. ;
N220 X67.5 ;
N230 G98 X135. ;
```

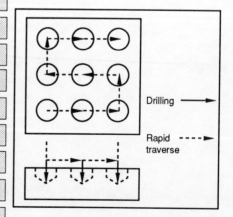

2.2.2 Subprograms
M98 P____ L__ ;

A subprogram is a sequence of operations which can be programmed by the user and called from within a program or another subprogram. It is used to reduce programming time on repetitive machine operations. A subprogram is stored in the CNC controller memory the same as a program. A subprogram is called by M98 P (subprogram #) and may be called many times within the same program. The address L specifies the number of times to repeat the subprogram. If L is not specified, the subprogram is carried out once.

A subprogram is like a canned cycle in that it is used for repetitive operations. However subprogram moves are fixed and are usually used for a specific program while canned cycles pass values which control location and depth.

Block format (example)

```
N210 M98 P1010 L3 ;
```

N100 Sequence number
M98 Call subprogram
P   Subprogram number
L   Repeat subprogram 3 times

```
N70 M99 ;
```

N70 Sequence number
M99 Return to the block in the main
     program after which the subprogram
     was called

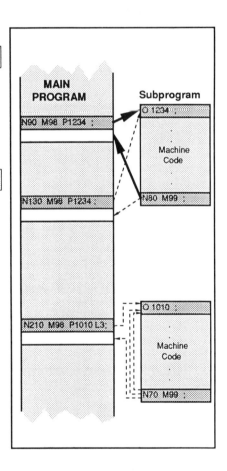

Example of Subprogram

The following program uses a subprogram to drill a set of holes, similar to the canned cycle example.

Main Program Part
N70 G90 G00 X0. Y0. Z50. ;
N30 M98 P1019 L3 ;

Subprogram
O 1019 ;
G91 ;
N20 G99 G81 X50. Y0. Z-20. R-40. L2 F100.;
N30 G98 X50. ;
N40 G00 X-150. Y50. ;
N50 M99 ;

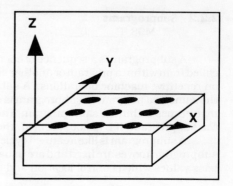

1. Call Subprogram

   N90 M98 P1019 L3 ;

   N90   Sequence number
   M98   Call subprogram
   P   Subprogram number
   L   Repeat three times

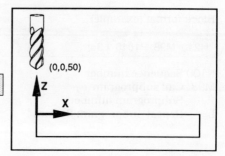

2. Subprogram Part

   N20 G99 G81 X50. Y0. Z-20. R-40. L2 F100. ;

   N20   Sequence number
   G99   R point return level
   G81   Drilling cycle
   X, Y X and Y distance to hole position
   Z   Z-distance from R point to Z point
   R   Z-distance from initial level to R point
   L   Repeat two times
   F   Feedrate

   N30 G98 X50. ;

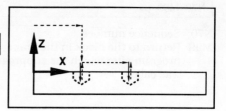

   N30   Sequence number
   G98   Initial point return level
   X   X distance to hole position

   N40 G00 X-150. Y50. ;

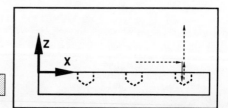

   N40   Sequence number
   G00   Positioning
   X, Y X and Y distance to start next row of holes

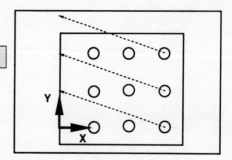

2.2.3 Custom Macros
```
G65 P___ L_ A_ B_ C_ ... X_ Y_ Z_ ;
G66 P___ L_ A_ B_ C_ ... X_ Y_ Z_ ;
G67 ;
```

A custom macro is a sequence of operations, similar to a subprogram. The difference between the two is that a custom macro is passed variables from within the main program and is therefore more powerful than a subprogram. Some custom macros, depending upon the macro number, can be programmed by the user and some are predefined and stored in the CNC controller memory.

Variables

In macro programming, a variable is formed by placing a number after the symbol '#'. That is, #110 is variable 110. A variable can be placed after an address so that the address is filled with the value of the variable.

A variable which has been undefined is called <vacant>. The variable #0 is always defined as <vacant>.

There are three different types of variables: local, common and system.

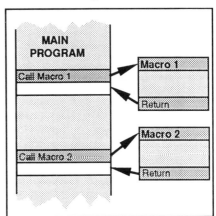

Local Variables   [#1 - #33]

A local variable is a variable used locally in a macro. It does not keep its value throughout the program. Local variables are used for argument transfer, as they are vacant in their initial status.

Common Variables   [#100 - #149]
                    [#500 - #509]

Common variables are the same throughout the entire program. They can be freely determined by the user. Variables #100 through #149 are cleared when the power is turned off, whereas variables #500 through #509 are saved in memory.

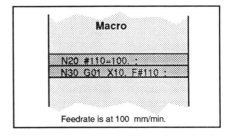

Feedrate is at 100 mm/min.

System Variables [#1000 - #5105]

System variables cannot be freely defined by the user as their locations are reserved in memory.

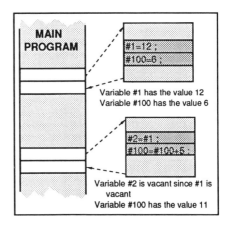

Variable #1 has the value 12
Variable #100 has the value 6

Variable #2 is vacant since #1 is vacant
Variable #100 has the value 11

G65 Custom Macro Simple Call

G65 P___ L_ A_ B_ C_ ... X_ Y_ Z_ ;
G65 P___ L_ A_ B_ C_ I_ J_ K_ I_ J_ K_ ... ;

A custom macro with a macro number defined by address P can be called from within a program by G65. Macro numbers, or P values, range from 9000 to 9899. Address A through Z (omitting G, L, O, N, and P) can be used for the transfer of variables, as can variables A, B, C and ten sets of I, J and K. Addresses not needed can be omitted.

Block format (argument assigment I example)

```
N80 G65 P9050 L2 A30. C10. J18. W2. ;
```

N80 Sequence number
G65 Custom macro simple call
P   Macro number
L   Repeat two times
A   Transfer value 30 to variable #1
C   Transfer value 10 to variable #3
J   Transfer value 18 to variable #5
W   Transfer value 2 to variable #23

Argument Assigment I

Address	Variable	Address	Variable
A	#1	Q	#17
B	#2	R	#18
C	#3	S	#19
D	#7	T	#20
E	#8	U	#21
F	#9	V	#22
H	#11	W	#23
I	#4	X	#24
J	#5	Y	#25
K	#6	Z	#26
M	#13		

Block format (argument assigment II example)

```
N90 G65 P9051 B5. I3. K7. I11. K4. K8. J12. ;
```

N90 Sequence number
G65 Custom macro simple call
P   Macro number
B   Transfer value 5 to variable #2
I   Transfer value 3 to variable #4
K   Transfer value 7 to variable #6
I   Transfer value 11 to variable #7
K   Transfer value 4 to variable #9
K   Transfer value 8 to variable #12
J   Transfer value 12 to variable #5

Argument Assigment II

Address	Variable	Address	Variable
A	#1	K_5	#18
B	#2	I_6	#19
C	#3	J_6	#20
I_1	#4	K_6	#21
J_1	#5	I_7	#22
K_1	#6	J_7	#23
I_2	#7	K_7	#24
J_2	#8	I_8	#25
K_2	#9	J_8	#26
I_3	#10	K_8	#27
J_3	#11	I_9	#28
K_3	#12	J_9	#29
I_4	#13	K_9	#30
J_4	#14	I_{10}	#31
K_4	#15	J_{10}	#32
I_5	#16	K_{10}	#33
J_5	#17		

Block format (argument assignment I and II examples)

```
N100 G65 P9052 C20. J2. J3. E54. ;
```

N100 Sequence number
C   Transfer value 20 to variable #3
J   Transfer value 2 to variable #5
J   Transfer value 3 to variable #8
E   Transfer value 54 to variable #8

If two addresses are assigned to the same variable, only the latter one is used. Therefore variable #8 has the value of 54.

Sec. 2.2]   **Advanced Machine Codes**   65

Example of Custom Macro Simple Call

The following program uses a macro to drill a set of holes similar to the canned cycle and subprogram examples.

Main Program Part
N50 G90 X0. Y0. ;
N60 G00 Z20. ;
N70 G65 P9090 L3 X50. Y0. Z-10. R-15.
    T1000 F100 A3 ;

Macro
O 9090 ;
N10 G91 ;
N20 G99 G82 X#24 Y#25 Z#26 R#18 P#20
    F#9 L#1 ;
N30 G00 X-[3*#24] Y#24 ;
N40 G90 M99 ;
%

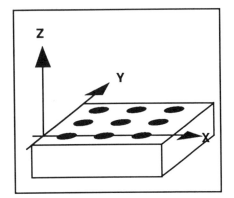

1. Call Custom Macro

   N70 G65 P9090 L3 X50. Y0. Z-10. R-15.

   T1000 F100 A3 ;

   N70 Sequence number
   G65 Custom macro simple call
   P   Macro number
   L   Repeat macro three times
   X   X distance to hole position (variable #24)
   Y   Y distance to hole position (variable #25)
   Z   Z distance from point R to hole bottom
        (variable #26)
   R   Z distance from initial level to point R
        (variable #18)
   T   Dwell (variable #20)
   F   Feedrate (variable #9)
   A   Repeat canned cycle (variable #1)

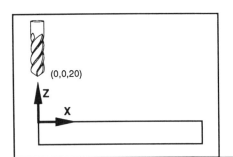

2. Custom Macro Part

   N20 G99 G82 X#24 Y#25 Z#26 R#18 P#20

   F#9 L#1 ;

   N20   Sequence number
   G99   R point return level
   G82   Drilling cycle, counter boring
   X, Y X and Y distance to hole position
   Z   Z distance from R point to hole bottom
   R   Z distance from initial level to R point
   P   Dwell at hole bottom
   F   Feedrate
   L   Repeat canned cycle 3 times

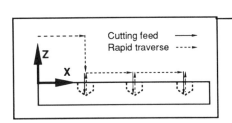

```
N30  G00  X-[ 3*#24 ]  Y#24 ;
```

N30   Sequence number
G00   Positioning
X, Y X and Y distance to start new row of
      holes (-150,50)

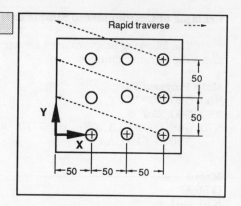

Sec. 2.2]   **Advanced Machine Codes**   67

G66, G67 Custom Macro Modal Calls
    G66 P___ L__ A__ B__ C__ ... X__ Y__ Z__ ;
    G66 P___ L__ A__ B__ C__ I__ J__ K__ I__ J__ K__ ... ;
    Custom macro modal call
    G67 ;
    Custom macro modal cancellation

A custom macro with a macro number defined by address P is called after every block in the main program in which motion occurred. G66 starts the modal call and G67 is used to cancel the modal call. G67 can use the same two types of addresses as G65.

Block format (custom macro modal call example)

    N110 G66 P9053 L2 X10. Y14. Z-10. F100. ;

N110 Sequence number
G66 Custom macro modal call
L   Repeat two times after each block with motion
X   Transfer value 10 to variable #24
Y   Transfer value 14 to variable #25
Z   Transfer value -10 to variable #26
F   Transfer value 100 to variable #9

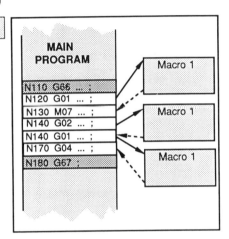

Block format (custom macro modal cancel example)

    N160 G67 ;

N160 Sequence number
G67 Custom macro modal cancellation

Example of Custom Macro Modal Call

The following program drills a set of holes similar to the custom macro simple call example.

Main Program Part
N50 G90 G00 X0. Y0. ;
N60 G00 Z20. ;
N70 G66 P9081 L3 X50. R-15. Z-10. F100. ;
N80 G00 X50. Y30. ;
N90 G00 X50. Y80. ;
N100 G00 X50. Y130. ;
N110 G67 ;

Macro
O9081 ;
N10 G91 ;
N20 G00 Z#18 ;
N30 G01 Z#26 F#9 ;
N40 G00 Z-[#18 + #26] ;
N50 G00 X#24 ;
N60 G90 M99 ;
%

1. Custom Macro Information

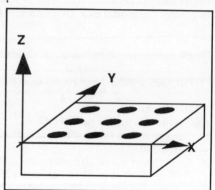

| N70 G66 P9081 L3 X50. R-15. Z-10. F100. ; |

N70 Sequence number
G66 Custom macro modal call
P   Macro number
L   Repeat macro three times
X   X distance to next hole position (variable #24)
R   Z distance to R point (variable #18)
Z   Z distance to hole bottom (variable #26)
F   Feedrate (variable #9)

2. Call Custom Macro

| N80 G00 X50. Y30. ; |

| N90 G00 X50. Y80. ; |

| N100 G00 X50. Y130. ; |

After each one of these block, the custom macro is called.

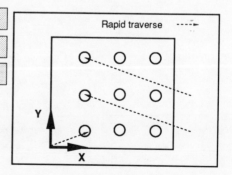

Sec. 2.2] **Advanced Machine Codes** 69

3. Custom Macro Part

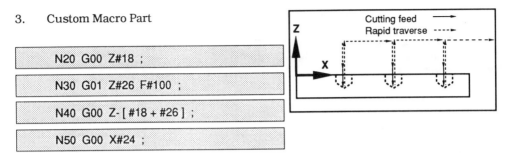

```
N20 G00 Z#18 ;
N30 G01 Z#26 F#100 ;
N40 G00 Z-[ #18 + #26 ] ;
N50 G00 X#24 ;
```

The above blocks perform the same operations as the canned cycle G82.

Manual Part Programming [Ch. 2]

Example of Macro and Arthimetic Operations

The following program uses a simple macro call to drill holes evenly spaced along a circular

Main Program Part
N50 G90 G00 X0. Y0. ;
N60 G00 Z50. ;
N70 G65 P9100 X0. Y0. R-10. Z-15. D60.
    A20. K18. T1000 ;

Macro
O 9100 ;
N10 #101 = #24 ;
N20 #102 = #25 ;
N30 #30 = 1. ;
N40 WHILE [#30 LE ABS [#6]] DO 1 ;
  N50 #31 = #1 + 360. * [#30 - 1] / #6 ;
  N60 #101 = #24 + [#7 / 2.] * COS [#31] ;
  N70 #102 = #25 + [#7 / 2.] * SIN [#31] ;
  N80 G90 G00 X#101 Y#102 ;
  N90 G91 G98 G82 X0. Y0. Z#26 R#18 P#20 ;
  N100 #30 = #30 + 1. ;
N110 END 1 ;
N120 G00 Z- #18 ;
N130 G90 M99 ;
%

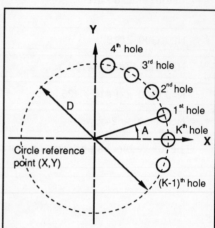

1. Call Custom Macro

| N70 G65 P9100 X0. Y0. R-10. Z-15. D60. A20. |
| K18. T1000 ; |

N70 Sequence number
G65 Custom macro simple call
P   Macro number
X   X coordinate of circle centre (variable #24)
Y   Y coordinate of circle centre (variable #25)
R   Z distance from initial level to R point (variable #18)
Z   Z distance from R point to hole bottom (variable #26)
D   Diameter of circle (variable #7)
A   Start angle in degrees (variable #1)
K   Number of holes (variables #6)
T   Dwell (variable #20)

2. Custom Macro Part

N10 #101 = #24 ;   Store X value of circle centre
N20 #102 = #25 ;   Store Y value of circle centre
N30 #30 = 1. ;   Hole counter
N40 WHILE [#30 LE ABS [#6]] DO 1 ;   While loop, repeats until drilled all holes
  N50 #31 = #1 + 360. * [#30 - 1] / #6 ;   Calculate angle above the horizontal
  N60 #101 = #24 + [#7 / 2.] * COS [#31] ; X coordinates of hole position
  N70 #102 = #25 + [#7 / 2.] * SIN [#31] ; Y coordinates of hole position
  N80 G90 G00 X#101 Y#102 ;   Drill hole
  N90 G91 G98 G82 X0. Y0. Z#26
       R#18 P#20 ;
  N100 #30 = #30 + 1. ;   Counter increased
N110 END 1 ;   End of while loop

Chapter 3
Computer-Assisted Programming

Computer assisted programming methods are much faster and more reliable than manual programming techniques. There are a variety of forms of computer assisted programming. The common feature of these programs is that the part and machining paths are not defined directly with G-code but through English-like statements or through interactive graphic instructions.

In this chapter three types of computer assisted programming methods are discussed. The first method is called APT (Automatically Programmed Tools) and is presented in Section 3.1. It is the original computer assisted method of part programming and has been widely used over the last two decades. It consists of a series of English-like geometry and motion statements, which are used to define the workpiece and machine tool path. An extensive (usually mainframe) computer program is then used to convert the APT instructions into G-code programs.

This is followed, in Sections 3.2 and 3.3, by two examples of interactive computer graphics-based programs, which are supplied by machine tool manufacturers. The programs, called Symbolic FAPT for lathes and Conversational Programming for mills, run directly on the controllers of CNC machine tools. They are fairly comprehensive and enable the part figure, blank shape, and machine tool path to be specified graphically at the control console. They allow for input of workpiece material, tool shape, surface roughness etc., and enable feeds and speeds to be automatically generated and manually overridden.

The third method, given in Section 3.4, is an interactive, PC-based graphics package. The program, called SmartCAM, is typical of graphics-based CAM packages in which the machine tool path is developed from part definition generated with CAD packages such as AutoCAD or Personal Designer.

3.1 APT

APT uses English-like language statements to define part shape and tool motion as well as machine tool dependent data, such as feedrates and spindle speeds. This data is contained in an APT part-program. An APT processor program is used to read these statements, interpret the meanings, and perform all the necessary calculations in order to generate a series of cutter location points that define the cutter path. The APT processor is a computer program which runs on a mainframe computer, possibly at a central site with time-sharing facilities. The generalized APT output is converted to the particular format G-code required by the CNC machine using a postprocessor program. Although G-codes are fairly well standardized, differences do exist between machine suppliers and CNC machines can be supplied with a variety of subsets of available codes. A description of postprocessor programs is given in Chapter 5.

APT is a three-dimensional system which can be used to define complex geometrical shapes and to control up to five axes CNC machines. A major advantage of APT is that it has developed into an accepted standard for machine tools. There are many versions of the APT language available, each with particular benefits and characteristics. The prime disadvantage of APT is that it is uses English-like commands to define geometry instead of the much more convenient graphical methods. Other disadvantages are that it requires extensive computing capability and can have a slow response (particularly with programmers who use the system intermittently and require multiple overnight runs on time-shared facilities).

The four types of statements in the APT language are:
- geometry statements which define primitive elements such as points, lines, circles, planes, cones and spheres
- motion statements which describe the tool path in relation to the part geometry
- postprocessor statements which give specific machine tool code information as well as feeds and speeds
- auxiliary statements which give part and tool tolerances.

APT part programs usually list the part and the postprocessor reference number followed by the program statements as follows:
   PARTNO___
   MACHIN/___
   Geometry statements
   Motion statements and machine tool commands
   FINI.

3.1.1 Geometry Statements
The general form of geometry statements is:

symbol = geometry type / descriptive data

For example a point, named P1, can be defined at coordinates X = 100 mm, Y = 200 mm and Z = 300 mm by the statement:
   P1 = POINT/100.0, 200.0, 300.0
A line, named L1, can be defined between two points, P1 and P2, by the statement:
   L1 = LINE/P1, P2
or through the point, P1, and parallel to another line, L2, by the statement:
   L1 = LINE/P1, PAREL, L2

A plane, named PL1, can be defined between three points, P1, P2 and P3, by the statement:
    PL1 = PLANE/P1, P2, P3
or through the point, P1, and parallel to another plane, PL2, by the statement:
    PL1 = PLANE/P1, PAREL, PL2
A circle, named C1, can be specified by a centre position, P1, and a radius of 100 mm by the statement:
    C1 = CIRCLE/CENTER, P1, RADIUS, 100.0
or by a centre position, P1, and tangent to line, L1, by the statement:
    C1 = CIRCLE/CENTER, P1, TANTO, L1.

3.1.2 Motion Statements

Motion statements use the defined geometry primitives in order to define tool movement. The general form of motion statements is:

    motion statement/descriptive data

The instruction to go from an initial starting point, P1, and from which all other points are referenced, is given by the statement:
    FROM/P1
The instruction to go to the point, P2, is given by the statement:
    GOTO/P2
or to go to the absolute position X = 100 mm, Y = 200 mm, and Z = 300 mm is given by the statement:
    GOTO/100.0, 200.0, 300.0
The instruction to move by an incremental (or delta) amount from the present position is given by the statement:
    GODLTA/100.0, 200.0, 300.0

Contouring commands are used to direct the tool along two intersecting surfaces. The **drive surface** guides the side of the cutter, the **part surface** defines the position of the bottom of the cutter, and the **check surface** defines the limit of current tool motion. Modifier words, such as TO, ON, PAST or TANTO, are used to govern the position of the tool in relation to the check surface, as shown in Figure 3.1. Motion statements, GOLFT (go to the left), GOFWD (go forward) and GOUP (go up), are also used to control the cutter motion.

The instruction for the tool to move forward, with the drive surface, S1, on the left hand side, and past the check surface, S2, is given by the statement:
    GOLFT/S1, PAST, S2.

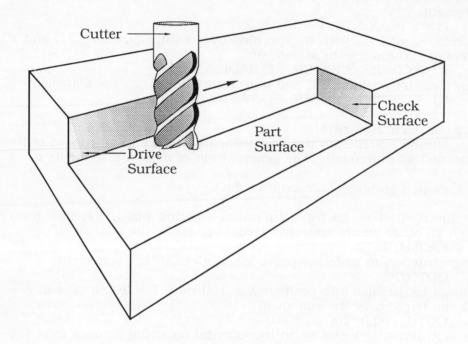

Figure 3.1 APT contouring surfaces

3.1.3 Postprocessor and Auxiliary Statements

Statements which specify machine tool related functions, such as those covered by F-, S-, T-, and M-codes, are defined in the postprocessor statements. For example, the instruction to set the feedrate at 100 mm per minute is given by the statement:
  FEDRAT/100, MPM
The instruction to set the spindle at 1500 rpm in a clockwise direction is given by the statement:
  SPINDL/1500, RPM, CLW
The instruction to set the coolant on is given by the statement:
  COOLNT/ON

Other auxiliary statements are used to define cutter size, part number, and curve tolerance. The instruction to define a 50 mm diameter cutter is given by the statement:
  CUTTER/50.0

Sec. 3.1]　　　　　　　　　　　**Apt**　　　　　　　　　　　　75

3.1.4 Examples of APT Program

An APT program for the milling example given in Section 2.1.5, and shown in the following figure is given below. It is divded into five subprograms as follows:

1) APT facing part program
2) APT deep side cutting part program
3) APT track pocketing part program
4) APT side cutting part program
5) APT hole pattern part program

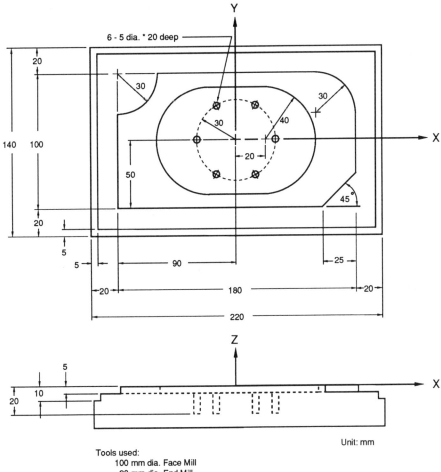

Unit: mm

Tools used:
　100 mm dia. Face Mill
　20 mm dia. End Mill
　5 mm dia. Drill

1) APT Facing Part Program

```
PARTNO      FACING

REMARK      GEOMETRIC DEFINITION
REMARK      POINT TO POINT
            PROGRAMING
SETPT =  POINT/-170, -70, 100
P1       =  POINT/-170, -70, 0
P2       =  POINT/170, -70, 0
P3       =  POINT/170, 0, 0
P4       =  POINT/-170, 0, 0
P5       =  POINT/-170, 70, 0
P6       =  POINT/170, 70, 0

REMARK       MOTION COMMAND
             CUTTER/100
             SPINDL/1200,CLW
             FEDRAT/500
             FROM/SETPT
             GOTO/P1
             GOTO/P2
             GOTO/P3
             GOTO/P4
```

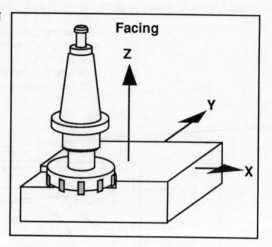

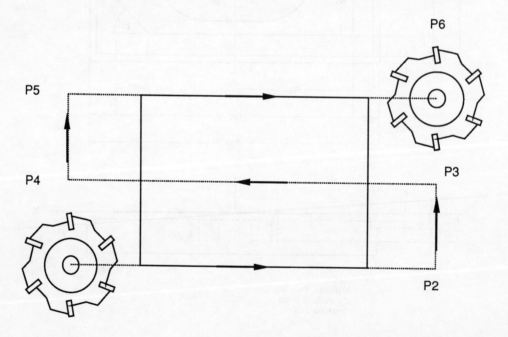

2) APT Deep Side Cutting Program

```
PARTNO      DEEP SIDE CUTTING

REMARK      GEOMETRIC DEFINITION
REMARK      CONTINUOUS PART
            PROGRAMING
SETPT =  POINT/-120, -75, 100
PA       =  POINT/-105, -65
PB       =  POINT/105, -65
PC       =  POINT/105, 65
PD       =  POINT/-105, 65
BASLIN   =  LINE/PA,PB
RITSID=  LINE/PB,PC
TOPLIN   =  LINE/PC,PD
LFTSID   =  LINE/PD,PA
XYPLN =  PLANE/PA,PB,PC
PSURF =  PLANE PARLEL,
         XYPLN,ZSMALL,10

REMARK      MOTION COMMAND
            CUTTER/20
            SPINDL/1200,CLW
            FEDRAT/500
            FROM/SETPT
            GO   /TO,BASLIN,TO,PSURF,
             ON,LFTSID
            GO FWD/BASLIN,PAST,RITSID
            GO LFT/RITSID,PAST,TOPLIN
            GO LFT/TOPLIN,PAST,LFTLIN
```

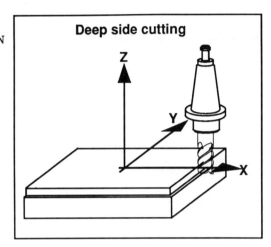

Deep side cutting

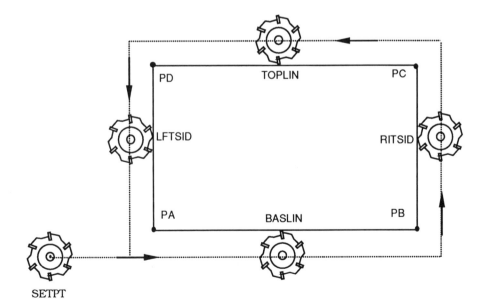

3) APT Track Pockecting Program

PARTNO   POCKET

REMARK   GEOMETRIC DEFINITION
SETPT   = POINT/-20,-.95,100
S1   = POINT/-20,-.95,-5
S2   = POINT/20,-.95,-5
S3   = POINT/-20,.95,-5
S4   = POINT/-20,-10.64,-5
S5   = POINT/20,-10.64,-5
S6   = POINT/-20,-30,-5
S7   = POINT/20.-30,-5
C1=CIRCLE/20,0,-5,.95
L1=LINE/20,0,-5,20,.95,-5
C2=CIRCLE/-20,0,-5,.95
L2=LINE/-20,0,-5,S1
C3=CIRCLE/20,0,-5,10.64
L3=LINE/20,0,-5,20,10.64,-5
C4=CIRCLE/-20,0,-5,10.64
L4=LINE/-20,0,-5,-20,-10.64,-5
C5=CIRCLE/20,0,-5,30
L5=LINE/20,0,-5,20,30,-5
C6=CIRCLE/-20,0,-5,30
L6=LINE/-20,0,-5,-20,-30,-5

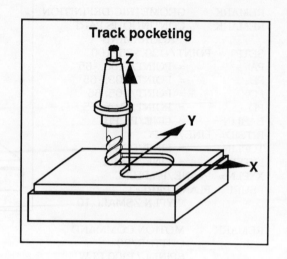

Track pocketing

REMARK   MOTION COMMAND
    CUTTER/20
    SPINDL/1200,CLW
    INTOL/.1
    OUTTOL/.1
    FEDRAT/500.00
    FROM/SETPT
    GOTO/S1
    GOTO/S2
    AUTOPS
    DNTCUT
        GODLTA/0,-.01,0
        GO/ON,C1
    CUT
    INDIRV/1,0,0
    TLON,GOFWD/C1,ON,L1
    GOTO/S3
    AUTOPS
    DNTCUT
        GODLTA/-0,.01,0
        GO/ON,C2
    CUT
    INDIRV/-1,-0,0
    TLON,GOFWD/C2,ON,L2
    GOTO/S4
    AUTOPS
    DNTCUT
        GODLTA/-0,.01,0
        GO/ON,C4
    CUT
    INDIRV/-1,-0,0
    TLON,GOFWD/C4,ON,L4
    GOTO/S6
    GOTO/S7
    AUTOPS
    DNTCUT
        GODLTA/-0,-.01,0
        GO/ON,C5
    CUT
    INDIRV/1,-0,0
    TLON,GOFWD/C5,ON,L5
    GOTO/-20,30,-5
    AUTOPS
    DNTCUT
        GODLTA/-0,.01,0
        GO/ON,C6
    CUT
    INDIRV/-1,-0,0
    TLON,GOFWD/C6,ON,L6
    GOTO/SETPT
    FINI

Sec. 3.1] Apt 79

4) APT Side Cutting Program

PARTNO   SIDE CUT

REMARK   GEOMETRIC DEFINITION
REMARK   CONTINUOUS PART
          PROGRAMING
PS   = POINT/-110, -60, 100
P1   = POINT/-90, -50
P2   = POINT/65, -50
P3   = POINT/90,-25
P4   = POINT/90,20
P5   = POINT/60,50
P6   = POINT/-60,50
P7   = POINT/-90,20
CTR1   = POINT/60,20
CTR2   = POINT/-90,50
CIRC1   = CIRCLE/CENTER,CTR1,
          RADIUS,30
CIRC2   = CIRCLE/CENTER,CTR2,
          RADIUS,30
L1   = LINE/P1,P2
L2   = LINE/P2,P3
L3   = LINE/P3,P4
L5   = LINE/P5,P6
L7   = LINE/P7,P1

XYPLANE = PLANE/P1,P2,P3
ZSURF   = PLANE/PARLEL,XYPLANE,
          ZSMALL,5

REMARK   MOTION COMMAND
          CUTTER/20
          SPINDL/1200,CLW
          FEDRAT/500
          FROM/SETPT

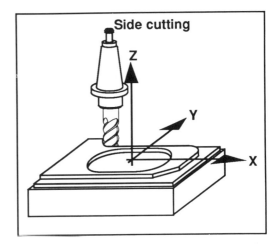

GO   /TO,L1,TO,ZSURF,ON,L7
GO FWD/L1,PAST,L2
GO LFT/L2,PAST,L3
GO LFT/L3,TANTO,CIRC1
GO FWD/CIRC1,TANTO,L5
GO FWD/L5,TANTO,CIRC2
GO FWD/CIRC2,PAST,L7
GO FWD/L7,PAST,L1
FEDRAT/2000
GOTO/SETPT
STOP
FINI

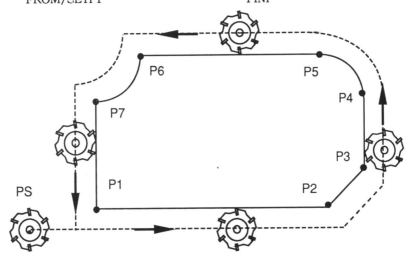

5) APT Drilling Program

PARTNO   DRILL SIX HOLES

CUTTER/5
SPINDL/2000,CLW
TOOLNO/small mill
FROM/0.0,0.0,20
CYCLE/DRILL,20.00,MMPR,500.00,
       TRAV,5.0
GOTO/30,0,0.0
GOTO/15,25.98,0.0
GOTO/-15,25.98,0.0
GOTO/-30,-0,0.0
GOTO/-15,-25.98,0.0
GOTO/15,-25.98,0.0
CYCLE/OFF
GOTO/0.0,0.0,20
STOP
FINI

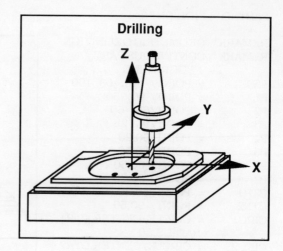

Drilling

3.2 SYMBOLIC FAPT

Symbolic FAPT is produced by FANUC for use with the lathe controller. It is a computer assisted programming system which allows part programs to be made in response to questions and instructions on a graphic CRT display.

Each process, from design drawings to G-code programming, is made throught the interactive operation with the graphic CRT display.

Symbolic FAPT allows for selection of part shape, material, component shape, machine home position and machining process. G-codes are then automatically generated. The FAPT menu selections are given below and discussed in greate depth in Section 3.2.1. A FAPT programming example is given in Section 3.2.2.

The initial menu selections are given as follows.

1. Blank and Part (Drawing & Blank)

- Selection of material
- Setting of standard surface roughness
- Selection of drawing format
- Specification of blank figure

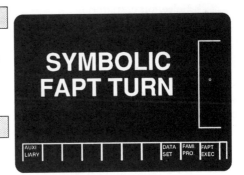

2. Blank and Part (Part Figure)

- Plotting of program coordinate axes and blank figure
- Part figure input
- Input, modification of part figure data
- Blank figure input for special blank

3. Home Position & Index Position

- Setting of home position
- Setting of index position

4. Definition of Machining

- Kind of process
- Specification of multiprocess
- Modification and change of process
- Output of single process
- Setting of data for machining
- Definition of machining figure

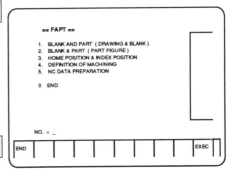

5. NC Data Preparation

- NC data prepared
- Tool path drawn

Computer Assisted Programming [Ch.3]

3.2.1 Execution of Symbolic FAPT

The five-step procedure of Symbolic FAPT given below is normally followed sequentially although any section can be chosen.

1. Blank and Part (Drawing & Blank)

☞ 1 INPUT Blank and Part (Drawing & Blank)

- Selection of material

  Selection is based on the material to be used.

  Example

  MN = 4

  MN Material selection (aluminum)

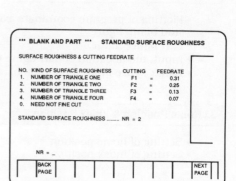

- Setting of standard surface roughness

  Selection is based according to the number of triangle ▽ marks attached to machining drawings. The lower the number of triangles, the rougher the surface.

  Example

  NR = 2

  NR Surface roughness (▽▽)

- Selection of drawing format

  Selection is based on the orientation of the machine coordinate system.

  Example

  DF = 1

  DF Drawing format

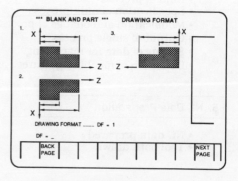

Sec. 3.2] Symbolic FAPT

- Specification of blank figure

Selection is based on the geometry of the blank.

Blank Figure .....BF -round, hollow or special figure
Blank Size .....L -length of blank
D -diameter of blank
DO -inner diameter of hollow cylinder
Base Line .....ZP -where origin exists in blank
Thickness .....TX -surplus thickness in special figure (X component)
TZ -surplus thickness in special figure (Z component)

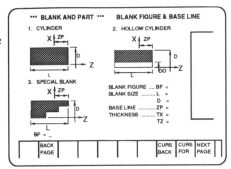

Example

BF = ⎡1⎤   L = ⎡147⎤   D = ⎡51⎤   ZP = ⎡21⎤

BF Cylinder type
L Cylinder length
D Cylinder diameter
ZP Distance of origin from the front of blank

2. Blank and Part (Part Figure)

  ⎡2⎤ ⎡INPUT⎤ Blank and Part

- Plotting of program coordinate axes and blank figure

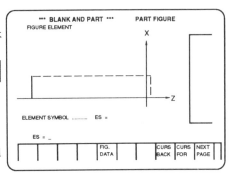

The blank above the X axis is plotted. The solid line indicates the position of the chuck.

- Part figure input

The part figure is input using the symbol keys sequentially along the profile of the part. The ten arrow keys for direction input and the keys indicating chamfering, rounding, threading, grooving and necking are used to

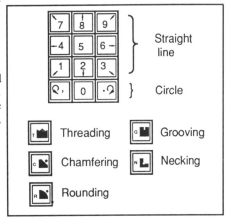

Symbol key example

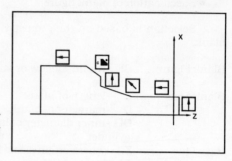

• Input and modification of part figure data

The numerical data required after selecting any of the symbol keys can be input by arthimetic operations and as incremental values. Editing of data is also possible.

- Arithimetic operation and function calculation

An arithmetic operation can be used to input dimensions. Addition, subtraction, multiplication, division and optional functions such as sine, cosine, tangent can be used. Angles are measured in degrees.

Symbol	Function	Example
[+]	Addition	15 [+] 3.8
[-]	Sub.	23.6 [-] 12.7
[•]	Mult.	6.1 [•] 2.32
[/]	Division	34.9 [/] 3.4
[T]	tan	[T] 23
[S]	sin	[S] 12
[C]	cos	[C] 67
[A][S]	arc sin	[A][S] 0.26
[A][C]	arc cos	[A][C] 0.44
[A][T]	arc tan	[A][T] 0.37
[R]	sqrt.	[R] 173
[P]	power	[P] [(] 2.6 , 5.2 [)]

- Input of incremental value

Dimensions can be input by specifying them as incremental values. The letter [I] is selected after the dimension to specify it as incremental. Distances in the X axis direction are then radius values.

Incremental
50.4 [I]

- Change of figure element symbol

Changes can be made to the part figure by inserting, deleting or changing any of the

- Change of numerical data

Symbolic FAPT asks for numerical data after the input of a symbol key. If a numerical value was input incorrectly, the Cursor Forward and Cursor Backward soft keys are used to return to the incorrect value.

Press [R▶] with [◣] flickering will insert

Press [DEL] with [↑] flickering will delete

Sec. 3.2] Symbolic FAPT

- Blank figure input for special blank

When the blank figure is of a special shape (BF = 3) and values for TX and TZ were not input, and after the part figure was input, the system will ask for more information about the shape.

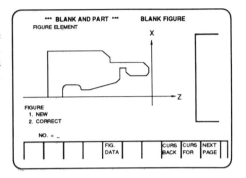

3. Home Position and Index Position

 3 INPUT Home Position and Index Position

- Setting of home position

The machine home position is used for indicating the relation between the program coordinate system and the machining coordinate system.

Example

DXH = 100   ZH = 100

DXH X diameter distance to home position
ZH Z distance to home position

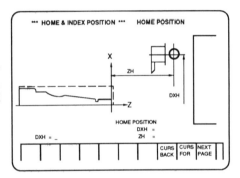

- Setting of index position

The index position indicates the position for a tool change and is usually given the same coordinates as the home position.

Example

DXH = 100   ZH = 100

DXH X diameter distance to index position
ZH Z distance to index position

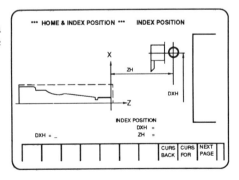

4. Definition of Machining

 4 INPUT Definition of Machining

• Kind of machining process

Selection is based on the kind of machining required. More than one machining process is possible. After selecting a process, tool data and the machining area are input.

Example

PROC. = 3

PROC Machining process

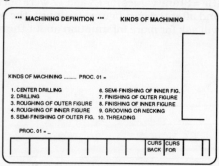

• Setting of data for machining

- Tool data (T code)

A suitable tool is chosen for the machining process. If the tool number and tool offset values are registered in the tool table in the CNC controller memory, only the tool identification number need be specified.

Example

ID = 1   TN = 1   TM = 1

ID Identification number
TN Tool selection number
TM Tool offset number

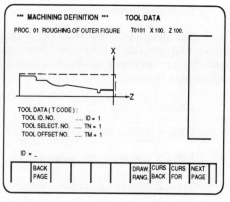

- Machining start position

The start position for the machining process is required. This position is usually programmed the same as the home and index positions.

Example

DXO = 100   DZ = 100

DXO X diameter coordinate of start position
DZ   Z coordinate of start position

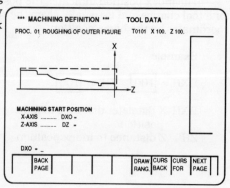

Sec. 3.2]   **Symbolic FAPT**   87

- Cutting conditions

The cutting conditions such as depth of cut, spindle speed and feedrate for the machining process are required.

Example

CX = [2]   CZ= [2]   TX = [1]   TZ = [1]

D = [2]   U = [2]   V= [1500]   F = [1.5]

CX   Clearance quantity in X axis
CZ   Clearance quantity in Z axis
TX   Finish allowance in X axis
TZ   Finish allowance in Z axis
D   Depth of cut
U   Return amount quantity

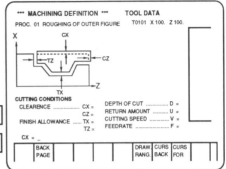

• Designation of machining area

- Cutting direction

The cutting direction of the machining process is required. When the cutting direction differs in the end face and outer diameter, the end face should be cut first.

Example

CD = [↓]

CD   Cutting direction

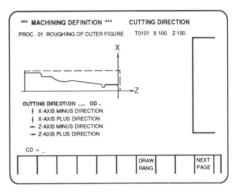

- Cutting area definitions

The cursor (■) displayed along the part figure can be moved with the Cursor Forward and Cursor Backward soft keys to specify the cutting area.

After specifying the cutting area, another cutting area can be chosen if it is to be done with the same tool. This is used when the cutting directions in the end face and the outer diameter are different.

Example

CN = [1]

CN   Choice to machine another area with the same tool

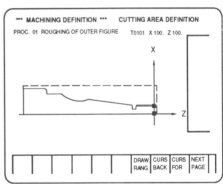

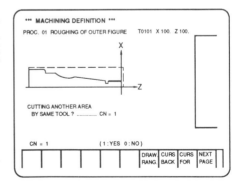

● Specification of multiprocess

Another machining process can be selected after all the information is input for the first one. When more than one machining process is needed, the input order determines the machining order.

Example

PROC. = 7

PROC Machining process

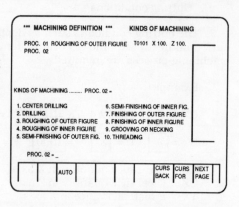

● Modification and change of process

The machining process can be altered by cancelling them all or by editing a single process.

Example

NO. = 2

NO Machining plan correction

When NEW is selected, all processes are cancelled and new ones can be specified. After choosing CORRECTION, the machining process order can be changed by inserting or deleting a process, or by using the Cursor Forward and Cursor Backward soft keys.

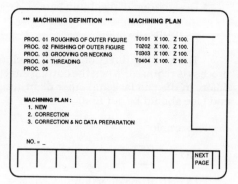

● Output of single process

By selecting CORRECTION & NC DATA PREPARATION, all the defined processes are displayed. Select the single process to be output by Cursor Forward and Next page soft keys. By this function, only one process NC data can be output with corrections, after all processes are defined, if necessary.

Example

NO. = 3

NO. Machining plan correction and NC data preparation

Sec. 3.2]   **Symbolic FAPT**   89

5. NC Data Preparation

   5   INPUT   NC Data Preparation

- NC data preparation

  A program number is entered. To prepare NC data according to the machining process order select either the Start or Register soft key. Both will prepare NC data and draw the tool path on the screen but the NC data is registered in the machining memory with the Register soft key.

  Example

  O = 30 Register

  O   Program number

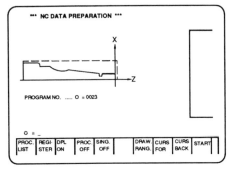

- Tool path drawn

  The tool path is drawn after NC data is prepared. The NC data is displayed when the DPL soft key is ON and only the tool path is shown when OFF.

  DPL ON

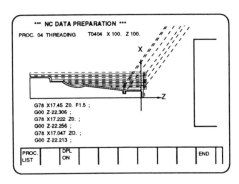

  DPL OFF

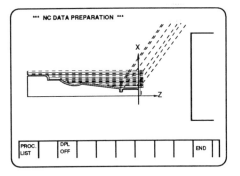

3.2.2 Example of Symbolic FAPT

The sequence of operations using the Symbolic FAPT method to program and machine the automobile stub shaft axle is given below.

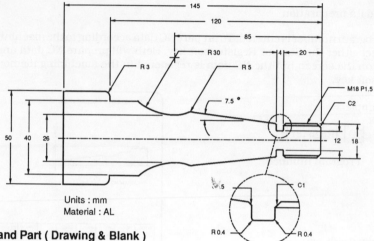

Units : mm
Material : AL

1. Blank and Part (Drawing & Blank)

 ☐1 │INPUT│ Blank and Part (Drawing & Blank)

MN = ☐4 NR = ☐2 DF = ☐1 BF = ☐1
L = ☐147 D = ☐51 ZP = ☐2

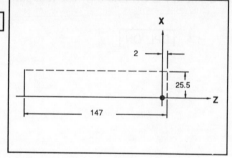

2. Blank and Part (Part Figure)

 ☐2 │INPUT│ Blank and Part (Drawing & Blank)

ES = ☐↑   Start Point
       SDX = ☐0.   SZ = ☐0.   PE = ☐0
       DX = ☐12.
ES = ☐C   C = ☐2.
ES = ☐←   Z = ☐-22.

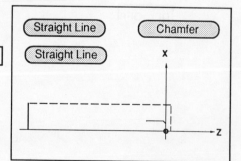

Sec. 3.2] Symbolic FAPT

ES = [T▲] EE = [0] LD = [1.5] NT = [1.]
ES = [G▲] EE = [1] DN = [1] WT = [4.]
DT = [3.]

Each corner of a groove can be rounded, chamfered or left unmachined.

ES = [C▲] C = [1.]
ES = [R▲] R = [0.4]
ES = [R▲] R = [0.4]
ES = [C▲] C = [0.5]

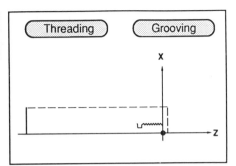

ES = [◣] TN = [1] A = [7.5]
ES = [R▲] R = [5.]
ES = [◯] TL = [1] TN = [1] R = [30.]
CDX = [86.] CZ = [85.]
ES = [◯] TL = [1] R = [30.] CDX = [86.]
CZ = [85.]

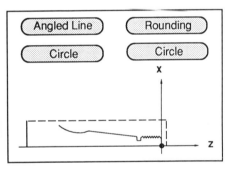

ES = [←] Z = [-122.]
ES = [↑] D = [40.]
ES = [R▲] R = [3.]
ES = [←] Z = [-145.]

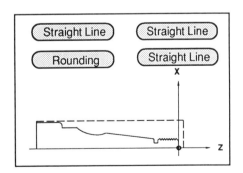

3.3 CONVERSATIONAL AUTOMATIC PROGRAMMING

Conversational AutomaticProgramming is produced by FANUC for use with their milling machine controllers. It is a computer aided programming system which allows part programs to be made in response to questions and instructions on a graphic CRT display.

Each process from design drawings to G-code programming is made through the interactive operation with the graphic CRT display.

Conversational mode allows for selection of part shape, material, machine home position and machining process. G-codes are then automatically generated. The menu selection are given belowed and discussed in greater depth in section 3.3.1 and a programming example is given in section 3.3.2.

1. Initial Set
- Selection of material, home position, coolant, etc.

2. Hole
- Drilling, boring, tapping, etc.

3. Facing
- Square and circle facing

4. Side cutting
- Circle outside, square outside, track outside etc.

5. Pocketing
- Circle, square, and track pocket

6. CNC Language
- Enter commands similar to NC format (F, G, M, S and T codes)

7. Contour
- Programming a shape that is made of lines at angles, arcs, etc.

8. Hole Pattern
- Point, line at angle, square, bolt hole circle, arc, etc.

9. Cycle End
- End of machining

10. Conversion
- Convert CNC statements into conversational display

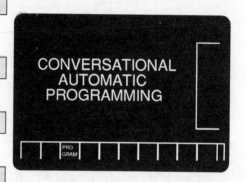

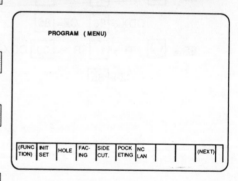

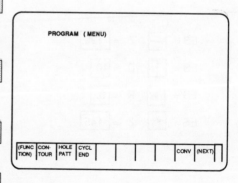

Sec. 3.3]   **Conversational Automatic Programming**   93

3.3.1 Execution of Conversational Programming

Some of the Conversational Automatic Programming procedures are given below.

3. Facing

 [Facing]

The type of facing is selected, and depending on the choice morel information is requested such as depth and length of cut.

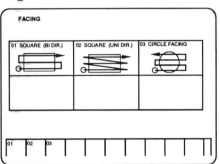

Example

[01]   Square (Bi Direction)

4. Side cutting

 [Side Cutting]

The type of side cutting is selected, and depending on the chosen path more information is requested such as the starting point and cutting width.

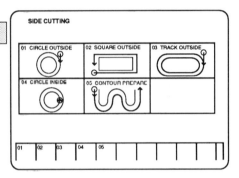

Example

[03]   Track Outside

8. Hole Pattern

 [Hole Pattern]

The type of hole pattern is selected, and depending on the chosen sequence more information is requested such as hole depth and the spacing between holes.

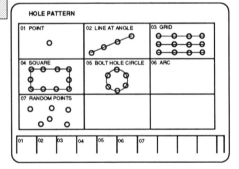

Example

[04]   Square

6. CNC Language

 [CNC Language]

Commands similar to CNC format are chosen and depending on the function more information is requested such as start and end points, radius, tool ID., spindle speed and feedrate.

Example

[03]   Arc

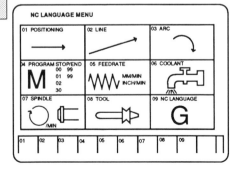

94 **Computer Assisted Programming** [Ch.3

3.3.2 Example of CNC Conversational Programming:

The sequences of operations using Conversational Programming method to program and machine the shape is shown below.

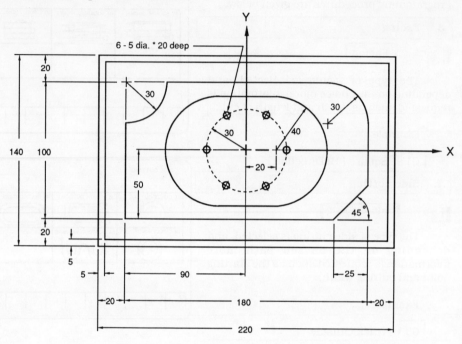

O0002;
ZI01 (INITIAL SET) W7. M9. U1. I100. H1. X500 Y250 Z400;

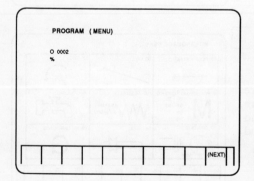
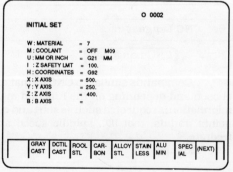

ZF01(SQUARE (BI DIR.))B1. U220. V140. X-100. Y-70. Z0. W50. T1001. Q8. H14. D24. M3. S1400. F1000. K50. C10.;

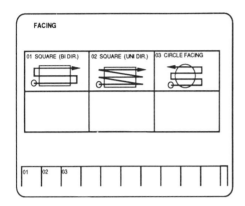
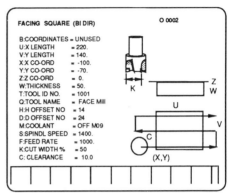

ZS02(SQUARE OUTSIDE) B1. U210. V130. X-150. Y-65. Z-10. W10. T1002. Q7. H15. D25. M3. S1400. F1000. C10.;
ZE99(END OF CYCLE);

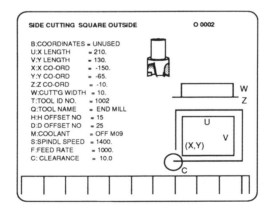

ZP03(TRACK POCKET)B1. A40. X20. Y0. U-20. V0. R10. Z-5 W5. E250. T1002. Q7. H15. D25. M3. S1400. F1000. J800. C01. K50.;
ZE99(END OF CYCLE);

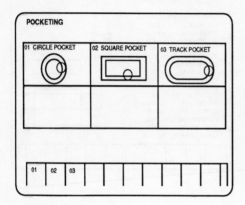
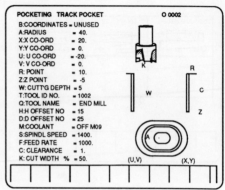

Sec. 3.3]   **Conversational Automatic Programming**   97

3.3.3 Example of Conversational Programming

The sequence of operations using Conversational Programming method to program and machine the shape is shown below.

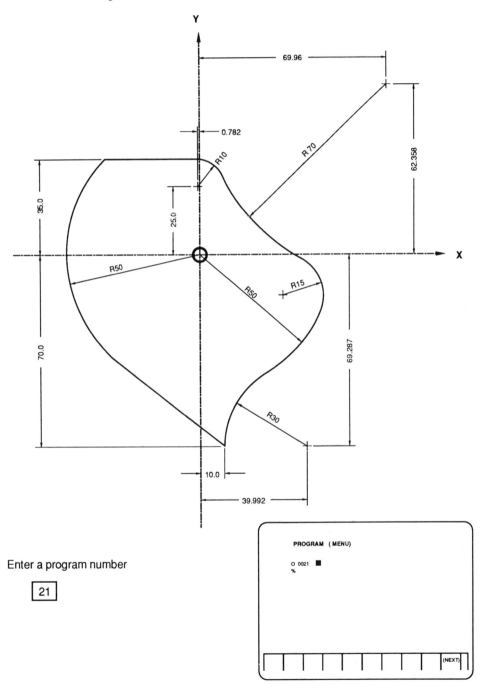

Enter a program number

21

1. Initial Set

 INITIAL SET 1

W = CARBON STL   M = M09
U = G21   I = 20.   H = G92
X = 0.   Y = 0.   Z = 100.

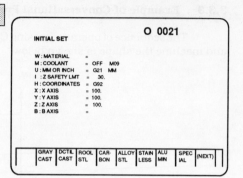

4. Side Cutting

 SIDE CUTTING 5

B = UNUSED   Z = -10.   R = 3.
I = 20.   V = 5.   C = 0.   W = 10.
A = UNUSED   T = 1000   Q = AUTO
H = 2   D = 26   M = M09   S = 795
F = 238   J = 150

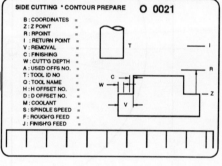

7. Contour

 CONTOUR
APPROACH

P = CIRCLE TANGENT   X = -50.
Y = 0.   D = LEFT   I = 10.   J =
R = 10.   C = LEFT   F = 500   E = 80

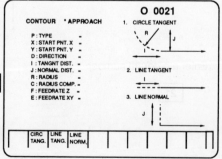

Sec. 3.3] Conversational Automatic Programming

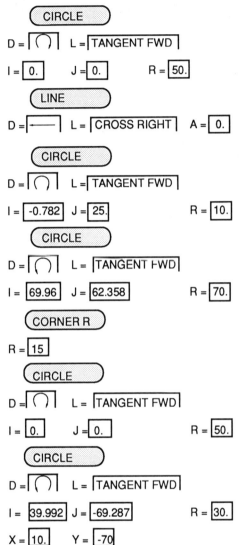
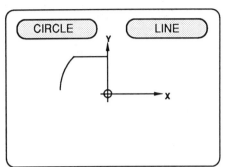
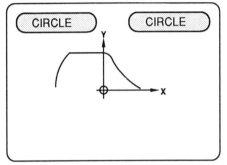
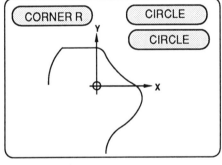

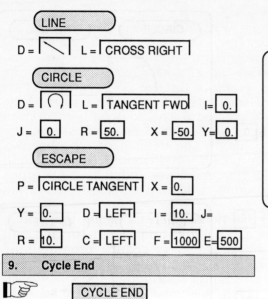
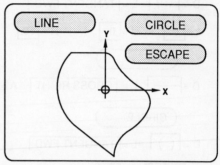

9. **Cycle End**

☞ CYCLE END

3.4 SMARTCAM

SmartCAM is a PC-based, two-dimensional, menu-driven, graphical part programming system. The part may be defined by selecting an element of the part geometry, such as a line or arc, from an on-screen menu and entering the endpoint coordinates of the element in response to system prompts for the X, Y values. Part geometry is displayed as each element is defined; any errors in the definition can be interactively corrected. The tool path is derived directly from this definition and can be displayed for verification. The part may, more easily and conveniently, be defined using other PC-based CAD packages, such as AutoCAD or Personal Designer, and ported directly to SmartCAM using a CAD drawing database file called a DXF file. Each machining operation is defined seperately, on layers within the CAD package

Within SmartCAM special functions (such as facing tool paths, pocketing and island avoidance for milling, and rough turning and threading cycles for turning) are provided to reduce programming time. The program does not have the capabilities of the integrated CAD/CAM programs described in Chapter 4 although for a PC-based program it is very effective. It can be used for a range of CNC equipment including milling machines, lathes, electric discharge machines (EDM), and flame cutting machines.

The main menu with two sub-menus, called Job Plan and Shape, are shown in Table 3.1. A further two sub-menus of the Shape module, called View and Tool Path, are given in Table 3.2. The purpose of the Job Plan module is to provide a data base of tool information and cutting parameters. The Shape module is used for part definition and to generate CNC code. Within Shape, part definition elements may be created or modified, and finishing and roughing tool paths can be generated and displayed. As shown in Table 3.2, the View sub-menu allows the user to adjust the graphics display with windows, viewpoint, zoom etc. and the Tool Path sub-menu to enter or change tool path information such as operation sequence, direction of cut, starting position, tool-path profile, lead-in and lead-out etc. Some of the machining capabilities include:

- Wall Offset option, which defines a tool path that is parallel or concentric to previously defined lines or arcs. It can be used to leave a stock allowance on rectangular and circular profiles or to leave material for a roughing pass.
- Roughing option, which is used to Pocket or Face with a mill, Turn or Face with a lathe, and Clear an area with a punch. With the Pocket option, in which a cavity is machined in a block of material, the tool path starts at the centre of the cavity and works out to the walls. Both Pocket and Face options provide a choice of Linear, Spiral, or Zig Zag tool paths, Island avoidance and Stock allowance.
- Code option, which generates the G-code for a defined tool path. If standardized 'start' and 'end' procedures are required, the information to generate these tasks is stored in

the Template file and is read and incorporated in the part program as the code is generated.

Table 3.1
SmartCAM Menus

System Menu
Job Plan
Shape
Edit Plus
Utilities
CAM Connection
Tape-to-Shape
Drafting System
Leave SmartCam

Job Plan sub-menu
New	Print
Job Info	Files
Tools	Read
Edit	Save
Insert	List Dir
Delete	Del Fill
Move	Leave
Cam	Layers
Edit	Delete

Shape sub-menu
New
Edit Shape
View
Group
Tool Path
Roughing
Code
Files

**Table 3.2
SmartCAM Menus**

View, sub-menu of Shape
Window
Zoom
Pan
Full
Base
Chg Base
Last Window
View 3D
Isometric
Show Path
Window
Name Window
Input Window
Get Window
Window Clear
Output
Adjust
Screen
Printer
Digitizer
List Dir
Colors

Tool Path, sub-menu of Shape
Change
Offset
Tool
Layer
Z-Position
Depth
Hole Op
Rapid Clear
Follow
Sequence
Tool Sort
Rev Order
Prof Start
Lead In/Out
Wall Offset
Explode
Roughing
Pocket(mill)
Face (lathe)

3.4.1 Example of SmartCAM Programming

A SmartCAM program for the milling example given in Section 2.1.5 is given below. Three sets of geometry for track pocketing, side cutting and hole pattern are defined within SmartCAM. The relevant cutter paths are produced automatically as each section is defined. The geometry can also be defined through AutoCAD, and transferred to SmartCAM as a DXF file. The principle operations are:

1. select JOB PLAN to create a job plan
2. select SHAPE and sub-menus NEW and EDIT SHAPE to create the part geometry
3. select SHAPE sub-menu VIEW to display the tool paths
4. select SHAPE sub-menu CODE to generate the CNC code
5. select EDIT PLUS to list the CNC code

1.   JOB PLAN

In the NEW menu the filename, metric units, milling operation, drawing number, part number, machining operations (track pocketing, side cutting and hole making), and part material (aluminum) are selected. The end-mill and drill are defined as numbers 1 and 2 respectively and relevant cutter data, such as diameter, length, cutting speed and feed etc., are recorded. The screen work area, or window limits, required in order to display all the geometry and the tool paths for the part is set as:

    Min X: -130   Min Y: -100
    Max X: 130   Max Y: 100

2.   SHAPE
Insert a tool change point
Select EDIT SHAPE/Insert/Point
End X: 150
End Y: 130
Select Quit

Select the milling tool number 1
Select EDIT SHAPE/Update/Feature Chg/Tool
Select a geometry element associated with tool #1
Select Element 1
Select Tool Number 1
Select Clear Z: 30
Select Prof Top: 0
Select Z Level -5
Select Quit

- Define and display side cutting profile

Select Start Point   Figure 3.2(a)
Enter End X: -90
Enter End Y: -50
Select Line

Enter End X: -90   Figure 3.2(a)
Enter End Y: 50

Select Arc (intersection CCW)   Figure 3.2(b)
Enter Radius: 30
Enter Centre X: -90

Select Line (intersection)   Figure 3.2(b)
Enter End X: 90
Enter End Y: 50

Select Line   Figure 3.2(c)
Enter End X: 90
Enter End Y: -50

Select Line   Figure 3.2(c)
Enter End X: -90
Enter End Y: -50

Select Chamfer   Figure 3.2(d)
Angle from 1st element: 45
Size parallel 1st element: 25

Select Blend   Figure 3.2(d)
Select the line
Enter Radius: 30

Select EDIT SHAPE/Update/Feature Chg/Offset   Figure 3.2(e)
Select a geometry element associated with tool #1
Select Right
Select Quit

• Define and display track pocket

Select EDIT SHAPE/Insert   Figure 3.3(a)
Select Start Point
Enter End X: 0
Enter End Y: -40

Select Line   Figure 3.3(a)
Enter End X: 20
Enter End Y: -40

Select Arc (Tangent)   Figure 3.3(a)
Enter Radius: 40
Enter End X: 20

Select Line (Tangent)   Figure 3.3(b)
Enter End X: -20

Enter End Y: 40

Select Arc (Tangent)   Figure 3.3(b)
Enter Radius: 40
Enter End X: -20

Select Line (Tangent)   Figure 3.3(b)
Enter End X: 0
Enter End Y: -40

Select EDIT SHAPE/Update/Feature Chg/Offset   Figure 3.3(c)
Select a track pocket element
Select Left
Select Quit

Rough the pocket   Figure 3.3(c)

Select MAIN/Roughing/Pocket
Select an element on the pocket boundary
Select Spiral
Select Continue
Enter Finish Allowance: 1
Enter Width of Cut: 10
Enter Final Depth: -10
Select GO

• Define and display hole pattern

Insert a tool change point
Select EDIT SHAPE/Insert/Point
End X: 150
End Y: 130
Select Quit

Select the drill, tool number 2
Select EDIT SHAPE/Update/Feature Chg/Tool
Select the tool change point
Select Number 2

Select Start Point   Figure 3.4
Enter End X: 30
Enter End Y: 0

Select EDIT SHAPE/Group/Copy   Figure 3.4
Select Rotate Copy
Enter Pivot Point X: 0
Enter Pivot Point Y: 0
Enter Rotation Angle: 60
Enter Number of Copies: 5

Select EDIT SHAPE/Update/Feature Chg   Figure 3.4
Select Hole OP
Select the holes associated with the drill (tool #2)
Select Clear Z: 30
Select Prof Top: 0
Select Depth Hole/Full Depth
Select Quit

3.   SHAPE/Main
Select VIEW to display tool path

4.   SHAPE/Main
Select CODE to generate the CNC code
Select Quit
Select Leave

5.   EDIT PLUS
To list the NC code:
Select Files/Read
Enter Filename

A 3-D isometric plot of the part is shown in Figure 3.5.

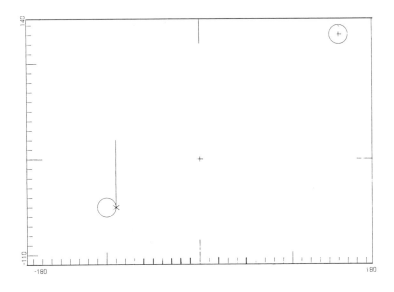

Figure 3.2(a) SmartCAM sidecutting

108   **Computer-Assisted Programming**   [Ch.3

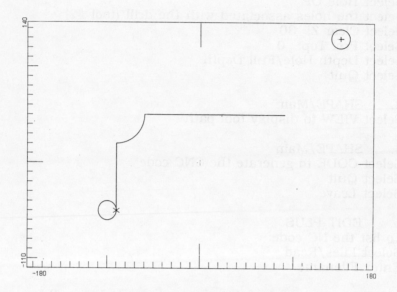

Figure 3.2(b) SmartCAM sidecutting

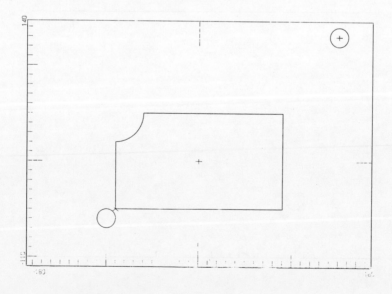

Figure 3.2(c) SmartCAM sidecutting

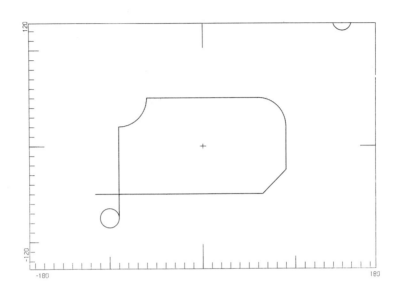

Figure 3.2(d) SmartCAM sidecutting

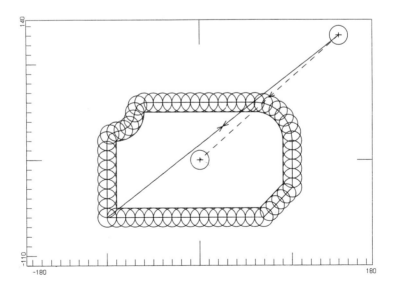

Figure 3.2(e) SmartCAM sidecutting

110 Computer-Assisted Programming [Ch.3]

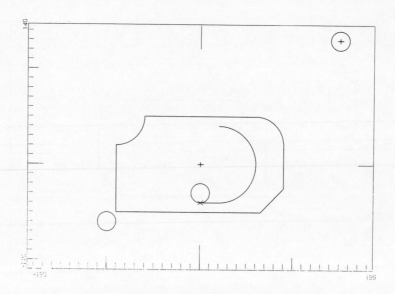

Figure 3.3(a) SmartCAM trackpocket

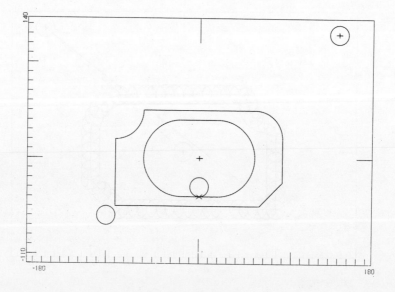

Figure 3.3(b) SmartCAM trackpocket

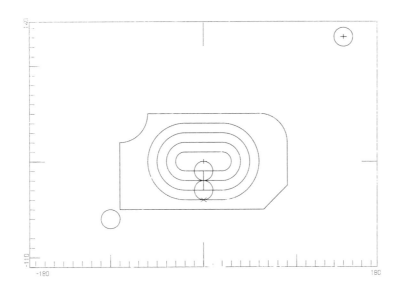

Figure 3.3(c) SmartCAM trackpocket

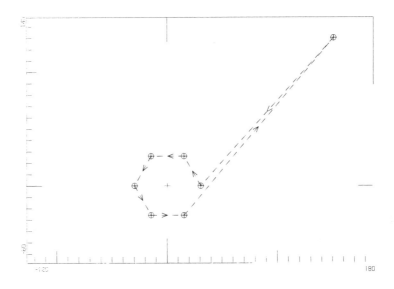

Figure 3.4 SmartCAM hole pattern

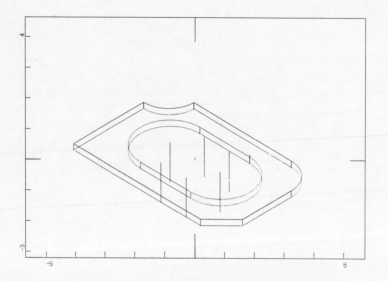

Figure 3.5 SmartCAM isometric view

Chapter 4
Integrated CAD/CAM Programming

Integrated computer aided design and manufacturing programs give the most advanced means of defining and programming CNC machine tools. These programs form an integrated approach to part definition (CAD) and machining definition (CAM) together with drafting, engineering analysis and data base management. They enable a direct link to be formed between design and manufacture and utilize an integrated data base (design and drafting specifications are utilized for manufacturing). The integrated data base can also be used to assist in process planning, production planning and control, inventory control and computer integrated manufacturing.

There are a number of excellent integrated CAD/CAM packages such as Anvil-5000, MEDUSA, Computervision CADDS4 and Intergraph. The systems are relatively expensive and run on mid-size workstations. However, with the advance in computer technology some of the systems (for example Anvil-5000 and CV Personal Designer) are available, in a restricted form, on PC's. In fact, with time, the computer assisted programming methods and the integrated programming methods are growing closer together.

With integrated CAD/CAM programs the part can often be defined in wire frame or solids modelling mode, it can be analysed through finite element programs, and the manufacturing processes can be interactively determined. Modules are available for virtually all types of CNC machining applications, including milling, turning, drilling, punching, flame cutting etc. All tool information is defined and graphically verified through tool management facilities. Multiple colour tool path displays are generated with cutting time analysis and total. Output from the machining modules is given as cutter location files, usually in APT IV CLFile format. This does not imply that APT has been used in the definition process but merely that the output file is listed in this particular format. As with all other previous methods the CLfile requires postprocessing in order to produce a suitable, machine specific, G-code file.

Although current programs have a wide range of capabilities, the goal of being able to automatically determine the optimum manufacturing processes and machining specifications from unambiguous design data is still a long way away. Integrated CAD/CAM programs do not have artificial intelligence, or the ability to make reasoned judgements, in order to automatically convert design specifications to part programs. Some progress has been made with solids modelling definition and analysis although the difficulty of recognising manufacturing operations from the CAD definition still remains; for example, what constitutes a hole and how it should be made. This difficulty has lead to another approach which attempts to define parts from a manufacturing perspective. In this approach, called parametric feature representation, the object is defined in terms of subsequent manufacturing processes such as holes, faces, pockets etc. The next generation of programs which incorporate some form of artificial intelligence for optimum machining will indeed be interesting.

In this chapter the operation of a widely used integrated CAD/CAM program, called Anvil-5000, is discussed. Emphasis is placed on the capabilities of the program to define parts and cutter paths rather than the specifics of the program.

4.1 ANVIL-5000

Anvil-5000 is a fully integrated system for computer-aided design, drafting, engineering, manufacturing, information management and data base support. The main menu selection, shown in Table 4.1, includes:

- System modals, which enable system functions, or defaults (such as colour, font, graphical level, decimal place and design tolerance) to be established. The system modals remain in effect until modified by the user.
- Data Base Management, which allows libraries of parts, patterns and templates to be defined. It also allows shared data bases for design, analysis and manufacturing.
- Display Control, which enables single or multiple views (up to eight) to be displayed as well as a variety of orthographic, perspective or auxiliary views. It also enables the user to zoom, or magnify, any region of the design.
- Basic Geometry, which allows entities such as points, lines, arcs and other curves (including conics, splines, offset curves, strings, and polygons) to form the basic building blocks of a wire frame model.
- Extended Geometry, which allows 3-D curves and surfaces (including planes, surfaces of revolution, ruled surfaces, curve driven surfaces, and Coon's and Bezier surfaces) to be defined.
- Drafting, which allows a variety of dimensions, symbols, labels, character sets, cross-hatching etc. to be inserted.
- Analysis, which allows curve, section or volume data to be obtained. For example, curve curvature, second moments of section, surface area and moments of inertia can all be obtained.

Table 4.1
Anvil-5000 Main Menu

1. System Modals
2. Blank/Unblank
3. Delete
4. File/Terminate
5. Special Functions
6. Data Base Management
7. Input/Output/Regeneration
8. Display Control
9. Point
10. Line
11. Arc Circle
12. Other Curves
13. Manipulation
14. Data Verify
15. Extended Geometry
16. Drafting
17. Machining
18. Analysis
19. Entity Control

4.1.1 Part Definition

Components are defined either as three-dimensional wire frame, or solids model, objects. In wire frame mode the geometry is generated from basic geometric elements including points, lines and arcs, called from the main menu. Points can be entered, as shown in the sub-menu given in Table 4.2, in a variety of ways including screen position with a mouse, numerical coordinates, polar position and the junction of two lines. Lines and circles can be similarly entered in many ways. A line, for example, may be defined between two screen positions, between two previously defined points, parallel to a line and tangent to another line, or as the intersection of two planes, while circles defined as centre point and radius, through three points, or inscribed within three curves. The resulting geometry can be displayed in multiple views, as shown in Figure 4.1, and manipulated by translation, rotation, scaling, mirroring, trimming, clipping, stretching and duplication. A single isometric wire frame view is shown in Figure 4.2 and in Figure 4.3 with hidden line removal. In many cases the use of standardized families of parts can be used to reduce the design and development.

Table 4.2
Anvil-5000 Point Sub-menu

1. Screen Position
2. Enter Coordinates
3. Polar
4. Delta
5. Vectored
6. Arc Centre
7. On a Curve
8. Curve End
9. Intersection of Two Curves
10. Spherical
11. Bearing/Distance
12. Curve Normal
13. Multiple Points
14. Surface Points

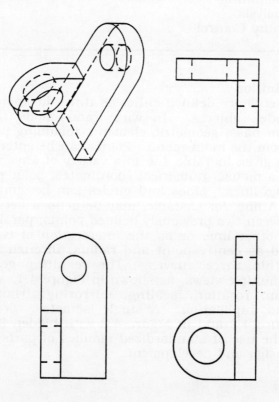

Figure 4.1   Multiple views of a component part

Sec. 4.1]   **Anvil-5000**   117

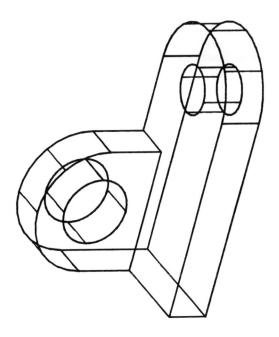

Figure 4.2 Single isometric wire frame view

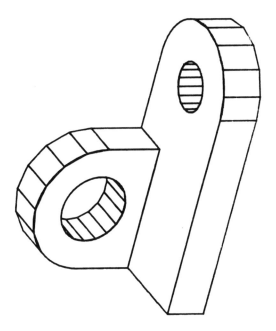

Figure 4.3 Isometric view with hidden line removal

In solid modelling, the component geometry is generated by combinations of simple shapes (called primitives) such as cylinders, cones, boxes, wedges or spheres. The solids modelling option is accessed through the Extended Geometry module of the main menu and given in Table 4.3. The primitives are combined through a boolean approach (with union, intersect and difference) to give a unique definition of the object. Solids can be displayed in a fully shaded form or in a wire frame form (with or without hidden lines removed). A simple example of part definition using the solids modelling option, with a cylinder and a cone, is given in Figure 4.4. A more intricate view of a centrifuge desalination unit with full exploded sub-system definition is given in Figures 4.5 and 4.6. Solid models may also be used for analysis of mass properties, bills of materials, etc. as well as CNC machining. Although the solids option has these advantages, it is less user friendly and is much slower in operation than the regular wire frame approach.

Table 4.3
Anvil-5000 Omnisolids Sub-menu

1. Modals
2. Parallelepiped
3. Cylinder
4. Axial Sweep
5. Rotational Sweep
6. Warped Surface Normal Offset
7. Warped Surface Projected to Plane
8. General Surface Enclosure
9. Fillet Solid
10. Composite Solid
11. Utilities

Figure 4.4 Simple solids model with the union of two primitives

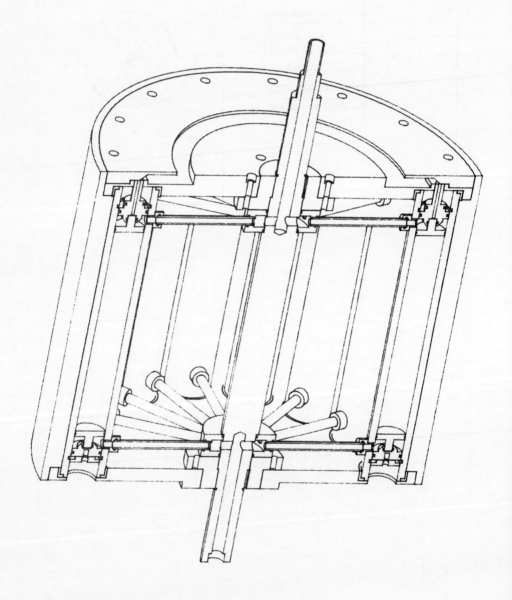

Figure 4.5 Solids model of a centrifuge desalination unit

Figure 4.6 Exploded sub-system definition of the desalination unit

4.1.2 Tool Path Definition

The numerically controlled machining option is integrated within Anvil-5000 and accessed through the main menu, as shown in Table 4.1. The machining module, shown in Table 4.4, provides automated techniques for tool path generation. The tool is controlled in an interactive point-to-point operation mode, and both part and tool path are displayed, as shown in Figure 4.7. The sub-menu for milling is given in Table 4.5. There are a range of subroutines operations, such as profiling, flange cutting, pocketing, area clearance and with selection of lace and non-lace cutting patterns. Tool paths can be collected together and manipulated with rotation, scaling etc. They can also be inspected, interactively edited and analysed for the approximate cutting times prior to postprocessing. Tool types, with offset values, cutting speeds, etc. are organized through the tool management facility.

Table 4.4
Anvil-5000 Machining Sub-menu

1. Modals
2. Tool Management
3. Milling
4. Turning
5. Drilling
6. Punching
7. Verification
8. Modification
9. Output
10. Utilities

Table 4.5
Anvil-5000 Milling Sub-menu

1. Modals
2. Create a Tool Path
3. Create Based on Existing Path
4. Move a Tool Path in Composite
5. Blank a Tool Path
6. Delete a Tool Path
7. Rename a Tool Path
8. Verify a Tool Path
9. List Paths Within Composite
10. Go to Display and Edit

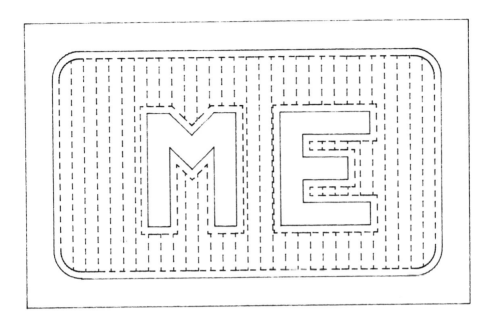

Figure 4.7 Tool path (broken line) and part (solid line) for the letters M E

4.1.3 Example of Anvil-5000
The sequence of operations using Anvil-5000 to define the milling example given in Section 2.1.5, generate tool paths, and output a CLFile is shown below. Listings of sections of the CLFile for facing, deep side cutting and side cutting operations are given in Tables 4.6, 4.7 and 4.8.

1. Create a 2-D part
   - use the point and line functions to create the outside rectangle
   - use the offset curve function to create the inside rectangle.
   - use the centre and radius function to create the centre circle and divide into six segments to generate six points.
   - create a 5 mm circle on each of the points in the centre circle
   - use the circle and line functions to create a pocket around the centre circle
   - use the line, circle, chamfer, fillet and trim functions to create a rectangle outside the pocket

2. Create a 3-D part
   - project the first inside rectangle to a depth of $Z = 5$ mm
   - project the second inside rectangle to a depth of $Z = 10$ mm

- project the pocket to a depth of Z = 10 mm
- project the outside rectangle to a depth of Z = -20 mm
- project the six 5 mm holes to a depth of Z = -10 mm

3. Display in multiple views (Figure 4.8)
   - use display control function to display part in four views
   - use input and output function to create a plot file

4. Create toolpath
   - use machining function to define tools and create a toolpath composite called **mill**
   - define tool 1 = 100 mm diameter face-mill
     tool 2 = 20 mm diameter end-mill
     tool 3 = 5 mm diameter drill
   - define five toolpaths inside the composite **mill**
     a) facing: using tool 1 (Figure 4.9)
     b) deep side cutting: using tool 2 (Figure 4.10)
     c) track pocketing: using tool 2 (Figure 4.11)
     d) side cutting: using tool 2 (Figure 4.12)
     e) drill six holes: using tool 3 (Figure 4.13)

5. Cutter Location Output
   - use machining function to output the CLFile

Table 4.6
Anvil-5000 Facing Operation

```
SPINDL/1200.00,CLW
FROM/0.00,0.00,100.00
RAPID
GOTO/-110.00,-20.00,100.00
FEDRAT/508.00
GOTO/-110.00,-20.00,-2.00
FEDRAT/254.00
GOTO/60.00,-20.00,-2.00
GOTO/60.00,20.00,-2.00
GOTO/-60.00,20.00,-2.00
GOTO/-60.00,-70.00,-2.00
GOTO/-110.00,-70.00,-2.00
FEDRAT/254.00
GOTO/-110.00,70.00,-2.00
GOTO/110.00,70.00,-2.00
GOTO/110.00,-70.00,-2.00
GOTO/-110.00,-70.00,-2.00
FEDRAT/508.00
GOTO/-110.00,-70.00,100.00
RAPID
GOTO/0.00,0.00,100.00
SPINDL/OFF
```

Table 4.7
Anvil-5000 Deep Side Cutting Operation

```
SPINDL/1200.00,CLW
FROM/0.00,0.00,100.00
RAPID
GOTO/-105.00,-75.00,100.00
FEDRAT/508.00
GOTO/-105.00,-75.00,-10.00
FEDRAT/254.00
GOTO/115.00,-75.00,-10.00
GOTO/115.00,75.00,-10.00
GOTO/-115.00,75.00,-10.00
GOTO/-115.00,-65.00,-10.00
FEDRAT/508.00
GOTO/-115.00,-65.00,100.00
RAPID
GOTO/0.00,0.00,100.00
SPINDL/OFF
```

Table 4.8
Anvil-5000 Side Cutting Operation

```
SPINDL/1200.00,CLW
FROM/0.00,0.00,100.00
RAPID
GOTO/-90.00,-60.00,100.00
FEDRAT/508.00
GOTO/-90.00,-60.00,-5.00
FEDRAT/254.00
GOTO/69.14,-60.00,-5.00
GOTO/100.00,-29.14,-5.00
GOTO/100.00,20.00,-5.00
SURFACE/60.00,20.00,-5.00,0.00,0.00,1.00,40.00
GOTO/99.90,24.00,-5.00
GOTO/98.31,31.84,-5.00
GOTO/95.20,39.21,-5.00
GOTO/90.69,45.82,-5.00
GOTO/84.95,51.39,-5.00
GOTO/78.22,55.72,-5.00
GOTO/70.77,58.63,-5.00
GOTO/62.89,60.00,-5.00
GOTO/60.00,60.00,-5.00
GOTO/-70.00,60.00,-5.00
GOTO/-70.00,50.00,-5.00
SURFACE/-90.00,50.00,-5.00,0.00,0.00,-1.00,20.00
GOTO/-70.10,47.17,-5.00
GOTO/-71.68,41.74,-5.00
GOTO/-74.70,36.96,-5.00
GOTO/-78.94,33.21,-5.00
GOTO/-84.06,30.80,-5.00
GOTO/-89.64,29.90,-5.00
GOTO/-90.00,30.00,-5.00
GOTO/-100.00,30.00,-5.00
GOTO/-100.00,-50.00,-5.00
FEDRAT/508.00
GOTO/-100.00,-50.00,100.00
RAPID
GOTO/0.00,0.00,100.00
SPINDL/OFF
```

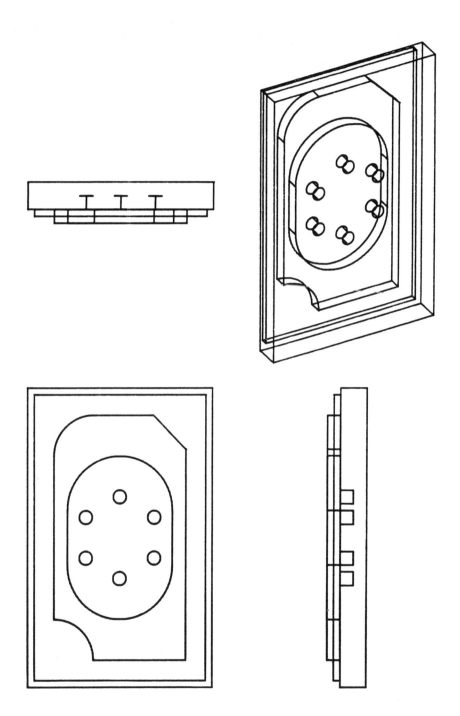

Figure 4.8 Multiple view display of the Anvil-5000 defined part

128   **Integrated CAD/CAM Programming**   [Ch.4

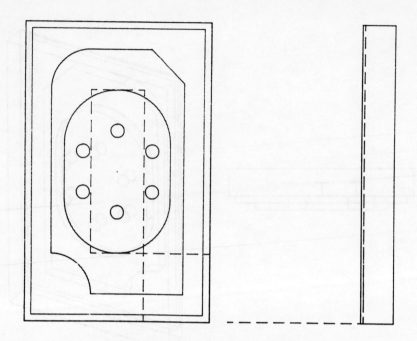

Figure 4.9 Facing Operation

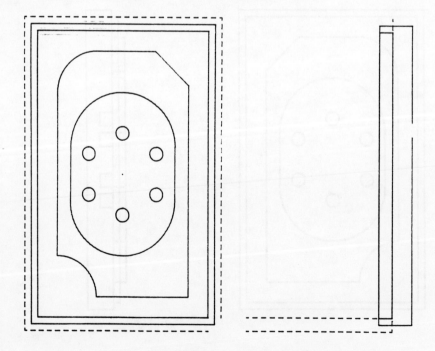

Figure 4.10 Deep Side Cutting Operation

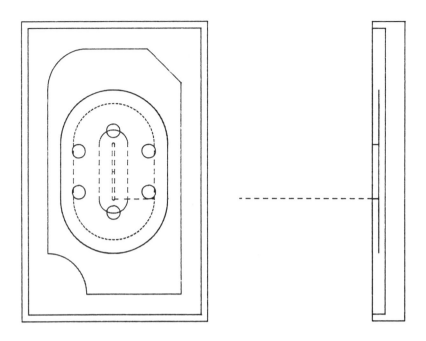

Figure 4.11 Track Pocketing Operation

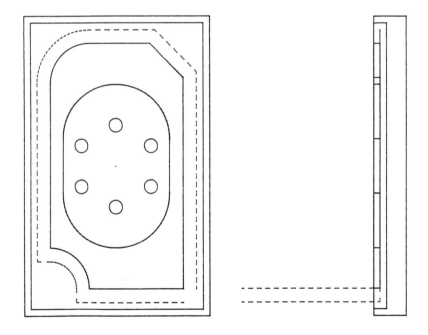

Figure 4.12 Side Cutting Operation

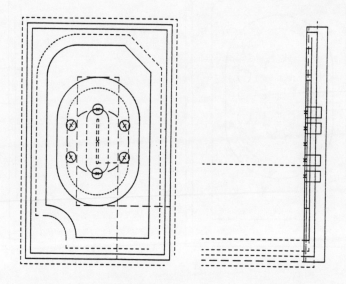

Figure 4.13 Drilling and Complete Cutter Path

Chapter 5
Direct Numerically Controlled Machining

Transmission and storage of part programs to numerically controlled machine tools has been accomplished through the medium of punched paper tape. Usually, the part program is manually punched on paper tape and then taken physically to the machine and loaded into a paper tape reader. While punched paper tape has proved a reliable medium, it is one of the last examples of the punched hollerith card system that evolved from the early French silk pattern weaving machines.

With the advent of low-cost computers and the rapid data transfer between computer peripherals, downloading of part programs into the CNC machine controller is now becoming more common. The digital transfer of data through a standard RS-232 C serial link provides a flexible means of communication that is reliable and can handle large amounts of data. This type of communication, through coded electrical pulses, is used with all computer peripherals, such as plotters, printers and digitizers; and in a sense the CNC machine is being treated as a computer peripheral.

Direct numerically controlled machining, DNC, is defined as the direct transmission of part programs from an external computer to the controller memory of a machine tool. Most modern CNC machine tools are supplied with a tape reader as well as RS-232 communication port connections. Thus, part programs may be downloaded to the controller memory of the CNC machine. A central computer can then be used to control a number of separate machine tools and form the basic programming link for flexible manufacturing cells, FMS, and integrated manufacturing systems. With FMS a number of CNC machine tools may also be connected with automated materials handling systems. The potential framework then exists for a single computer to control machine loading, scheduling and all production planning and control operations. If the number of machines is large, small dedicated satellite computers can be used at each machine and the host computer used to store and dispatch programs as required. Communication programs resident in the host computer can be used to detect when programs are complete and provide information on machine utilization and pieces produced etc. This two-way transfer of data is an essential feature of DNC.

An alternative means for DNC transmission is for the part program to be downloaded a block, or section of program, at a time. This is achieved by directing the digital data to a buffer memory located behind the tape-reader (BTR). In this approach, the data is transmitted to the controller as though it were reading paper tape, although it is actually taking ASCII code data from a computer. The buffer memory (which is typically 4 Kbytes, or the equivalent of 4000 characters of part program, or 100 blocks of part program, or 10 m of paper tape) is usually filled with data, and refilled so that it remains between the 1.2-4 Kbyte level. This process can be considered as analogous to filling and refilling a bath of water with the bath plug removed. Thus, the machine tool is being run from a remote computer with a few blocks of instructions time delay. This can effectively become real time control of the machine tool if the part program is fed through to the buffer memory in single blocks of G-code.

The advantage of the BTR approach is that longer and more detailed part programs can be easily handled without the need for expensive CNC controller memory. Also, with the increased use of computer-assisted programming, part programs exist within remote computer memory and are in a conveniently coded form for direct transmission to the controller of the machine tool. Using this approach, the size of the part program is virtually unlimited. This feature is particularly useful with curved surface machining where programs require large arrays of surface points and can be the equivalent of many kilometres of paper tape in length. The full editing features of a host computer can be used to add or delete complete sections of these long part programs, even during the machining process.

In this chapter the means of establishing a DNC remote buffer link is described together with the listing of an assembly language data transmission program. In Section 5.2 an approach for developing postprocessor programs for converting cutter location files obtained from CAD/CAM packages, in APT IV code, to G-code for CNC machine tools is given.

5.1 DNC REMOTE BUFFER LINK

5.1.1 Data Transmission

A typical control layout for a numerically controlled machine with a remote buffer is given in Figure 5.1. The CNC machine can be programmed directly from the machine controller, the paper tape reader, or the remote computer with the buffer memory. Part program data is stored in the remote computer in ASCII code. Alphanumeric characters are represented by a combination of seven bits with the eighth bit as a parity check feature. A bit in this case is an on/off switch representing a binary '0' or '1'. The even parity version of the ASCII format is the same as the ISO paper tape format, shown previously in Figure 1.9.

Sec.5.1] **DNC Remote Buffer Link** 133

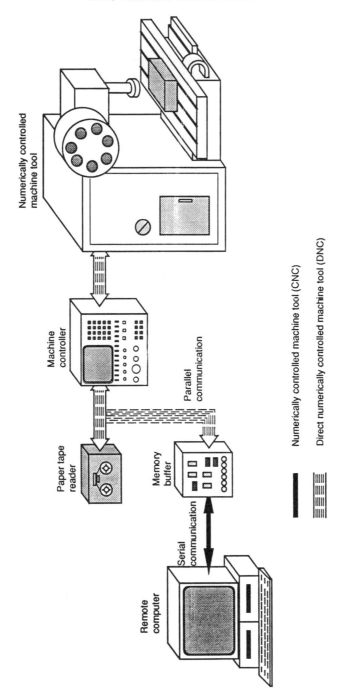

Figure 5.1 Schematic for a numerically controlled machine with a remote memory

Data is transferred between electronic devices in ASCII code with each '0' bit being represented by a +3 volt signal and each '1' bit by a zero volt signal. Transmission of characters can occur one bit at a time on a single line connection (serial communication) or with eight bits, representing a single character, sent simultaneously down eight lines (parallel communications). CNC machines normally have an RS-232 C serial connection. Serial transmission of the character 'A', together with start bits and stop bits, is shown in Figure 5.2. The RS-232 C serial link actually has 25 separate communication lines, each associated with a particular response. In this application only nine of the lines are required and the pin connections for these lines are shown in Figure 5.3. Data is transmitted on pin number 3 and the other pins are used for 'hand shaking' protocols between the devices. As shown schematically in Figure 5.4, electrical signals are sent and received on the other pins in a specified sequence to ensure that the two devices are ready to send and receive data.

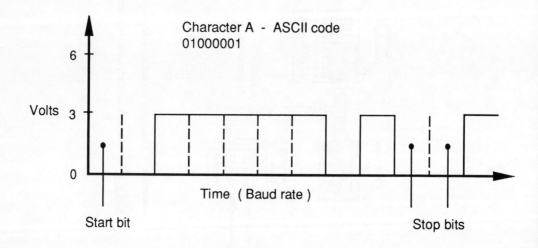

Figure 5.2 Analogue display of serial transmission of the character 'A'

Sec.5.1]   DNC Remote Buffer Link   135

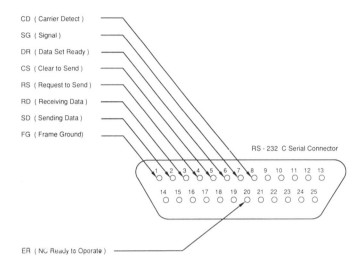

Figure 5.3 RS-232 C serial pin connections

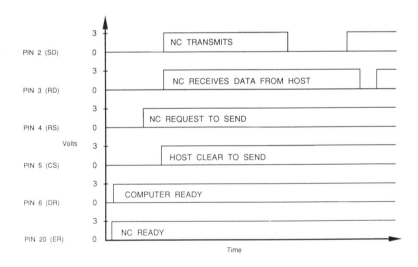

Figure 5.4 Schematic representation of electrical signal transmissions between devices

5.1.2 Transmission Program

Transmission of data from the host computer to the machine tool via the remote buffer is controlled by a software program. The program reads the G-code file (equivalent to the part program that would have gone to the punched paper tape), ascertains that communication lines and devices are open, and transmits the data, as required by the remote buffer and the machine tool. Computers which operate in a multi-user environment have asynchronous input controllers (AIC) that monitor the transmission lines at the operating system level and are able to treat the DNC link as a regular terminal. An input/output RS-232 C port is assigned for the DNC machine and the characteristics of baudrate, etc. are set according to those of the DNC machine. On the IBM-PCs, with a single-user environment, the software program is required to check the status of the transmission lines and to regulate data flow to the remote buffer.

The flowchart of a computer program for data transmission from an IBM-PC to a remote buffer is given in Figure 5.5 and the program listing is given in Appendix 1. As the program is required to operate at a low level, it is written in assembly language. The program is divided into three main parts, namely, initialization of input/output ports, opening and reading the part program file, and monitoring and transmitting the data as characters. As the G-code file is read and transmitted, a character at a time, checks are made for special characters such as a ';' sign (which indicates a new block) or a '%' sign (which indicates the end of program) and appropriate action taken.

With some controllers the tool path can be passed from the computer and displayed on the machine controller before the part is machined. Also, programs can be passed from the machine controller and the paper tape reader back to the host computer.

5.1.3 Transmission Problems

The process described above simply fills the remote buffer but does not ensure that data is passed in complete blocks or lines of program. In a multi-user environment where it is possible for the buffer to be starved of input data, the incomplete blocks of program can cause unexpected and dangerous consequences. For example, if the following block of program,

G00X10.0Y10.0Z10.0

is partially stored at the end of the buffer file as

G00X10.0Y10.0Z1

and is starved of other input data, the effect is for the machine to operate at a depth of $Z = 1$ mm; that is, the cutter undercuts the workpiece by 9 mm!

Experience has shown that the difficulty can be overcome by making the transmission program a non-swapable process within the computer, by giving the transmission program a suitably high priority or by implementing a protocol that transmits program data in blocks. In the latter case, the data block would be transmitted to the remote buffer as,

G00X10.0Y10.0Z10.0;

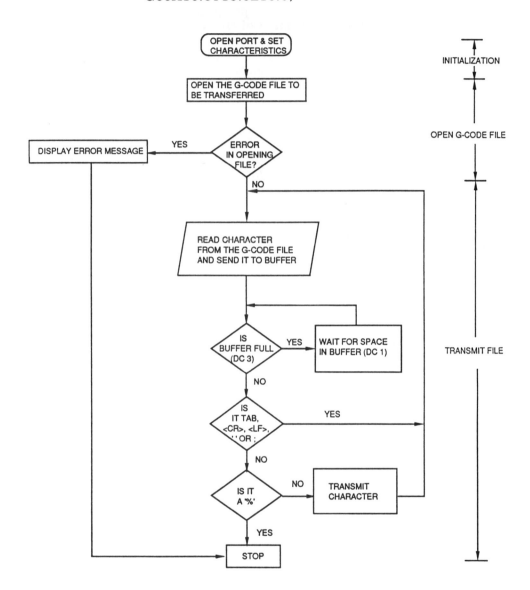

Figure 5.5 Flowchart of the computer program to control data transmission between the IBM-PC and the machine tool

5.2 POSTPROCESSORS

A postprocessor is a computer program which converts the APT format CLfile output from CAD/CAM programs to machine dependent G-code format commands. Because of the variety of CNC controllers, these programs are frequently written on an individual basis for particular machines and installations. The following sections illustrate the basic features of a simple postprocessor program and enable modifications to be made for particular installations.

Typical CLfile commands, in APT IV format, are given in Table 5.1. They consist of instructions, such as GOTO, CIRCLE, RAPID, etc., and are used to define the full tool path. A listing of common G-code commands is given in Chapter 2. The function of a postprocessor program is to establish a one-to-one correlation between the CLfile and G-code commands. A few equivalent CLfile and G-code commands are given in Table 5.2. For example, the CLfile command 'RAPID' is equivalent to "G00" in G-code. Other commonly used equivalences are absolute or incremental programming, linear and circular interpolation, cutter compensation, and units of measurement.

Table 5.1
CLfile Commands in APT IV Format

Command	Parameter	Description
CIRCLE	Radius, centre, and start point	Centre point, radius and start point of circle
COOLNT	On/Off	Turns coolant on or off
FEDRAT	Value	Set maximum feedrate to value
FINI	-	Finish program
FROM	X, Y, Z	Starting position of tool
GOTO	X, Y, Z	Destination position
INTOL	Value	Inside tolerance
OUTTOL	Value	Outside tolerance
RAPID	-	Go fast
SPINDL	Speed (rpm), CW/CCW	Spindle speed and direction of rotation
STOP	-	End of program
TOOLNO	Tool parameters like	Description of tool
UNITS	Inch/mm	Units of length

Table 5.2
Equivalent CLfile and G-code Commands

APT Codes	G-codes
CIRCLE	G02 and G03
COOLNT	M07 and M08
FEDRAT	G01
FINI	%
FROM	G92, G54, G55 etc.
RAPID	G00
SCALE	G58
SPINDL	M03 and M05
STOP	M30 and M02
TOOLNO 1-20	T 1-20
UNITS	G20 and G21

The flowchart of a computer postprocessor program is given in Figure 5.6 and the program listing is given in Appendix 2. The program is written in C programming language and is divided into two parts. In the first part, input and output files are opened and CLfile commands are read. In the second part, CLfile commands are converted to G-codes.

The first command in a CLfile is "PARTNO", which is converted into a dummy file number O0111. Subsequent CLfile commands have blanks in the first six characters and command variables in the next four characters. The blanks are ignored and the character string is used to uniquely identify an APT command. Once the command has been identified, certain flags are checked and a unique correspondence is determined.

Some CLfile commands can have multiple meanings. The "GOTO" command, for example, can be translated into "G00" or "G01". Therefore, once the command is identified as "GOTO", a check is made on the flag called "previous-command". If "previous-command" equals "RAPI", then "GOTO" is translated into "G00", or else it is translated into "G01" with a feedrate specified from a previous command. The "CIRCL" command can also be translated into "G02" or "G03". The "CIRCL" command gives the starting point and the radius of the circle and is followed by a set of "GOTO" commands that give a series of intermediate positions on the circle. Once the command is identified as "CIRCL", the first "GOTO" point is used to determine the direction of the circle and the following commands are processed until the end point of the circle is found.

Many other features can be added to a postprocessor program according to the needs of the installation. For example, a graphic display of the tool path (as shown in Chapter 9), or machining time calculations can be readily incorporated.

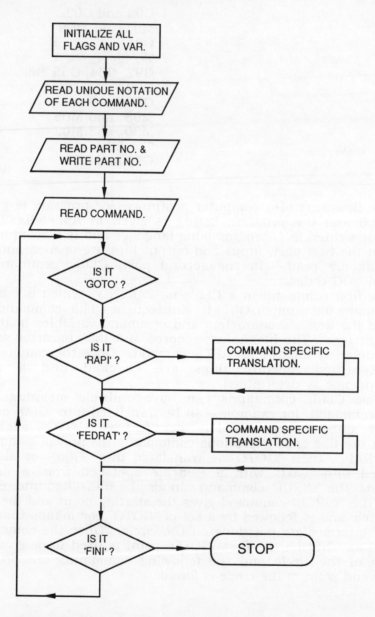

Figure 5.6 Flowchart of the postprocessor program

Sec.5.2] **Postprocessors** 141

5.2.1 Example of Program Conversion

The CLfile, generated through Anvil-5000, for machining a letter 'R', and the equivalent G-code generated through the postprocessor program, are given below.

```
PARTNOTOOLPATHR
    UNITS/MM
    INTOL/0.03
    OUTOL/0.03
    FROM/35.00,-40.00,0.00
    RAPID
    GOTO/35.00,-40.00,25.40
    RAPID
    GOTO/35.00,-0.00,25.40
```

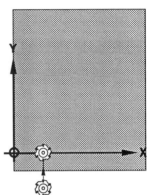

```
    FEDRAT/254.00
    GOTO/35.00,-0.00,11.63
    GOTO/35.00,145.00,11.63
    GOTO/85.00,35.00,11.63
```

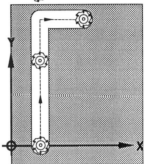

```
    CIRCLE/85.00,100.00,11.63,0.00,0.00,1.00,45.00
    GOTO/86.56,145.00,11.63
          90.95,144.64,11.63
          95.28,143.84,11.63
            .
            .
            .
          112.68,64.48,11.63
          109.03,61.95,11.63
          107.73,61.16,11.63
    GOTO/135.53,0.00,11.63
    GOTO/93.56,0.00,11.63
```

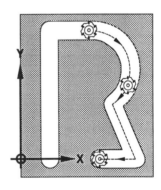

```
    GOTO/75.00,40.83,11.63
    GOTO/75.00,0.00,11.63
    GOTO/35.00,-0.00,11.63
    RAPID
    GOTO/35.00,-0.00,25.40
    RAPID
    GOTO/35.00,-40.00,25.40
    STOP
    FINI
```

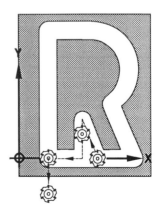

142   **Direct Numerically Controlled Machining**   [Ch.5]

G-code for the letter R.

```
O0111;
G21;
G92   X 35.000    Y -40.000    Z 0.000;
G90;
G00   X 35.000    Y -40.000    Z 25.399
G00   X 35.000    Y   0.000    Z 25.399;
```

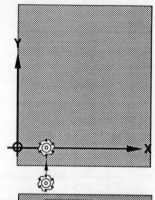

```
F 254.000 ;
G01   X 35.000    Y   0.000    Z 11.630;
G01   X 35.000    Y 145.000    Z 11.630;
G01   X 85.000    Y 145.000    Z 11.630;
```

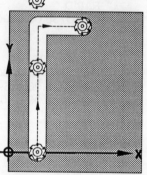

```
G02   X 107.729   Y  61.160    Z 11.630   R 45.00;
G01   X 135.529   Y   0.000    Z 11.630;
G01   X  95.559   Y   0.000    Z 11.630;
```

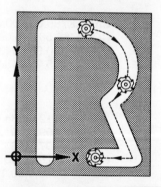

```
G01   X 75.000    Y  40.830    Z 11.630;
G01   X 75.000    Y   0.000    Z 11.630;
G01   X 35.000    Y   0.000    Z 11.630;
G00   X 35.000    Y   0.000    Z 25.399;
G00   X 35.000    Y -40.000    Z 25.399;
M05;
M30;
%
```

Chapter 6
General Curved Surface Machining

Historically, the definition and machining of curved surfaces has presented more challenging problems than the better established approaches, outlined in Chapters 3 to 5, for planar surfaces. This broad area covers many aspects of engineering and includes, for example, car, ship, aeroplane, telephone, computer cabinet housing definition as well anatomical shapes, such as limbs, sockets and faces. The need in these applications can be for a mock-up model, injection moulding dies, patterns for sandcasting or simply a physical model of a measured shape.

Curved surfaces can be defined in a variety of ways, such as existing models, parallel or orthogonal sections, tables of offsets, curved boundary lines, random surface data, laser scanned surface data or CAD definition through mathematical models. Many of these methods do not define the full surface but merely define a few points within the surface. Also, if the data are obtained by some form of measurement, they will probably be sparsely distributed and will contain irregularities. Smoothing of input data is then required, followed by creation of a mathematical model of the surface and generation of a regular array of surface points for machining. In the case of laser scanning, the measured cloud data is very dense, somewhat inconsistent, and requires data compression and filtering. In the definition of these surfaces, there are often specific restrictions imposed, such as smoothness of form (or avoidance of unnecessary undulations) when defining car body shapes, for example.

The basic approach for defining and machining all of the above examples is as follows:
- Data input, which gives some form of full, or partial, surface definition
- Surface modelling, which defines the full, or partial (at sufficient resolution for machining purposes), surface definition
- Surface orientation, which aligns the surface, in relation to the machine tool, for tool access and part location
- Surface normal calculations, which define the normal vectors at all surface vertices
- Cutter location calculations, which define the cutter offset positions at all surface vertices

- G-code generation, which defines the part program when used with additional machine data such as feedrate and spindle speed.

An essential fact to remember with curved surface machining is that no matter what type of surface definition or sophisticated CNC machine tool is involved, at some point the individual point-to-point tool movements across the surface will need to be defined. The spacing of such movements will determine the number of points visited by the tool on the surface and hence the smoothness of the finished surface. A large number of points will result in a smoother surface although the machining time will be longer. There is usually a resolution which gives a compromise between the machining time and the subsequent finishing operations. A suitable resolution for machining might be 0.1 to 2.0 mm spacing. Another fact to remember is that doubly curved surfaces cannot in general be generated exactly (there is a continuing and variable geometry mismatch between the tool and the curved surface) and so surface machining cusps are inevitable.

In this chapter the definition and machining of curved surfaces is discussed. The positioning of a variety of shaped milling cutters in relation to arbitrarily curved surfaces is considered and equations derived. The description and listing of a series of C-based computer programs are given for all aspects of surface definition and machining.

6.1 DATA INPUT

Data to describe a surface can be obtained from a variety of sources. A prime objective of the data input phase is to convert the data, in whatever form it is presented, into a suitably formatted data file at sufficient resolution for machining purposes. In the following work, and the associated computer programs, the surface data are organized in lines and vertices, as shown in Figure 6.1. A line consists of an ordered array of surface points or vertices, and the surface is defined by a number of ordered lines. For machining, the surface is defined by a rectangular matrix of data points. This does not mean that the surface need be rectangular but that there should be an equal number of vertices per line. The three coordinate components (X, Y, and Z) of vertex number [3] on line number [2] are expressed in the form X[2][3], Y[2][3] and Z[2][3].

Sec.6.1]   **Data Input**   145

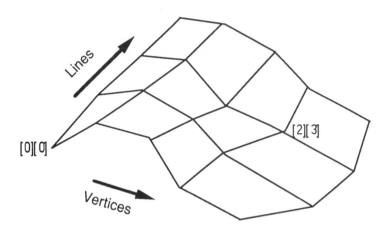

Figure 6.1: Surface definition using an array of lines and vertices

The data file format for describing surfaces is given in Table 6.1. The first line in the file lists the number of lines on the surface. For each line the number of vertices on that line is listed. The X, Y and Z coordinate components of each vertex occupies one line in the file.

Table 6.1
Data File Format

```
4                                      Number of lines in file
4                                      Number of vertices in line
    0.000000   0.000000    6.945513    X, Y, and Z coordinates
   11.112825   0.000000    3.195067
   22.225651   0.000000    3.438101
   30.004660   0.000000    9.168039
4                                      Number of vertices in line
    0.000000   0.000000    6.945513    X, Y, and Z coordinates
   11.112825   7.396981    5.695435
   22.225651   6.250916    6.806497
   30.004660   0.000000    9.168039
4                                      Number of vertices in line
    0.000000   0.000000   11.078021    X, Y, and Z coordinates
   11.112825   7.396981   10.904495
   22.225651   6.667675   12.085170
   30.004660   0.000000   13.543784
4                                      Number of vertices in line
    0.000000   0.000000   11.668512    X, Y, and Z coordinates
   11.112825   7.396981   11.598998
   22.225651   6.756888   13.213834
   30.004660   0.000000   15.002296
```

6.2 SURFACE MODELLING

A popular method for fitting curves and surfaces, in computer aided manufacturing applications, is through the use of fourth order B-spline approximations. These B-splines generate a shape that has continuity in position, slope and curvature. They have a variation diminishing property which forces the curve or surface to lie within the convex hull of the enclosing polygon, thereby smoothing the data vertices, as shown in Figure 6.2. B-splines also have the property that changes to control vertices only affect local shape and not the complete curve or surface.

In the curve and surface fitting programs given below, parametric B-spline approximation curves are defined as:

$$R(t) = \sum_{Vertex=0}^{NVertices} P_{Vertex} B_{Vertex,4}(t) \tag{1}$$

where $R(t)$ is the parametric B-spline approximation, P is the control polygon (i.e. data vertices) and $B(t)$ is the basis function value. Parametric B-spline approximation surfaces are defined as:

$$R(u,v) = \sum_{Line=0}^{NLines} \sum_{Vertex=0}^{NVertices} Q_{Line,Vertex} B(u)_{Vertex,4} B(v)_{Line,4} \tag{2}$$

where $R(u,v)$ is the parametric B-spline approximation, Q is the control polyhedron (i.e. data vertices) and $B(u)$ and $B(v)$ are the normalized basis function values.

The basis function values are calculated as follows:

$$B_{Vertex,Order}(t) = \left\{\frac{t - t_{Vertex}}{t_{Vertex+Order-1} - t_{Vertex}}\right\} B_{Vertex,Order-1}(t) + \left\{\frac{t_{Vertex+Order} - t}{t_{Vertex+Order} - t_{Vertex+1}}\right\} B_{Vertex+1,Order-1}(t) \tag{3}$$

The initial conditions for order one are:

$$B_{Vertex,1}(t) = 1 \quad \text{if } t_{Vertex} \leq t < t_{Vertex+1}$$
$$B_{Vertex,1}(t) = 0 \quad \text{otherwise} \tag{4}$$

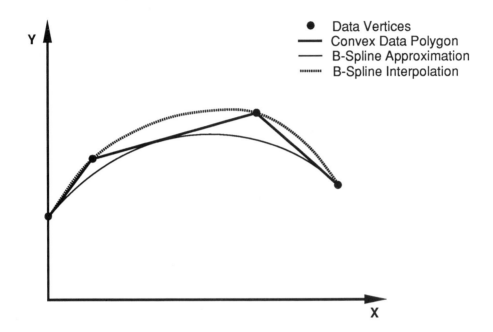

Figure 6.2 B-spline approximation with enclosing polygon

In order to understand the formation and addition of basis functions to produce a composite B-splines curve a series of divided difference calculations, based on Equation (2), is shown in Table 6.2. The ordered data vertices which form the control polygon are given in Table 6.2 as 0,1,2,...6,7,8. Multiple data vertices are added at the end regions and the corresponding knot points established as t_0, t_1,...t_{13}, t_{14}. The resulting basis functions for the first, second, third and fourth order divided differences are calculated for both the first and sixth data intervals and the final basis functions are plotted in Figure 6.3. It can be seen that the basis functions in the first data interval, where there are multiple knots (t_0, t_1, t_2, t_3) are not symmetric while the interior steady state regions are formed from similar symmetric shaped basis functions.

Table 6.2
Basis Function by Divided Difference

Data Points	Knot Points	Basis Function Order			
		k = 1	k = 2	k = 3	k = 4
0	t_0	$B_{0,1} = 0$	$B_{0,2} = 0$	$B_{0,3} = 0$	$B_{0,4} = (1-t)^3$
0	t_1	$B_{1,1} = 0$	$B_{1,2} = 0$	$B_{1,3} = (1-t)^2$	$B_{1,4} = 3t - \frac{9}{2}t^2 + \frac{7}{4}t^3$
0	t_2	$B_{2,1} = 0$	$B_{2,2} = (1-t)$	$B_{2,3} = 2t - \frac{3}{2}t^2$	$B_{2,4} = \frac{3}{2}t^2 - \frac{11}{12}t^3$
0	t_3	$B_{3,1} = 1$	$B_{3,2} = t$	$B_{3,3} = \frac{t^2}{2}$	$B_{3,4} = \frac{t^3}{6}$
1	t_4	$B_{4,1} = 0$	$B_{4,2} = 0$	$B_{4,3} = 0$	$B_{4,4} = 0$
2	t_5	$B_{5,1} = 0$	$B_{5,2} = 0$	$B_{5,3} = 0$	$B_{5,4} = \frac{(1-t)^3}{6}$
3	t_6	$B_{6,1} = 0$	$B_{6,2} = 0$	$B_{6,3} = \frac{(1-t)^2}{2}$	$B_{6,4} = \frac{1}{6}(4 - 6t^2 + 3t^3)$
4	t_7	$B_{7,1} = 0$	$B_{7,2} = (1-t)$	$B_{7,3} = \frac{1}{2} + t - t^2$	$B_{7,4} = \frac{1}{6}(1 - 3t + 3t^2 - 3t^3)$
5	t_8	$B_{8,1} = 1$	$B_{8,2} = t$	$B_{8,3} = \frac{t^2}{2}$	$B_{8,4} = \frac{t^3}{6}$
6	t_9	$B_{9,1} = 0$	$B_{9,2} = 0$	$B_{9,3} = 0$	$B_{9,4} = 0$
7	t_{10}	$B_{10,1} = 0$	$B_{10,2} = 0$	$B_{10,3} = 0$	$B_{10,4} = 0$
8	t_{11}	$B_{11,1} = 0$	$B_{11,2} = 0$	$B_{11,3} = 0$	$B_{11,4} = 0$
8	t_{12}	$B_{12,1} = 0$	$B_{12,2} = 0$	$B_{12,3} = 0$	$B_{12,4} = 0$
8	t_{13}	$B_{13,1} = 0$	$B_{13,2} = 0$	$B_{13,3} = 0$	$B_{13,4} = 0$
8	t_{14}	$B_{14,1} = 0$	$B_{14,2} = 0$	$B_{14,3} = 0$	$B_{14,4} = 0$

Sec.6.2] **Surface Modelling** 149

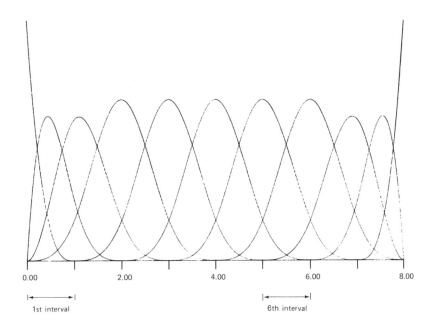

Figure 6.3 Basis functions developed from the divided difference formulation

The direct evaluation of B-spline basis functions as divided differences can generate numerical errors. This can be avoided using the recursive B-spline curve algorithm given below. The recursive algorithm variables are defined as:

- **Basis** is the value of the basis function B
- **Order** is the order of the basis function
- P_{Vertex} are the values of the coordinate components of a vertex in the control polygon
- $R_{SplineVertex}$ are the values of the coordinate components of a vertex in the resulting B-spline
- **SplnVrtx** is the current vertex position in the resulting B-spline
- **Knot** is the current position in the knot vector
- **NKnots** is the number of knots in the knot vector
- **NVertices** is the number of vertices on the control polygon
- **T** is the current location within the parametric knot interval
- **TStart** is the start of the parametric knot interval
- **TEnd** is the end of the parametric knot interval

- **DeltaT** is the series of incremental divisions into which the parametric knot interval is divided.

Step 1: Add a vertex in each end interval of the control polygon P at one-third the length of the interval from the ends.

Step 2: Initialize the knot vector to 0 for the first three knots, to the value of the control vertex number (starting at 0, up to NVertices-1), and to the value of the last vertex number for the last three knots.

Step 3: Initialize the coordinate components of $R_{SplineVertex}$ to zero at SplnVrtx = 0.

Step 4: Start a **for** loop with   3 = Knot <= NVertices

   Step 4.1: Start a **for** loop with   0 = n < NKnots, initialise the basis function values for order one using Equation 4

   Step 4.2: Set parametric knot interval variables TStart = Knot, TEnd = Knot+1.0,   T = TStart

   Step 4.3: Step through the knot interval with **do while** T < TEnd

      Step 4.3.1: Start a **for** loop Order = 2, 3, 4
      Start a **for** loop   0 = n < NKnots-2

         Step 4.3.1.1 Calculate the basis function values $B_{n,Order}$ using equation 4

      Step 4.3.2: Start a **for** loop with   0 = Vertex < NVertices
      Calculate the B-spline vertex components as a summation $R_{SplineVertex} = R_{SplineVertex} + P_{Vertex}B_{Vertex,4}$

      Step 4.3.3: Increase T to T=T+DeltaT. Increase SplnVrtx by one and initialize the values of $R_{SplineVertex}$ to zero.

The recursive B-spline surface algorithm is essentially similar except the iterative procedure is carried out in both the u and v directions.

6.2.1 Compiling and Using the Programs

Programs for defining and machining curved surfaces are described in the following sections. They are written in C programming language and presented as short, self-contained and functioning programs with relevant descriptions. Some frequently used and general functions are included in a library. A knowledge of C-programming is assumed.

Sec.6.2]   **Surface Modelling**   151

Each executable program is the result of compiling one or more C source code files and linking these with the library file. Each source code file that calls library functions has to include the library header file. All files generating an executable program are located in a self-contained directory. A typical directory FITSURF containing the main source code file FITSURF1.C, supporting functions in FITSURF2.C and function declarations (prototypes) in FITSURF2.H is given as follows:

    \FITSURF
    FITSURF1.C
    FITSURF2.C
    FITSURF2.H

The library functions are contained in a directory called LIBRARY which also contains the prototyping and linkable library file as follows:

    \LIBRARY
    LIBRARY.C
    LIBRARY.H
    LIBRARY.LIB

The functions included in the LIBRARY are listed in Appendix 3 with brief descriptions given below:

- **open_input_file** opens a designated file for reading and returns a file pointer. If the operation is unsuccessful, an error message is printed and the program aborted.
- **open_output_file** opens a designated file for writing and returns a file pointer. If the operation is unsuccessful, an error message is printed and the program aborted.
- **beep** creates a sound of one thousand hertz for a time of one tenth of a second.
- **get_one_double** reads a value from the screen and returns it as a double precision value. If an invalid entry is encountered, a beep sound is made and further data entry requested.
- **check_parameters_two** checks if there are two strings in the DOS input parameter string. If this is not the case, the screen is cleared and an error message is displayed.
- **check_parameters_three** checks if there are three strings in the DOS input parameter string. If this is not the case, the screen is cleared and an error message is displayed.
- **allocate_memory_for_one_line** allocates the memory for a single line of X, Y and Z axis component data. If unsuccessful, an error message is displayed and the program is aborted.
- **read_file_allocate** opens, allocates memory and reads a complete, suitably formatted, data file. If the memory allocation is not successful, an error message is displayed and the program is aborted.

- **read_one_line_allocate** allocates memory and reads, from an open data file, one complete set of vertices defining a line on the surface. The number of vertices read from the file is returned.
- **allocate_two_d_float_array** allocates the memory for a rectangular matrix of single precision floating point values. The size of matrix is specified and the pointer to the allocated memory is returned.
- **write_file** opens an output data file and writes the number of lines, the number of vertices and the surface data in the format shown above.
- **write_one_line** writes to an open data file the number of vertices and the complete set of vertices defining a line on the surface.
- **free_two_d_float_array** deallocates the memory for a rectangular matrix of single precision floating point values.

6.2.2 B-spline Lines

The program, called FITLINE, for fitting a fourth order parametric B-spline approximation curve to a set of three-dimensional data vertices is given in Appendix 4. It consists of two source code files, FITLINE1.C and FITLINE2.C (both of which include function calls to the LIBRARY), and one header file. Once compiled FITLINE is run in DOS by typing:

>FITLINE 'Input Data File' 'Output Data File'

FITLINE1.C consists of a **main** function, which calls three other functions called **check_parameters_two** (a library function), **information** and **fit_3d_curves**. Brief descriptions of the two new functions are as follows:

- **information** clears the screen and then displays information about the program's purpose and the input and output data files on the screen.
- **fit_3d_curves** fits a B-spline curve to the input data and saves the fitted curve data to the output data file. The number of vertices in the input data and the parametric division of the input intervals, delta-T, are read and the number of vertices on the spline calculated. Coordinate components of the input data are defined as pointers and memory is dynamically allocated. A for-loop is used to read a complete line of sparse data vertices from the input file and the curve is fitted using
- **bspline_3d_curve** in FITLINE2.C. The fitted B-spline curve is written to the output file.

FITLINE2.C consists of a function **bspline_3d_curve**, which calls three functions **add_vertices_in_end_intervals**, **knot_vector** and **spline_3d**, which in turn calls **allocate_two_d_float_array** (a library function), **basis_order_1**, **basis_order_n** and **free_two_d_float_array** (a library function). Brief descriptions of the new functions are as follows:

- **bspline_3d_curve** allocates memory and calls the relevant functions to produce a B-spline curve.
- **add_vertices_in_end_intervals** calculates a new vertex in the first interval at one-third of the interval length from the start vertex and another vertex in the last interval at one-third of the interval length from the end vertex.
- **knot_vector** establishes values for the complete set of knots including multiple knots at the beginning and end of the data set.
- **spline_3d** calculates the coordinate components of the B-spline for all output vertices. It uses the modified (with end interval vertices) control polygon, the knot vector and the basis function values obtained from **basis_order_n** in order to calculate the B-spline vertex value.
- **basis_order_1** initializes the values of the first order basis function for each knot interval on the knot vector as given in Equation 6. A for-loop is used to step through all the knots and sets the basis function value to 1.0 if the active knot number is the same as the knot in the loop, otherwise the value is set to 0.0.
- **basis_order_n** calculates the second, third and fourth order basis function values for each knot interval along the knot vector. There are four non-zero basis function at each knot interval and these are used to calculate the values of the basis functions at incremental divisions within the parametric knot intervals.

6.2.3 B-spline Surfaces

The program, called FITSURF, for fitting a fourth order parametric B-spline approximation surface to a regular set of three-dimensional data vertices (i.e. rectangular matrix of vertices) is given in Appendix 5. It consists of two source code files, FITLSURF1.C and FITSURF2.C (both of which include function calls to the LIBRARY), and one header file. Operation of the programs is similar to the line fitting routines given above except calculations are made in the line and vertex directions (i.e. u and v parameters), as shown in Figure 6.1. The number of vertices in the line and vertex directions, together with the delta-T parameter, determines the fineness of the resulting surface grid and hence the roughness of the machined surface. While the delta-T parameter always has a value between zero and one, separate values may be selected in the line and vertex direction. This allows a high resolution to be selected in the direction of high surface curvature (rapidly changing surface shape) together with a low resolution in the direction of low curvature.

6.2.4 AutoCAD Surfaces

Surfaces may be defined in using commercial CAD packages and, once in the correct data file format (shown in Table 6.1), can be machined using the programs given in the following sections. Within AutoCAD, for example, there are five ways of specifying surfaces:
- General polygon meshes
- Ruled surfaces
- Tabulated surfaces
- Surfaces of revolution
- Edge-defined surface patches

Boundary lines can be defined through the regular series of geometry elements, such as line, arc, 3-D polyline etc., and resulting surfaces can be generated with quadratic and cubic B-splines, or Bezier patches. Other surfaces can be obtained using Autolisp, which is a lisp-based programming language encapsulated within AutoCAD. Examples of surfaces generated through AutoCAD are given in Figures 6.4 to 6.8. An example of parametric blending between two boundary lines (in this case a circle and a rectangle) using Autolisp within AutoCAD is given in Figure 6.9.

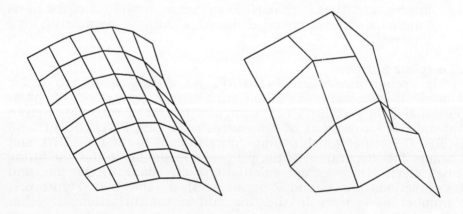

Figure 6.4 General polygon mesh - AutoCAD

Sec.6.2] **Surface Modelling** 155

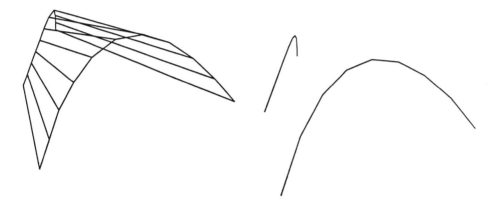

Figure 6.5 Ruled surface - AutoCAD

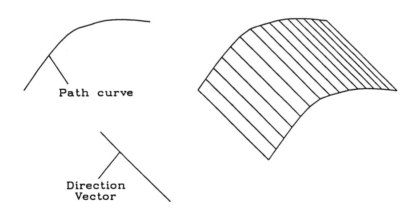

Figure 6.6 Tabulated surface - AutoCAD

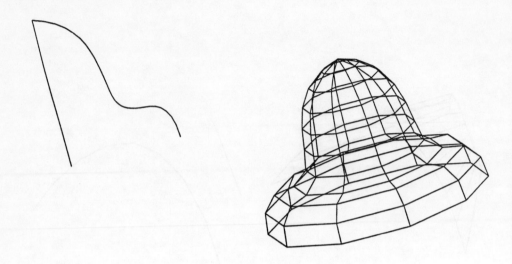

Figure 6.7 Surface of revolution - AutoCAD

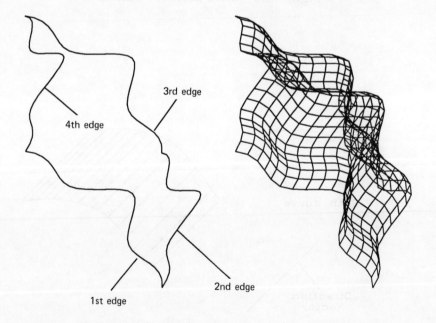

Figure 6.8 Edge-defined surface patches - AutoCAD

Sec.6.3] **Surface Orientation** 157

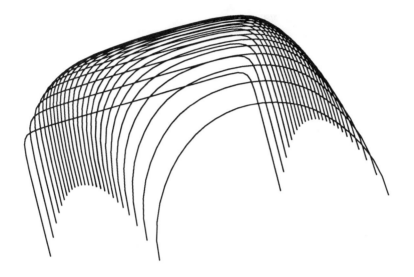

Figure 6.9 Parametric blending using Autolisp - AutoCAD

6.3 SURFACE ORIENTATION

Once the surface is defined, as a matrix of data vertices, it may require orientating in some way to the cutting tool and the machine tool axes. Homogeneous coordinate transformations may be used for this purpose. If a matrix of surface data points is represented by $[P_s]$ as follows:

$$[P_s] = \begin{bmatrix} X_{11}, & X_{12} & ...X_{21}...X_{NM} \\ Y_{11}, & Y_{12} & ...Y_{21} \; Y_{NM} \\ Z_{11}, & Z_{12} & ...Z_{21} \; Z_{NM} \\ 1, & 1 & ...1 \quad 1 \end{bmatrix} \qquad (5)$$

and the rotation, translation and scaling matrix by $[R]$, which is defined as:

$$[R] = \begin{bmatrix} a & b & c & T_x \\ e & f & g & T_Y \\ h & i & j & T_z \\ 0 & 0 & 0 & S \end{bmatrix} \qquad (6)$$

where a, b, ... j specify rotation, T_x, T_y, T_z specify translation and S specifies overall scaling, then the manipulated data set can be represented by $[P_M]$ as follows:

$$[P_M] = [R][P_s] \tag{7}$$

The matrix $[P_M]$, given as follows:

$$[P_M] = \begin{bmatrix} X_{M11}, & X_{M12} & ...X_{M21}, & ...X_{MNM} \\ Y_{M11}, & Y_{M12} & ...Y_{M21}, & ...Y_{MNM} \\ Z_{M11}, & Z_{M12} & ...Z_{M21}, & ...Z_{MNM} \\ S, & S, & ...S, & ...S \end{bmatrix} \tag{8}$$

is thus a description of the curved surface relative to the machine table.

For machining operations that require the X-Z plane lie on the machine table and that the tool approaches the workpiece from the Y-direction, the matrix $[R]$ is given as follows:

$$[R] = \begin{bmatrix} 1 & 0 & 0 & T_x \\ 0 & 0 & -1 & T_y \\ 0 & 1 & 0 & T_z \\ 0 & 0 & 0 & 1 \end{bmatrix} \tag{9}$$

The program, called ROTATE, for rotating a surface (defined by a set of data vertices) is given in Appendix 6. It consists of one source code file ROTATE.C (which includes function calls to the LIBRARY). The program applies the rotational components of the homogeneous coordinate transformations matrix (6) to a surface of data vertices.

ROTATE.C consists of a **main** function which calls four other functions, namely: **check_parameters_two**, **information**, **get_one_double_value** and **rotate_surface_file**. The function **get_one_double_value** is called three times to read the rotation angles about the X, Y and Z-axes. A brief description of the new function is as follows:

- **rotate_surface_file** opens the surface input file and the rotated surface output file. The surface is read one vertex at a time and the rotational equations are applied to each component of the vertex. The results are then written to the output file.

The program, called TRANSLAT, for translating a surface (defined by a set of data vertices) is given in Appendix 7. It consists of one source code file TRANSLAT.C (which includes function calls to the LIBRARY). The program applies the translational components of the homogeneous coordinate transformations matrix (6) to a surface of data vertices.

TRANSLAT.C consists of a **main** function which calls four other functions namely: **check_parameters_two**, **information**, **get_one_double_value** and **translate_surface_file**. The function **get_one_double_value** is called three times to read the translational components in the X, Y and Z direction. A brief descriptions of the new function is as follows:

- **translate_surface_file** opens the surface input file and the translated surface output file. The surface is read one vertex at a time and the translational equations are applied to each component of the vertex. The results are then written to the output file.

6.4 SURFACE NORMAL CALCULATION

In order to position a cutting tool relative to the correctly oriented surface vertices, the surface normal vectors are required at every vertex. The vector offsets to the tool centre can be calculated and the surface vertices and vector offsets added together to give the absolute position of the tool centre in space.

The surface normal unit vector, $\bar{n}$, at each vertex can be calculated, see Figure 6.10, as the cross-product of the two difference vectors, $\bar{D}_{Line}$ and $\bar{D}_{Vertex}$, formed from the four adjacent vertices so that:

$$\bar{n} = \frac{\bar{D}_{Line} \times \bar{D}_{Vertex}}{|\bar{D}_{Line} \times \bar{D}_{Vertex}|} \tag{10}$$

where

$$\bar{D}_{Line} = P_{[Line+1, Vertex]} - P_{[Line-1, Vertex]}$$
$$\bar{D}_{Vertex} = P_{[Line, Vertex+1]} - P_{[Line, Vertex-1]} \tag{11}$$

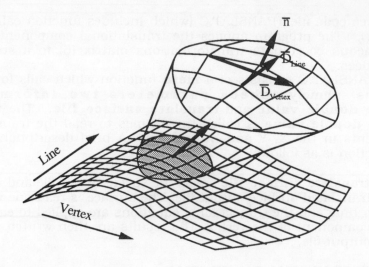

Figure 6.10 Surface normal vector at a surface vertex

A program, called SURFNORM, for calculating surface normal vectors to a regular matrix of data vertices is given in Appendix 8. It consists of one source code file SURFNORM.C (which includes function calls to the LIBRARY). Within the program a means of calculating vector normals (defined by Equations 10 and 11) at the boundaries of the surface has to made. The approach used in the program (and one which saves considerable complexity in handling special cases) is to extend the surface artificially by duplicating the four sets of end vertices. Figure 6.11 shows the memory representation of a surface defined by four lines and six vertices per line. The vertex numbers in the shaded centre field represent the original surface polyhedron vertex numbers.

	Vertex							
		0	0	1	2	3	4	5	5
Line	0	0,0	0,0	0,1	0,2	0,3	0,4	0,5	0,5
	0	0,0	0,0	0,1	0,2	0,3	0,4	0,5	0,5
	1	1,0	1,0	1,1	1,2	1,3	1,4	1,5	1,5
	2	2,0	2,0	2,1	2,2	2,3	2,4	2,5	2,5
	3	3,0	3,0	3,1	3,2	3,3	3,4	3,5	3,5
	3	3,0	3,0	3,1	3,2	3,3	3,4	3,5	3,5

Figure 6.11 Matrix of surface vertices in memory

SURFNORM.C consists of a **main** function which calls three other functions, namely: **check_parameters_two**, **information** and **make_normal_vectors_on_surface**. The new function **make_normal_vectors_on_surface** in turn calls the functions **read_one_line**, **copy_one_line**, **make_normal_vector_on_one_line**

Sec.6.4]   **Surface Normal Calculation**   161

and **write_one_line** (a library function). Brief descriptions of these functions, which are listed in Appendix 8, are given below:

- **make_normal_vectors_on_surface** makes the function calls to calculate the surface normal vectors on the surface. To conserve memory the data is loaded three lines at a time, as shown in Figure 6.12.

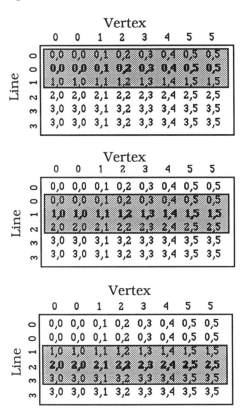

Figure 6.12 Data loading scheme to calculate the normal vectors on the line 0, 1 and 2

- **read_one_line** reads the coordinate components of one line of data vertices from the surface file into memory and arranges the data in the extended format.
- **copy_one_line** copies the X, Y and Z components from the specified source to the specified destination.
- **make_normal_vector_on_one_line** calculates the surface normal vectors for all vertices on a line. Three extended lines of surface vertices are loaded into memory. The normal vector for all vertices on the middle line are calculated, as given in Equations (10) and (11).

6.5 CUTTER LOCATION CALCULATION

A generalized shaped milling cutter can be represented, as shown in Figure 6.13. The shape is defined by a shank radius, R_1, and a fillet radius, R_2. A ball-mill is represented when the shank radius and the fillet radius are equal, ($R_1 = R_2$), and an end-mill is represented when the fillet radius is zero, as shown in Figure 6.14.

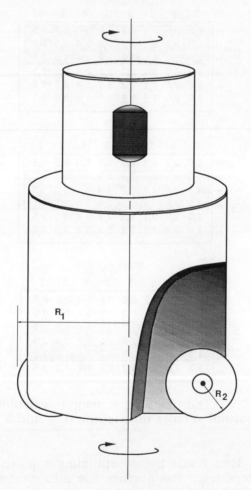

Figure 6.13 Generalized shaped milling cutter

In order to position the cutting tool relative to the surface polyhedron, the vector offset of the contact point on the tool to some arbitrary (but fixed) reference point on the tool is required. It is conventional to use the tool base centre (marked P_T in Figure 6.14) as

Sec.6.5] **Cutter Location Calculation** 163

the tool reference point. All subsequent programming is then referenced to this point on the cutting tool.

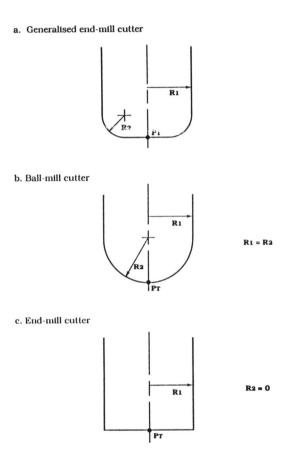

Figure 6.14 Milling cutter representation

The surface polyhedron with normal vector, $\bar{n}$, can be represented as shown in Figure 6.15. The resulting surface normal vector in the X-Y plane, $\bar{n}_{xy}$, is formed from two vector cross products as shown in Figure 6.16. The cross product $(\bar{k} \times \bar{n})$ gives a vector in the X-Y plane which is tangent to the surface normal (i.e. part of a horizontal contour line on the surface) while the cross product $((\bar{k} \times \bar{n}) \times \bar{k})$ gives a vector which is normal to the surface contour line.

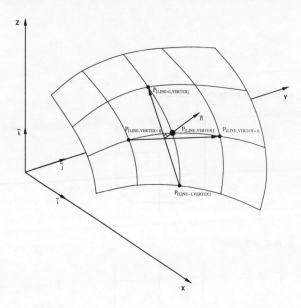

Figure 6.15 Surface polyhedron with normal vector

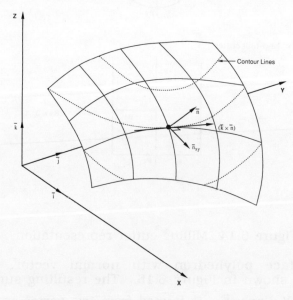

Figure 6.16 Surface polyhedron with normal vectors in the X-Y plane

Sec.6.5] **Cutter Location Calculation** 165

Thus the surface normal unit vector can be calculated as

$$\bar{n}_{xy} = \frac{(\bar{k} \times \bar{n}) \times \bar{k}}{|(\bar{k} \times \bar{n}) \times \bar{k}|} \quad (12)$$

This can be achieved numerically by projecting the surface normal vector $\bar{n}$ into the X-Y plane (i.e. taking the X and Y components of $\bar{n}$ and setting the z component to zero) and normalizing with respect to $\bar{n}_{x,y}$ as

$$\bar{n}_{x,y} = \left[\frac{n_x}{|n_{x,y}|}, \frac{n_y}{|n_{x,y}|}, 0 \right] \quad (13)$$

The absolute position of the reference point, P_T, for a generalised shape milling cutter is obtained by adding the surface vertex vectors and the tool offset vectors (or vector path from the cutter/surface contact point P_M and the reference point P_T), as shown in Figure 6.17.

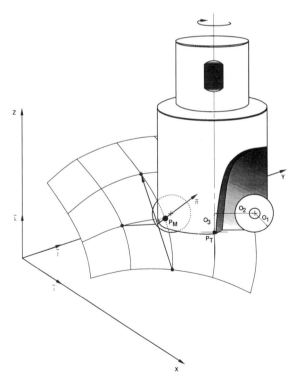

Figure 6.17 Generalized end-mill positioned at a surface vertex

The position of the tool reference point, P_T, may be given as,

$$[P_T] = [P_M] + [O_1] + [O_2] + [O_3] \tag{14}$$

Where the tool offset vectors $\overline{O}_1$, $\overline{O}_2$, and $\overline{O}_3$ are given by

$$\overline{O}_1 = \overline{n} R_2$$

$$\overline{O}_2 = \overline{n}_{xy}(R_1 - R_2)$$

$$\overline{O}_3 = \overline{k} R_2 \tag{15}$$

The offset for a ball-mill is thus given by

$$O_B = \overline{n} R_2 - \overline{k} R_2 \tag{16}$$

and the offset for an end-mill is given by

$$O_E = \overline{n}_{xy} R_1 \tag{17}$$

6.5.1 Cutter Offsets Program

The program, called OFFGEN, for calculating the generalized shaped cutter offsets for a regular matrix of data vertices is given in Appendix 9. It consists of one source code file OFFGEN.C (which includes calls to the LIBRARY). The offsets are automatically calculated for an end-mill by entering a fillet radius of zero and for a ball-mill by entering a fillet radius equal to a shank radius. The **main** function calls the four functions **check_parameters_two**, **information**, **get_one_double** and **calculate_offsets_general_mill** which in turn calls the function **check_for_equal**. The function **get_one_double** is called twice, to get the fillet and shank radii.

- **calculate_offsets_general_mill** opens an input file with the normal vector data and an output offsets file. Two for-loops lead through the surface. A single normal vector is read from the input file. The offset components and the surface normal vector in the X-Y plane are calculated, as shown in Equations (13) and (15). The function **check_for_equal** is called whenever the size of the input files has to be compared with the declared value (i.e. after reading the number of lines and the number of vertices on a line). The results of the offset calculation are written to the cutter location output file.

- **check_for_equal** compares two integer numbers. If the numbers are not equal, an error message is displayed and the program aborted.

6.5.2 Cutter Location Program

The program, called ADDOFFST, for calculating the cutter location file is given in Appendix 10. It consists of the source code file ADDOFFST.C (which includes calls to the LIBRARY). The **main** function calls the three functions **check_parameter_three**, **information** and **add_offsets_to_surface**.

- **add_offsets_to_surface** opens the surface vertex file, the cutter offsets file, and the cutter location file. While reading the surface vertex file and the cutter offsets file, calls to **check_for_equal** ensure that both files are the same size (i.e. the same number of lines and vertices). A vertex at a time is read from both files, added together and a cutter location value written to the cutter location file.

- **check_for_equal** compares two integer values. If they are not the same, an error message is posted and the program aborted.

6.6 G-CODE GENERATION

In order to machine the curved surface a selected cutter is moved across the surface, between adjacent vertices, in a series of straight line interpolation moves. The spacing of such movements, and hence the smoothness of the finished surface, is determined by the resolution of the surfaces vertices defined in Section 6.1 and 6.2. The program described below, called MAKEGCD, converts the geometrical information, in the cutter location file, into G-code instructions for the CNC machine tool. It creates a standard start and end sequence for the G-code file and interactively requests the following machining information:

- Clearance-over-zero - a safety height at which the cutter will travel, at rapid speed, to new machining locations.
- XStart, YStart and ZStart - cutter location at the beginning of the program.
- Feedrate - the velocity (units/minute) at which the cutter will travel when G01 instructions are executed.
- Spindle speed - rotational speed of the cutting tool spindle in revolutions per minute.
- Machining procedure - the machining path pattern would normally be defined in the order that the vertices are read from the cutter location file matrix (machine forward). In this case the cutter travels along one line of the path from the first to the last vertex, it then lifts to the defined clearance height and travels to the start of the next line. Within the program the machining path can also be defined in the reverse order (machine backward) and in an alternating forward and backward pattern (machine alternate). The latter pattern saves non-cutting motion, although it

introduces a lack of symmetry into the cutting process. In this interlaced pattern the cutter is always removing material. Other patterns can be introduced in the software merely by reordering the data as it is read from the cutter location file.
- Measurement system - inch and millimeter measurements are used to describe the surface.

The program, called MAKEGCD, for generating G-code for machining a three-dimensional arbitrary shaped surface is given in Appendix 11. The program reads a cutter location file and writes a G-code instruction file. It consists of one source code file MAKEDGCD.C (which includes calls to the LIBRARY). Once compiled the program is run in DOS by typing:

>MAKEGCD 'Input Data File' 'Output Data File'

MAKEGCD.C consists of a **main** function which calls the function **check_parameters_two**, **information** and **make_gcode_for_surface** which calls the functions **get_one_double**, **get_machining_procedure**, **get_measurement_system**, **header**, **make_gcode_forward**, **make_gcode_backward**, **make_gcode_alternating** and **ender**. Also used are the functions **make_gcode_forward_one_line**, **make_gcode_backward_one_line**, **gcode_begin_line_g00** and **read_one_line**. Brief descriptions of the new functions are given below:

- **make_gcode_for_surface** opens the input and output files and requests from the user the clearance-over-zero, XStart, YStart and ZStart, feedrate, spindle speed, machining procedure (**get_machining_procedure**) and measurement system (**get_measurement_system**). The header for the G-code file is produced by a call to the function **header**. A switch branches to the function for the desired machining procedure (**make_gcode_forward**, **make_gcode_backward**, **make_gcode_alternating**). Finally the call to the function **ender** completes the G-code file.

- **get_machining_procedure** requests the user to select the machining procedure and returns it to the calling function.

- **get_measurement_system** requests the user to select the measurement system and returns it to the calling function.

- **header** writes a standard Fanuc G-code start sequence to the output file. It writes a G92 with the start location of the cutter. Depending upon the selected measurement system the appropriate code, G20 or G21, is written. The next line (G90) specifies absolute programming. Finally the spindle speed (S) and the spindle on code (M03) are written.

- **make_gcode_forward** allocates the memory for the X, Y, Z coordinates large enough to hold one line, with up to 2000 vertices. Within a for-loop a line of cutter location coordinates is read by a call to **read_one_line** and the G-code is written by calling the function **make_gcode_forward_one_line**.

- **make_gcode_backward** allocates the memory for the X, Y, Z coordinates large enough to hold one line with up to 2000 vertices. Within a for-loop a line of cutter location coordinates is read by a call to **read_one_line** and the G-code is written by calling the function **make_gcode_backward_one_line**.

- **make_gcode_alternating** allocates the memory for the X, Y, Z coordinates large enough to hold one line with up to 2000 vertices. One line of cutter locations is read by a call to **read_one_line** and the G-code is written by a call to **make_gcode_forward_one_line**. Within a while-loop a line of cutter location coordinates is read by a call to **read_one_line** and the G-code written by alternate calls to the functions **make_gcode_backward_one_line** and **make_gcode_forward_one_line**.

- **read_one_line** reads one line of X,Y and Z coordinates from the cutter location file.

- **ender** writes a standard Fanuc G-code termination sequence to the output file. First the cutter is moved fast to the clearance over part height, then return to the start X and Y position and then moved to the start Z position. Finally the spindle is switched off (M05) and the program termination character (%) is written in the last line of the program.

- **make_gcode_forward_one_line** writes the G-code for one complete machining line in forward direction. Depending on the specified start condition it moves to the start vertex with G00 instructions (rapid) by calling the function **gcode_begin_line_g00** or with a G01 instructions. Within a for-loop that works through all vertices on the line, G01 instructions are written to the output file so that the cutter moves in straight line segments to the new cutter locations.

- **make_gcode_backward_one_line** writes the G-code for one complete machining line in reverse direction. Depending on the specified start condition it moves to the end vertex with G00 instructions (rapid) by calling the function **gcode_begin_line_g00** or with a G01 instructions. Within a for-loop that works through all vertices on the line in reverse manner, G01 instructions are written to the output file so that the cutter moves in straight line segments to the new cutter locations.

- **gcode_begin_line_g00** moves the cutter rapid (G00) to the clearance height, then moves it rapid (G00) to the new X and Y locations and then moves it under controlled feedrate (G01) down to the new Z coordinate location.

6.7 TOOLS FOR SURFACE DATA MANIPULATION

6.7.1 Inverting the Surface Normal Vector

Direction of the surface normal vectors is determined by the line and vertex vector cross-product, given in Equation 10. A component surface may be concave or convex; and the program SURFNORM (Section 6.4) has no means of knowing, for subsequent machining purposes, which direction is waste material and which direction is good material. Thus, after the surface normal vector is calculated, the direction has to be checked to see if it points to the correct side of the component surface; if this is not the case, the set of surface normals needs to be inverted.

The program, called INVNORM, for inverting the direction of the surface normal unit vectors on an arbitrary shaped surface is given in Appendix 12. The program reads a surface normal unit vector file and writes the inverted vectors to the output file. It consists of one source code file INVNORM.C (which includes calls to the LIBRARY). Once compiled the program is run by typing:

>INVNORM 'Input Data File' 'Output Data File'

INVNORM.C consists of a **main** function which calls the functions **check_parameters_two**, **information** and **invert_normal_vector**. A brief description of the new function is given below:

- **invert_normal_vector** opens the input and the output files. Within two nested for-loops the values of each surface normal unit vector in all three components are read. These values are multiplied by minus one and written to the output file.

6.7.2 Transposing the Surface Matrix

As shown in Section 6.1, a surface is represented by an ordered set of vertices and lines. Surface normal, cutter location and G-code calculations (which determines the direction of surface machining) are all similarly ordered. In order to machine in the orthogonal direction, it is necessary that the rows and colums of the surface matrix be interchanged. It is noted that machining in both directions across the surface produces a much smoother surface than simply increasing the density of points on the surface.

Sec.6.7]   **Tools for Surface Data Manipulation**   171

The program, called TRANSPOS, for transposing the surface matrix is given in Appendix 13. The program reads a surface file and writes the surface transposed to the output file. It consists of one source code file TRANSPOS.C (which includes calls to the LIBRARY). Once compiled the program is run by typing:

>TRANSPOS 'Input Data File' 'Output Data File'

TRANSPOS.C consists of a **main** function which calls the functions **check_parameters_two**, **information** and **transpose_surface** which in turn calls the functions **read_file_allocate** and **write_file_transposed**. Brief descriptions of the new functions are given below:

- **transpose_surface** calls the function **read_file_allocate** which allocates and reads a complete surface file. A call to **write_file_transposed** writes the file transposed to the output file.

- **write_file_transposed** opens the output file. Two nested for-loops work through the surface. Instead of having the outer loop counting the line and the inner loop counting the vertices, the outer loop counts the vertices and the inner loop counts the lines. Within the inner loop the coordinates of the surface vertices are written to the output file.

6.7.3 Reducing the Surface Resolution

During surface fitting the surface is generally fitted in a resolution necessary for the finishing cut. However, to advance from the raw material block to the finished part, a number of roughing cuts are usually necessary. It is advantageous to reduce the number of lines and vertices on the surface description to save machining time.

The program, called RSKIP, for reducing the surface resolution for roughing is given in Appendix 14. The program reads a surface file and writes the reduced resolution surface to the output file. It consists of one source code file RSKIP.C (which includes calls to the LIBRARY). Once compiled the program is run by typing:

>RSKIP 'Input Data File' 'Output Data File'

RSKIP.C consists of a **main** function which calls the functions **check_parameters_two**, **information** and **lines_and_vertices_to_skip** and **skip** which in turn calls the new functions **how_many_new_values** and **skip_this_one**. Brief descriptions of the new functions are given below:

- **lines_and_vertices_to_skip** reads from the screen the number of lines and the number of vertices to skip.

- **skip** opens the input and the output file. After reading the number of lines on the surface, a call to **how_many_new_values** determines how many new lines will be on the reduced resolution surface. A for-loop loops through all the lines. A call to **skip_this_one** determines if this line needs to be skipped. If this is the case, the vertices are read from the input file but not written. A call to **how_many_new_values** determines how many new vertices will be on the reduced resolution line. Within the following for-loop, the vertices on the line are read. A call to **skip_this_one** determines if the vertex is not to be skipped. The vertex coordinates are written to the output file.

- **how_many_new_values** determines how many values will remain if a certain number of values are skipped.

- **skip_this_one** determines if the current value needs to be skipped or not.

6.7.4 Adding Extensions to a Surface

Roughing an arbitrary shape surface from a rectangular block of material often requires large and irregular cuts. Plunge cutting, for example, where a cutter moves vertically down into the material, causes machining to take place on the underside of the cutter and can cause many end-mill cutters to encounter solid uncut material at the cutter centre. In these cases it is often advantageous to extend the geometry of the machined surface and to make sure that the start and/or end locations of each line of machining are outside the boundaries of the raw material. In the program given below each surface line is extended by a specified length with straight line interpolation.

The program, called RADDEND, for extending the surface in the line direction is given in Appendix 15. The program reads a surface file and writes the extended surface to the output file. It consists of one source code file RADDEND.C (which includes calls to the LIBRARY). Once compiled the program is run by typing:

>RADDEND 'Input Data File' 'Output Data File'

RADDEND.C consists of a **main** function which calls the functions **check_parameters_two**, **information**, **get_one_double** and **add_ends_to_surface** which in turn calls the function **make_new_vertex**. Brief descriptions of the new functions are given below:

- **add_ends_to_surface** opens the input and the output file. Two nested for-loops go through all the lines and vertices on the surface. The function **make_new_vertex** is called to create the new vertices at both ends of the lines. The extended surface is written to the output file.

- **make_new_vertex** calculates the new X and Y components for the new vertex that extends the surface. The calculation is carried out by using the straight line equations.

6.8 COMPARISON OF BALL-MILL AND END-MILL CUTTERS

Ball-mills rather than end-mills are often used for machining curved surfaces. Historically, the reasons are that ball-mills are easy to position in relation to curved surfaces and generate simple machining programs. However, when machining plane surfaces, end-mills are used as they match the geometry of the required surface. If a ball-mill were used to face a plane surface, it would require many more passes across the surface to generate the same surface finish as that produced with an end-mill.

The same argument holds when machining a wide class of smooth, low curvature surfaces such as those given in Chapter 7. The critical factor for rapid and efficient machining of all these surfaces is that the cutter shape should match the surface shape as closely as possible. The profile of an end-mill can be made to match that of a curved surface by inclining it correctly to the surface normal. As an end-mill is inclined to the surface normal, an elliptical profile is generated, as shown in Figure 6.18, and the effective radius of curvature, r_{eff}, on the cutter axis is given by

$$r_{eff} = \frac{R}{\sin\phi} \qquad (18)$$

where R is the cutter radius and ϕ is the angle of cutter inclination.

Thus, the effective radius of curvature of an end-mill varies from infinite to R as $0 \leq \phi \leq 90$ degrees. The effective radius of a ball-mill is restricted to the spherical radius of the cutter. For an inclination of five degrees, the effective cutter radius of an end-mill is approximately twelve times greater than a ball-mill of similar size, as shown in Figure 6.19.

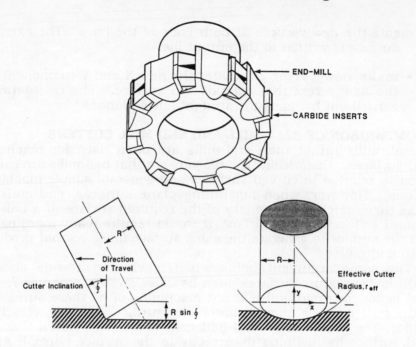

Figure 6.18 The effective radius of an end-mill cutter inclined to a plane surface

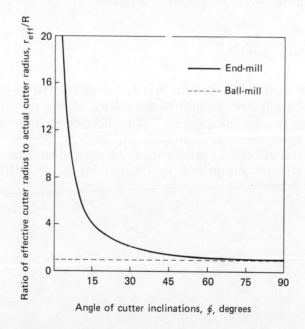

Figure 6.19 Ratio of effective cutter radius to actual cutter radius against angle of cutter inclination

Sec.6.8] **Comparison of Ball-mill and End-mill Cutters** 175

A cusp of material is formed between adjacent cutter paths across the surface, as shown in Figure 6.20. This surface roughness is a function of tool radius, end-mill cutter inclination, cross-feed length, and surface curvature.

The expression for a ball-mill cutter on a plane surface may be given as:

$$L = 2\sqrt{2hR - h^2} \qquad (19)$$

where L is the cross-feed length and h is the cusp height.

The expression for an inclined end-mill cutter on a plane surface may be given as:

$$h = \sin\phi\left\{R - \sqrt{(R^2 - L^2/4)}\right\} \qquad (20)$$

or

$$\frac{1}{L} = \frac{\sin\phi}{2\sqrt{h(2R\sin\phi - h)}} \qquad (21)$$

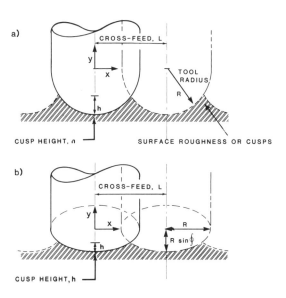

Figure 6.20 Cusp height resulting from ball-mill and end-mill cutters

Surface roughness resulting from ball-mill and inclined end-mill machining for four values of non-dimensionalised cross-feed, L, is shown in Figure 6.21. The results show that for the case of an end-mill of radius 50 mm, cross-feed distance 20 mm and cutter inclination 5 degrees, the resulting h/R ratio is 0.0018 or a cusp height of 0.088 mm. The resulting h/R ratio for the same radius ball-mill is 0.02 or a cusp height of 1 mm. The difference in surface finish is thus a factor of twelve to one. For a similar surface finish (i.e. a cusp height of say 0.25 mm), the number of machining passes across the surface, and hence the machining time, would be approximately six times greater with ball-mills than inclined end-mills.

Another important difference between ball-mills and end-mills is the speed of material removal or cutting speed. An inclined end-mill cuts, as shown in Figure 6.22, at the periphery of the cutter at a cutting speed, S_e, given by:

$$S_e = 2\pi RN, \qquad (22)$$

where N is the speed of cutter rotation. The material is removed at full and predetermined cutting speed, S_e, for all angles of cutter inclination.

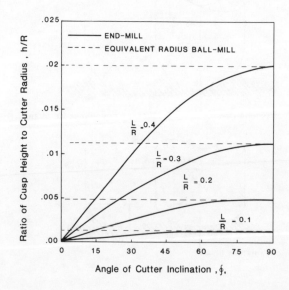

Figure 6.21 Ratio of cusp height to cutter radius against angle of cutter inclination

Sec.6.9] **Machining Through Circular Arc Interpolation** 177

A ball-mill cuts, as shown in Figure 6.22, along an arc that extends from the cutter axis to a point at radius, r, on the ball profile. In the vicinity of the cutter axis the ball-mill is rubbing the material away. The cutting speed, S_b, is given by:

$S_b \min = 0$

$S_b \max = 2\pi r N.$ (23)

An additional factor is that end-mills are readily obtainable in large sizes and with replaceable carbide inserts while ball-mills are usually available in relatively small sizes and made from high-speed steel. The ball-mills do not have the machining capability of carbide insert cutters and they require careful spherical resharpening. If an estimate is made for all these factors, it can be shown that a ball-mill can typically take 20 times longer than an inclined end-mill to machine a wide class of smooth, low curvature surfaces.

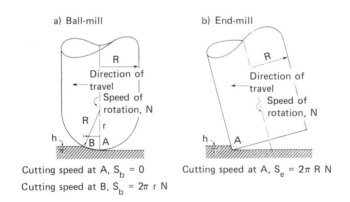

Figure 6.22 Cutting position and speed of ball-mills and end-mills

6.9 MACHINING THROUGH CIRCULAR ARC INTERPOLATION

In curved surface machining it is customary to represent and to generate the surface as a sequence of short **linearly** interpolated moves. Experimental measurements of such stop-start motion show that full cutting speed is achieved typically for only ten percent of the machining time; the remaining time is spent in accelerating, decelerating, or pausing between instructions. An algorithm is presented below which uses a recursive approach to maximize the number of **circularly** interpolated moves for surface generation. Reduction of machining time in the order of 5:1 and compression of the size of the cutter location file of 10:1 can frequently be achieved using this approach.

An example of the experimental velocity measurements with a single circular interpolated movement around a semi-circular path of 40 mm radius and a series of 30 linearly interpolated moves around the same semi-circle is given in Figure 6.23. The measurement configuration on the CNC machine tool is shown in Figure 6.24. It is clear from Figure 6.23 that with circular interpolation the Y-axis motion has a smooth velocity profile and the prescribed feedrate is maintained over the full path with no motion, while with linear interpolation the stop-start and pause motions are quite evident. The time to complete the circular interpolated motion was 2.8 s and 7.5 s for the linear interpolated motion, which is a time savings of 63 percent.

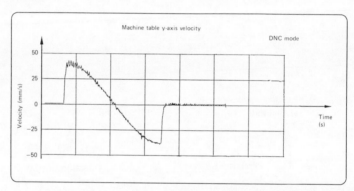

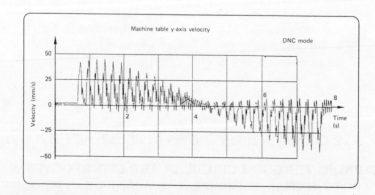

Figure 6.23 Circular and linear interpolation motions

Sec.6.9]   **Machining Through Circular Arc Interpolation**   179

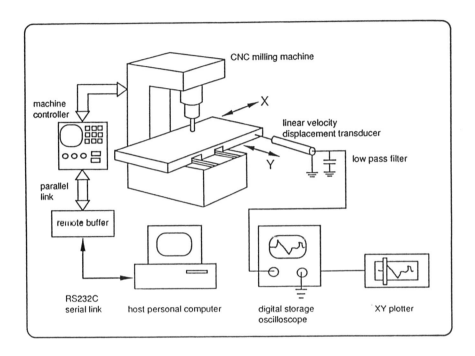

Figure 6.24 Velocity transducer mounted on a CNC machine

Segmenting a tool path, defined by a series of surface data points on a cross-sectional plane, into a sequence of arcs and lines can be accomplished in two stages. Firstly, circular arcs are fit to the data using a recursive testing process and a user-defined maximum allowable error between the fitted arc and the individual data points. Secondly, if the maximum allowable radius on the machine tool controller is exceeded and circular interpolation is not permissible, a straight-line section is fitted through consecutive points. The algorithm may be summarized as follows:

- an initial arc centre point is determined for a span through the entire data set (first, last and middle points) and the arc radius and centre calculated
- a test procedure is invoked to calculate the errors between the arc and the intermediate surface points, as illustrated in Figure 6.25, and given below as

$$\text{error} = |r - d| = \left| r - \left[(x_i - x_c)^2 + (z_i - z_c)^2 \right]^{1/2} \right|$$

(24)

- if the test fails (i.e. any error exceeds the specified maximum error tolerance), the span of the arc is decremented by one point and a new arc fitted through the first point, the new end point, and new mid-point
- if the test passes, the arc definition is retained and the appropriate G-code is generated
- a new arc is then fitted from the current arc end point to the final data point and the above steps repeated.

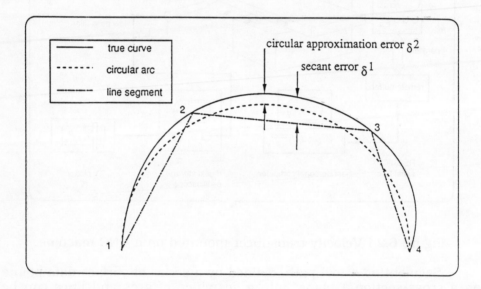

Figure 6.25 Circular and linear interpolation approximation errors

To determine the centre and radius of a circular arc passing through three data points, the equation of a circle of radius r and centre (h, k) is expressed in general form

$$X^2 + Z^2 + aX + bZ + c = 0 \tag{25}$$

where,

$$a = -2h$$
$$b = -2k \tag{26}$$
$$c = h^2 + k^2 - r^2$$

Sec.6.9]   **Machining Through Circular Arc Interpolation**   181

A system of three equations, in the variables a, b, and c, are created for the data points $\{(x_1, z_1), (x_2, z_2), (x_3, z_3)\}$, all of which lie on the arc.

$$\begin{bmatrix} x_1 & z_1 & 1 \\ x_2 & z_2 & 1 \\ x_3 & z_3 & 1 \end{bmatrix} \begin{bmatrix} a \\ b \\ c \end{bmatrix} = \begin{bmatrix} -\left(x_1^2 + z_1^2\right) \\ -\left(x_2^2 + z_2^2\right) \\ -\left(x_3^2 + z_3^2\right) \end{bmatrix} \quad (27)$$

The system is solved by forming the augmented matrix of the system and converting to echelon form through row operations. Back substitution is used to calculate the variables (a, b, c). In the computer implementation, Gaussian elimination with pivoting is employed over other methods to reduce the effect of roundoff errors in the calculation of the variables a, b, and c.

Examples of the segmentation of three data sets are shown in Figures 6.26, 6.27, and 6.28. In all three figures, the initial data set was segmented with an error tolerance of 0.102 mm in the fitting process and a maximum allowable radius of arc of 200.0 mm. In Figure 6.26, the data were obtained from a B-spline surface fitting algorithm (given in Section 6.2) and comprise but one line, from many hundreds, for a cross-sectional surface profile through a Francis turbine blade. In Figures 6.27 and 6.28, the data were generated from a laser scanner; Figure 6.27 shows a cross-section through a human torso and Figure 6.28 a cross-section through an automobile's automatic transmission gear shifter lever.

182 **General Curved Surface Machining** [Ch.6

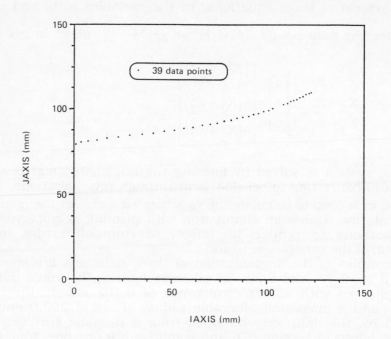

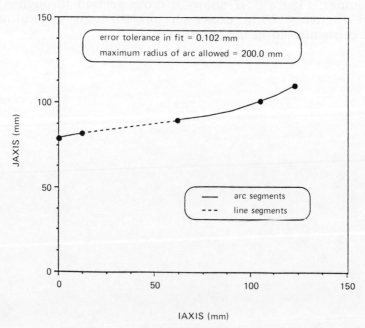

Figure 6.26 Circular interpolation of a turbine blade section

Sec.6.9] Machining Through Circular Arc Interpolation

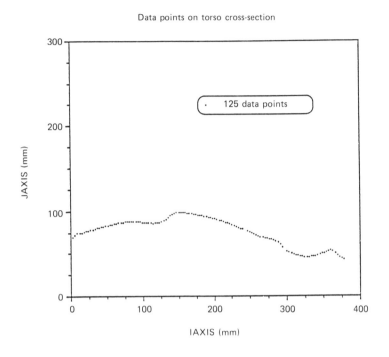

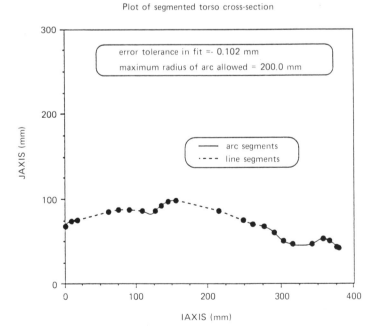

Figure 6.27 Circular interpolation of laser scanned human torso data

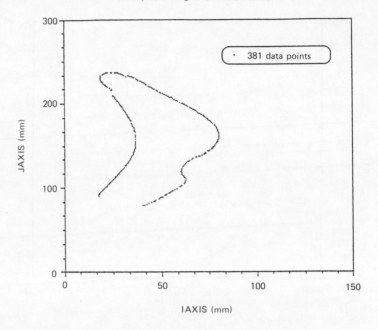
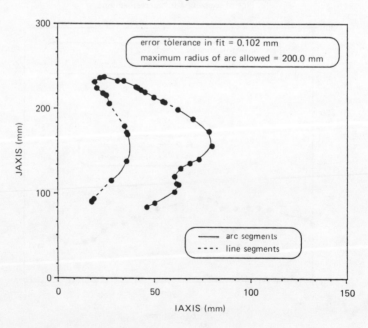

Figure 6.28 Circular interpolation of laser scanned gear lever data

Chapter 7
Curved Surface Machining: Case Studies

Case studies for a variety of curved surface machining applications are discussed. The approaches, techniques and programs are included in the discussion.

7.1 MARINE PROPELLERS

A marine propeller is a propulsion device which consists of a number of identically twisted blades spaced equally around a hub or boss. Propellers are normally manufactured by casting or forging. In both cases a full-scale pattern, normally made of wood or plastic, is required. Usually, the initial pattern is made from a number of separate blade sections mounted at the correct radial and angular positions relative to the axis of the propeller. To give the final blade shape, a skilled hand-blending operation between the sections is employed. For small diameter propellers, the initial pattern may be produced by drilling a series of holes of predetermined depth over the workpiece surface using a special pointed end-drill. Excess material is removed until the tapered holes gradually disappear and the blade shape remains. In the following section a method is presented for defining a marine propeller from a table of offsets and machining it using CNC machine tools.

7.1.1 Surface Definition

The blade shape is defined, in a table of offsets, by a series of aerofoil sections at specified radius ratios of the propeller, as shown in Figure 7.1. The face of each blade is part of a helicoidal-like surface which can be generated (for a true helicoidal surface) by a radial line which rotates about an axis and also moves along this axis at a distance directly proportional to the angular movement. The twist of the blade is defined by the helix, or pitch, angle, which may vary with the radius, while the inclination of the blade along the axis is given by the rake angle.

An aerofoil section can be defined in three-dimensional space by an arc length **distance s** (corresponding to the width or cord direction of the blade) measured along a helical path (pitch datum line) and an arc length **distance t** (corresponding to the thickness direction of the blade) measured along an orthogonal helical path, as shown in Figure 7.1. Where the face of the blade does not lie on a true helicoidal surface dimension, t_F and t_B in Figure 7.1 are needed to define the face and back surface respectively.

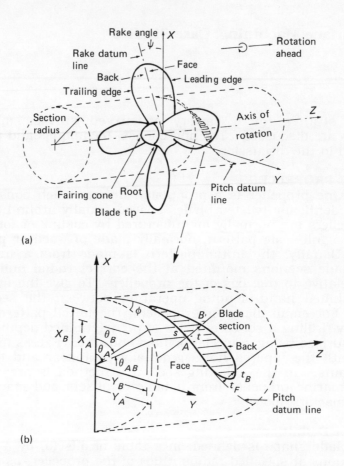

Figure 7.1 Marine propeller details

The rectangular coordinates of any point A on the pitch datum line in Figure 7.1 are given by:

$$X = r\cos\theta$$

$$Y = r\sin\theta$$

$$Z = \frac{P\theta}{2\pi}$$

(1)

Marine Propellers

where P is the pitch or axial distance corresponding to one complete revolution of the radial line r. It is defined by the expression

$$P = 2\pi r \tan \phi \quad (2)$$

Differentiating equations (1) with respect to the rotation of the radial line r (given by the angle θ_A) gives:

$$\frac{dX}{d\theta} = -r\sin\theta$$

$$\frac{dY}{d\theta} = r\cos\theta$$

$$\frac{dZ}{d\theta} = \frac{P}{2\pi} \quad (3)$$

It may be shown that

$$s = \int_{\theta=0}^{\theta=\theta_A} \sqrt{\left[\frac{dX^2}{d\theta} + \frac{dY^2}{d\theta} + \frac{dZ^2}{d\theta}\right]} d\theta$$

so that substituting for the gradients from equation (3) gives

$$s = \int_0^{\theta_A} \sqrt{\left[(-r\sin\theta)^2 + (r\cos\theta)^2 + (P/2\pi)^2\right]} d\theta$$

or solving and rearranging we have

$$\theta_A = \frac{s}{\sqrt{\left[r^2 + (P/2\pi)^2\right]}} \quad (4)$$

Substituting for θ_A in equations (1) gives

$$X_A = r\cos\frac{s}{\sqrt{\left[r^2 + (P/2\pi)^2\right]}}$$

$$Y_A = r\sin\frac{s}{\sqrt{\left[r^2 + (P/2\pi)^2\right]}} \quad (5)$$

and

$$Z'_A = \frac{Ps}{2\pi\sqrt{\left[r^2 + (P/2\pi)^2\right]}} \qquad (6)$$

However, because the blades are inclined at a rake angle ψ, the true Z coordinate is given by

$$Z_A = -r\tan\psi + \frac{Ps}{2\pi\sqrt{\left[r^2 + (P/2\pi)^2\right]}} \qquad (7)$$

For the case of the helix orthogonal and coaxial to the pitch datum line, the equivalent expression to equation (4) is

$$\theta_{AB} = \frac{-t}{\sqrt{\left[r^2 + (OP/2\pi)^2\right]}} \qquad (8)$$

where OP is the pitch for the orthogonal helix. The relationship between the orthogonal pitch datum line and the true pitch datum line may be seen from the expanded view in Figure 7.2 so that

$$\tan\phi = \frac{P}{2\pi r} = \frac{2\pi r}{OP}$$

or
$$OP = 4\pi^2 r^2 / P \qquad (9)$$

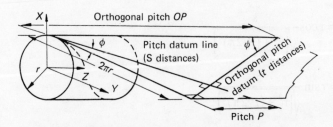

Figure 7.2 Expansion of the helical pitch datum line

Combining equations (8) and (9) gives

$$\theta_{AB} = \frac{-t}{\sqrt{\left[r^2 + (2\pi^2/P)^2\right]}} \qquad (10)$$

Rectangular coordinates of the point B, defined by a distance s along the pitch datum line and a distance t along the orthogonal pitch datum line are given by

$$X_B = r\cos(\theta_A - \theta_{AB})$$

$$Y_B = r\sin(\theta_A - \theta_{AB})$$

$$Z_B = Z_A + Z_{AB} = Z_A + (OP\theta_{AB}/2\pi) \qquad (11)$$

So by combining equations (4), (9), (10) and (11)

$$X_B = r\cos\left\{\frac{s}{\sqrt{(r^2 + g^2)}} + \frac{t}{\sqrt{\left[r^2 + (r^2/g)^2\right]}}\right\}$$

$$Y_B = r\sin\left\{\frac{s}{\sqrt{(r^2 + g^2)}} + \frac{t}{\sqrt{\left[r^2 + (r^2/g)^2\right]}}\right\}$$

$$Z_B = \frac{sg - rt}{\sqrt{(r^2 + g^2)}} - r\tan\phi \qquad (12)$$

where $g = P/2\pi$

From these equations a computer program may be written to read the propeller parameters, such as number of blades, overall diameter and rake angle together with the table of offsets, in order to define an array of rectangular coordinate points which define the propeller blade surface in three-dimensional space. The coordinates of the face and back surfaces of a single blade may be first calculated from the **s and t distances**. Other blades can be formed by homogeneous coordinate transformation of the single blade about the axis of rotation. The hub is generated by laying a series of equidistant, twisted space curves between the blades and along the envelope of revolution formed by the hub profile about the Z axis. In its simplest form, the hub may be taken as the frustum of a right circular cone. If the surface grid data definition is inadequate for machining purposes, a B-spline can be fit through the vertices and a finer mesh generated, as described in Chapter 6. An example of a four-bladed propeller is given in Figure 7.3.

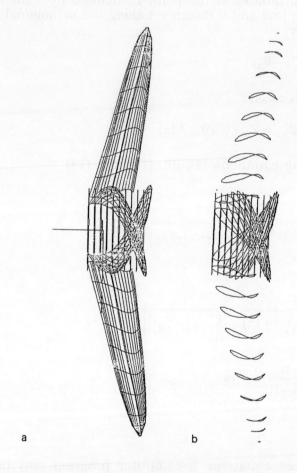

Figure 7.3 Computer graphic display of a four-bladed propeller

7.1.2 Surface Machining

With twisted multi-blade propellers, it is obviously not possible to see, nor is it possible to machine (with a three-axis machine), all parts of the surface from any single position. Machining of small four-bladed propellers of the type shown in Figure 7.4 has shown that three separate positions are required to gain access to the open parts of the face and back as well as between the blade roots and the interconnecting hub surfaces.

Sec.7.1] **Marine Propellers** 191

Figure 7.4 CNC machined four-bladed propeller

For location purposes, a circular rod of known length is attached to the centre of the block of workpiece material. Flats ground on the end of the bar serve as suitable location faces.

The back surface is generated by machining from the outer tip of the first blade in a series of decreasing circumferential paths. There is a limiting radius, called R*, at which further inward progression towards the root would have the effect of removing material from neighbouring blades. The back surface of neighbouring blades are similarly machined from outer tip to R*. If required, the surface finish can be improved by directing the milling cutter to travel over the blade surfaces in an orthogonal direction.

The face of the propeller is formed (down to the radius R*) in a similar way to the back. Precise location of the inverted workpiece is required to align face and back surfaces correctly. A cutter of small diameter is directed around the blade face and maximum hub profile leaving the partially completed blades mounted on a block of unmachined material at the hub. To machine the hub and the lower blade surfaces, the workpiece is mounted in a dividing head with the rod axis in a horizontal plane. The milling cutter is directed along a path that traverses down the back surface along the hub and outwards on the face until the blade surface is blended at radial R* position. Finally, the location rod can be removed by a turning operation at which time any suitable location holes and fairing cone-shape may be made. The size of propellers which may be machined is not limited by the method but only by the capacity of the CNC machine tool.

Examples of partially machined two-bladed propellers are given in Figures 7.5, 7.6 and 7.7.

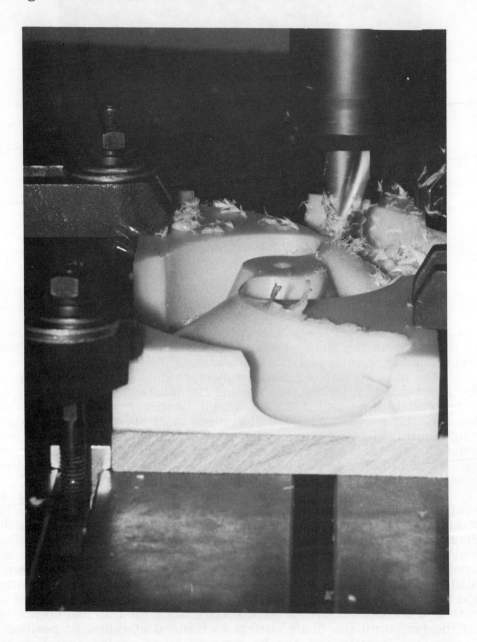

Figure 7.5 Machining the face surface of a two-bladed propeller

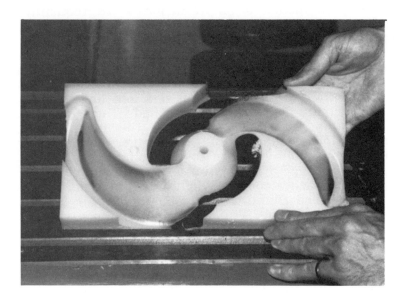

Figure 7.6 The face and hub surfaces of a two-bladed propeller

Figure 7.7 Machining the propeller profile to remove excess material

7.2 SHIP HULL MODELS

Scale replica models of ship hull forms are frequently required for hydrodynamic testing prior to vessel construction. Usually, a model maker produces a number of wooden sections of the waterline profiles of the hull, which are pinned and glued together. Removal of the excess material down to the waterline sections is done manually to produce a smooth shape. Data for the hull surface definition are taken from the naval architect's table of off-sets and lines plan, shown in Figure 7.8, which are neither faired nor lofted. Thus, the actual fairing for hydrodynamic models is done by the model maker when smoothing the model shape. This unspecified and arbitrary fairing gives no assurance of symmetry to the boat, nor is there any real knowledge of the shape definition between stationlines.

In this approach a computer lofting and manufacturing program for shipyards, called ShipCAM, is used to fair the lines plan and generate a suitable surface grid for machining on a CNC machining centre.

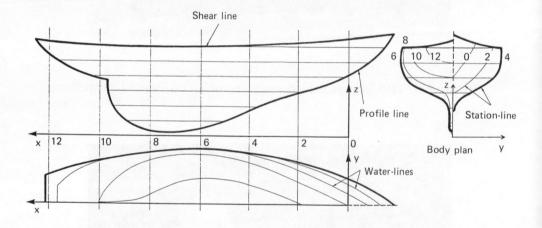

Figure 7.8 Lines plan showing stationlines and waterlines

7.2.1 Hull Definition

The surface is defined by a number of smoothed B-spline lines running along the entire length of the surface so that they evenly cover and represent the essential features of the surface. The B-spline vertices may be interactively adjusted through ShipCAM while observing curve curvature for line smoothness. Stationlines are fitted to these longitudinal lines and are viewed for consistency with the lines plan data. If the resulting surface does not represent the required surface form, then fine adjustments may be made to the longitudinal lines and the station lines redrawn. If a hard chine is present on the hull surface, then the chineline is taken as a boundary line and the surface treated as two separate patches. A B-spline surface fitted through the smoothed data vertices is shown in Figure 7.9 for a round bottom seiner and in Figure 7.10 for a double chine fishing boat.

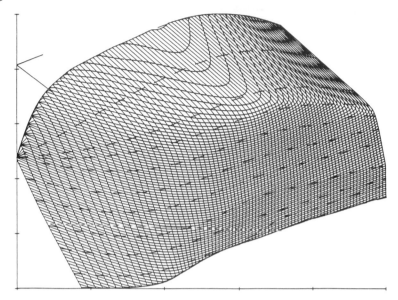

Figure 7.9 Perspective view of a round bottom seiner

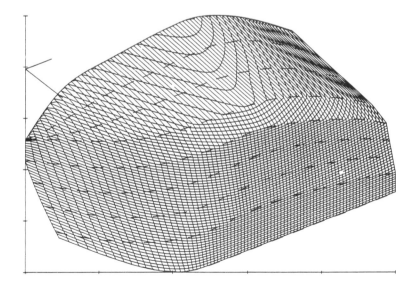

Figure 7.10 Perspective view of a double chine fishing boat

196   Curved Surface Machining: Case Studies   [Ch.7

7.2.2 Hull Machining

Ship hulls are normally formed in two halves. Materials such as kiln-dried spruce, cedar or jelutong wood can be used for model making. They have good dimensional stability, strength, machinability, cost and a fairly close grain to take a fine sanded finish. In order to conserve material and reduce machining time, each half is preformed from a series of waterline profile sections that are pinned and glued together, as shown in Figure 7.11.

Figure 7.11   Preformed and laminated hull sections

The plane of symmetry of the boat is used as the location plane for both of the half-hull preforms. The preform blanks are accurately aligned by a number of location dowels and secured onto a jig-plate in the orientation shown in Figure 7.12. The jig-plate, which has machined edges, can be accurately positioned and clamped to the milling machine table. This gives good alignment to the model and avoids the necessity of clamping directly on to the model surface. It also enables larger models to be accurately repositioned on the machine table during the machining process.

Machining is normally carried out with a 100 mm diameter carbide-tipped end-milling cutter, as shown in Figures 7.13 and 7.14. The cutter is directed along individual stationlines with offsets calculated as shown in Chapter 6. With three-axis machines the angle of inclination of the cutter to the surface normal, and hence the effective radius of cutting, is determined by the surface geometry, i.e.

Sec.7.3]   **Ship Hull Models**   197

there is no control over ϕ, the cutter inclination. However, for a typical ship hull model, it can be shown that the machining time is approximately 1/20th of that required for the more conventional spherical ball-mill cutter. With cutter movement along stationlines, the surface finish is much smoother than that obtained by machining along waterline in a 2 1/2 axis contour machining of the surface with the resulting square-cornered contour line ridges.

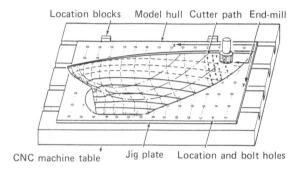

Figure 7.12 Location of the half hull form on the jig plate

Figure 7.13 Photograph of the hull form during machining

Figure 7.14 Photograph of the hull form during machining

Once the hull surface is complete, the edges of the shear and profile boundary lines can be machined with a 50 mm diameter end-mill cutter offset to the lines by the cutter radius. A slot is made along the shearline to allow for subsequent insertion of the keel. The matching half-hull shape is machined using the same program except that the Y-axis is mirrored on the machine console. The two machined hull forms, together with the keel, are aligned using dowel pins, assembled and glued together. The regular pattern of cusps of the machined hull surface are removed by sanding. A completed 2-m double chine fishing boat is shown in Figure 7.15.

A typical 2-m boat requires the equivalent of 5-10 km of paper tape. As the surface definition, display, cutter location and machining data are all calculated on a personal computer, the resulting files can be readily edited and are transmitted to the numerically controlled machine via the DNC link described in Chapter 5.

Figure 7.15 Completed 2-m double chine fishing boat

7.3 TURBINE BLADES

Turgo and francis turbine blades consist of a series of twisted blades mounted radially on a central rotor. Water impinges upon the blades to produce rotary motion, in much the same way as a waterwheel operates. A model of a single blade is required in order to develop an intricate sand casting mould with multiple blades (10 to 30) arranged around the periphery of the rotor. Data for francis and turgo turbines can be obtained by measuring existing manually produced patterns. These patterns are difficult to measure accurately, and the resulting data are sparse and irregular. An example of three-dimensional measured data from the internal surface of a turgo turbine model is shown in Figure 7.16. The data were taken to provide essential surface features without over constraining the surface and causing local undulations. The six lines (two of which are boundary lines) were individually smoothed such that the resulting set of orthogonal stationlines give a smoothly changing form, as shown in Figures 7.17(a) and 7.17(b). Internal and external surfaces with interconnecting boundaries and thickness constraints have to be smoothed concurrently.

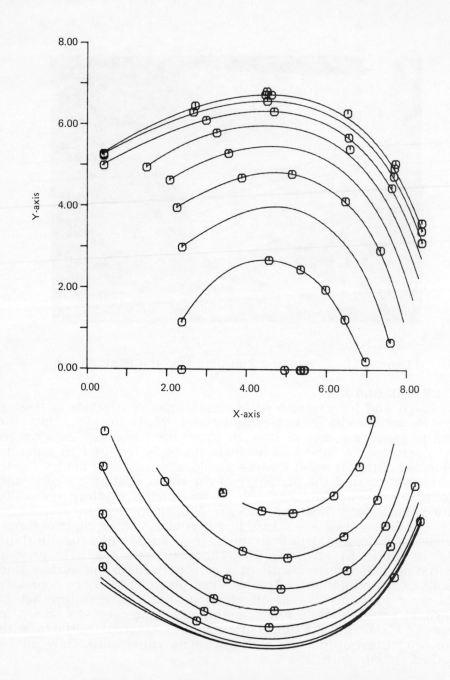

Figure 7.16 Measured internal surface of a turgo turbine (the two displayed sets are the fore and aft stationlines)

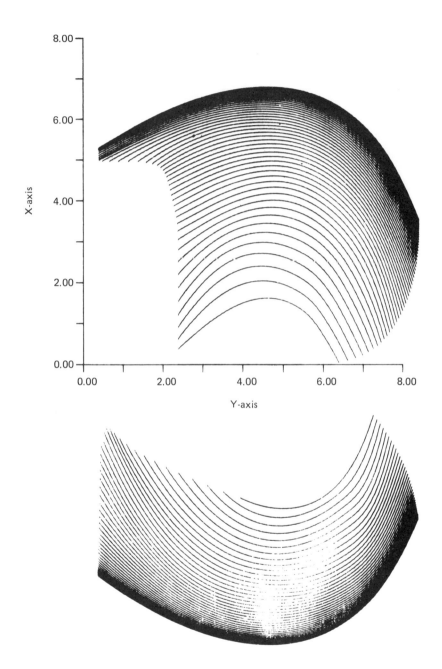

Figure 7.17(a) Section view of the internal surface of a turgo turbine fitted through six smoothed longitudinal lines

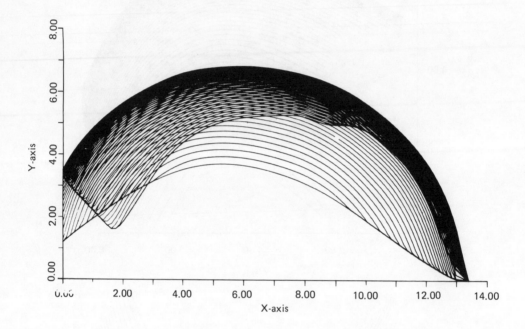

Figure 7.17(b) Longitudinal view of the fitted turgo turbine surface

Machining of the surfaces can be achieved with a ball-mill or end-mill (depending upon the size and curvature of the model) using the procedures given in Chapter 6. Two photographs of an original blade surface and a CNC machined model are given in Figure 7.18.

The advantages of CNC manufacture in this case are that models can be produced much more quickly, accurately and scaling of patterns for a variety of unit sizes can be readily achieved. Furthermore, the scaled patterns are identically shaped and give predictable power generation results from the completed turbine - a feature that is sadly lacking with manual methods.

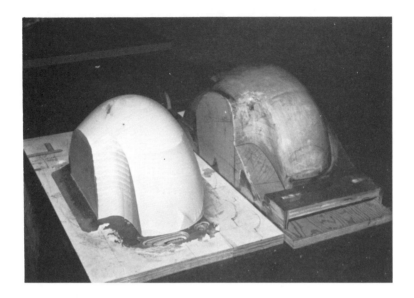

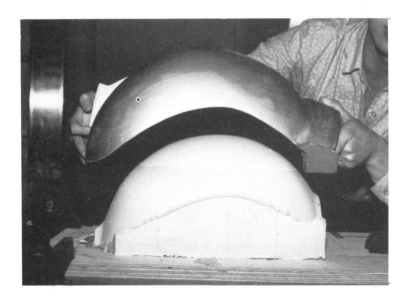

Figure 7.18 Original and CNC machined turgo turbine models

An example of a surface fitted for a lowhead turbine, using a similar approach, is given in Figure 7.19.

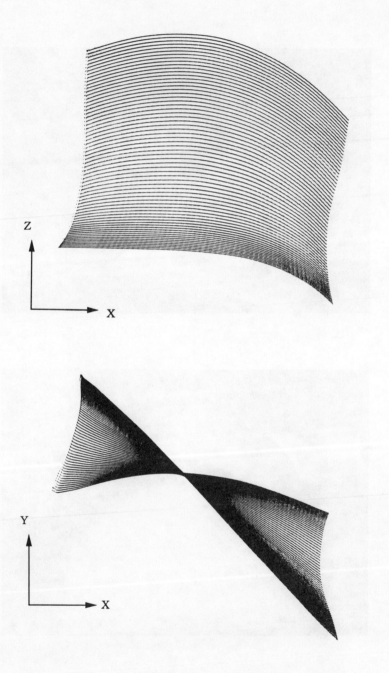

Figure 7.19 Fitted surface for a lowhead turbine

7.4 BIOMEDICAL APPLICATIONS

The replication of a variety of anatomical shapes can be achieved using CNC machining methods. The initial objective is to define the surface using **non-contact** methods, set up a suitable grid of surface data and machine the object using the techniques given in Chapter 6. Shadow moiré, photogrammetry and laser scanning are suitable non-contact methods and each has certain benefits. In this section three approaches of machining are presented; two utilizing moiré methods for surface definition and one utilizing laser scanning. The method in Section 7.4.2 illustrates the benefit of defining surfaces in terms of the subsequent machining operations.

7.4.1 Limb Shape Replication

Central perspective shadow moiré can be used to define limb shapes. The technique uses point source illumination in order to produce moiré contour fringes. An illustration of the shadow moiré fringes formed by natural sunlight casting a shadow of a picket fence on a snow drift is given in Figure 7.20. The same effect can be achieved, as illustrated in Figure 7.21, with an equispaced plane optical grating, a point source of light and a camera. The resulting shadow-pattern on the model is viewed through the grating from a position offset from the light source. A photograph of an apparatus for producing a shadow moiré fringe pattern on medical or die-making models is shown in Figure 7.22.

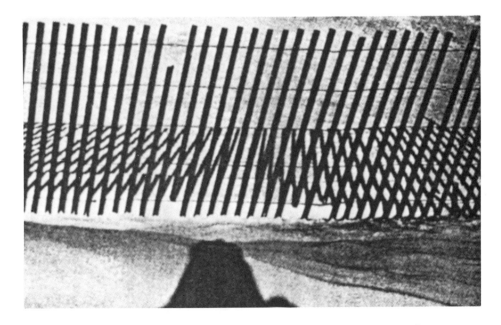

Figure 7.20 Shadow moiré pattern on a snow drift

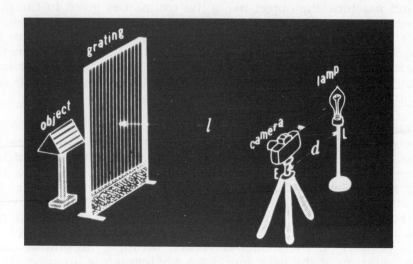

Figure 7.21 Schematic layout of moiré contourography

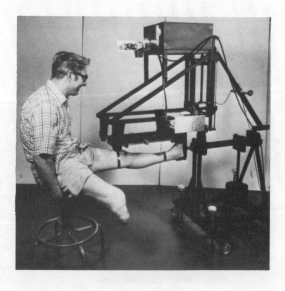

Figure 7.22 Shadow moiré apparatus

Sec.7.4] Biomedical Applications

The interval between successive fringes, which gives the height of object points from the grid plane, is not constant but is calculable from the fixed geometric positions of the lamp, camera, grid, and object. A three-dimensional perspective sketch of the shadow moiré system is given in Figure 7.23. The point source of light is located at S and the camera at E. Equispaced lines parallel to the x axis represent the centre lines of transparent spaces between the bars of the grating. A ray from S through any point Q on a transparent centre-line is scattered by the object surface at P, and passes through the centre-line of some other transparent gap at R to be visible at E as one point on a light fringe.

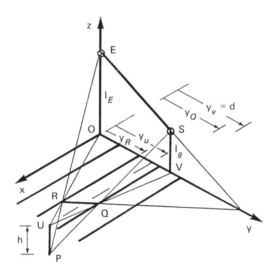

Figure 7.23 Line diagram showing geometric arrangement of grid, light source and camera

If the line PU is perpendicular to the grating and of magnitude h, then the geometry of the figure requires that:

$$\frac{h}{l_E} = \frac{RU}{OR} = \frac{y_U - y_R}{y_R} \tag{13}$$

and,
$$\frac{h}{l_S} = \frac{QU}{VQ} = \frac{y_Q - y_U}{y_V - y_Q} \tag{14}$$

also,
$$y_Q = y_R + N S_o \tag{15}$$

where, S_o = pitch of grating and N = integer value 0, 1, 2, 3, ...

For each value of N, the height h from the bright contour fringe on the object surface to the grid, has values of h_0, h_1, h_3... Thus, the integer N represents the fringe order and the first, or zero, bright fringe occurs at the grid plane. By combining equations (13), (14), and (15), the height of the nth <u>bright</u> fringe is given by:

$$h_n = \frac{N\, S_o\, \ell_E\, \ell_S}{\ell_E(d - N\, S_o) + y_R(\ell_S - \ell_E)} \qquad (16)$$

If the distances from the light source and camera to the grid are the same (i.e., $\ell_E = \ell_S$), then equation (16) reduces to:

$$h_n = \frac{\ell_S\, N}{(d/S_O - N)} \qquad (17)$$

and if $d \ggg S_o$, as is often the case, then:

$$h_n = \frac{\ell_S\, N\, S_o}{d} \qquad (18)$$

From Figure 7.24 it can be seen that the point P will appear to be located in the grating plane at point P'.

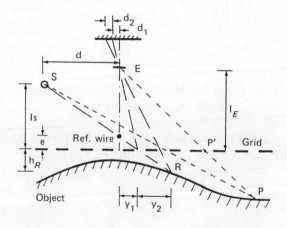

Figure 7.24 Perspective correction and fringe order determination

To correct this apparent shift of the true coordinates of P, the measured coordinates x^1, y^1 of P' requires corrections as follow:

$$x = x^1 M \left[1 + \frac{h_P}{\ell_E}\right] \qquad (19)$$

$$y = y^1 M \left[1 + \frac{h_P}{\ell_E}\right] \qquad (20)$$

where x^1 and y^1 are coordinate points of any fringe measured on the shadow moiré photograph and M is the photograph scale factor.

In order to calibrate the contour fringe order number, a reference wire is positioned above the grid and directly in line with the camera, as shown in Figure 7.24. A shadow of the central reference wire is seen on the grid and on the moiré fringes on the object surface as shown in Figure 7.25. From this geometry it can be shown that:

$$h_R = \frac{\ell_s y_2}{d + y_1} \qquad (21)$$

where,

$$y_1 = d_1 M \qquad (22)$$

And with due allowance being made for perspective correction:

$$y_2 = d_2 M \left[1 + \frac{h_R}{\ell_E}\right] \qquad (23)$$

Therefore, combining Equations (21), (22), and (23):

$$h_R = \frac{1}{\left\{\left[\dfrac{d + d_1 M}{d_2 \ell_s M}\right] - \dfrac{1}{\ell_E}\right\}} \qquad (24)$$

Thus, by measuring the distances d_1 and d_2 to any particular fringe on the shadow moiré photograph, a value of h_R and hence all fringe order numbers can be obtained.

Digitization of the fringe pattern and conversion of data, in line with the above equations, results in an array of surface points. No attempt is made to combine the data obtained from the three views around a leg shape, but instead each view is analysed separately and the model is machined in the same three angular positions. The overlap of each view gives a blending of the three separate surfaces to form the finished replica. An example of a CNC machined replica of a limb shape is given in Figure 7.26.

Figure 7.25 Shadow moiré fringe pattern formed on mannequin's leg

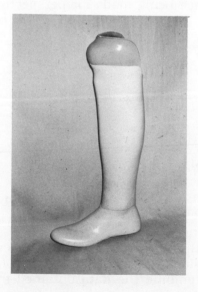

Figure 7.26 CNC machined limb shape

Sec.7.4]   **Biomedical Applications**   211

7.4.2 Shoe Lasts - Cylindrical Mill Turning

An shadow moiré approach, which is suited to cylindrical objects with compound curvature such as shoe lasts, is given below. In this approach the surface data is measured through a wrap-around moiré technique called pericontourography. Data for the cylindrical type object are generated in a form suitable for machining by quasi-helical mill-turning. In pericontourography the basic shadow moiré fringe pattern is formed on a rotating model and recorded, through a narrow vertical strip aperture, on a translating film, as shown in Figures 7.27 and 7.28. The image produced is in effect a map of contour lines (not of constant height) all around the object, as shown in Figure 7.29. Constant rates of rotation and translation are electronically synchronized through the camera controls.

Figure 7.27 Periphery camera with tubular stand, grid and rotary table

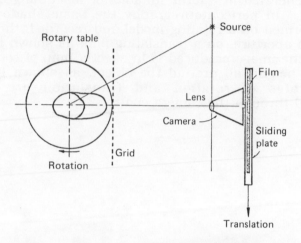

Figure 7.28 Basic elements of pericontourography

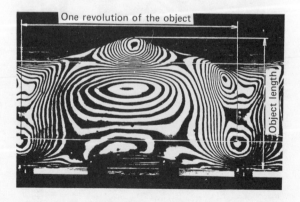

Figure 7.29 Pericontourograph of a shoe last

The photograph is digitized at the intersection of the fringe lines and a set of equally spaced grid lines in the rotational, or theta, direction as shown in Figure 7.30. The digitized reference points are used to calculate the surface points in cylindrical coordinates and to specify a quasi-helical cutter path suitable for subsequent machining.

The workpiece shape is produced by incrementally rotating the part on a fourth axis rotary table while traversing a spherical ended milling cutter along the part length, as shown in Figure 7.31. As surface shape is defined by radius at constant increments of angle of rotation, θ, and transverse feed position, z, the part rotation speed and longitudinal transverse speed are maintained at constant levels. Adjustment to the radial position of the cutting tool relative to the axis of rotation is all that is required.

Specifying the surface in terms of the subsequent machining process avoids the use of extensive surface fitting and reformatting routines. Machine control instructions are also simplified as the tool follows a continuous quasi-helical path around the workpiece and requires position control on one axis only.

With larger tool radii there is greater likelihood that some portions of the machined surface will be inadvertently removed, for example, when attempting to machine within a concavity where the radius of curvature of the tool is greater than the concave surface. For each machined vertex interference or undercutting, checks can be made at adjacent vertices and adjustments made, where required, to the radial tool position.

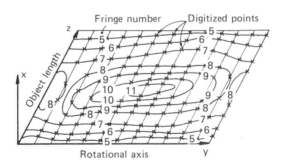

(a) z, y, fringe order number

Figure 7.30 Data transformations to define a quasi-helical cutter path

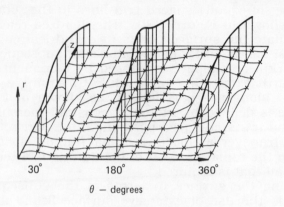

(b) r, θ, z at digitised points

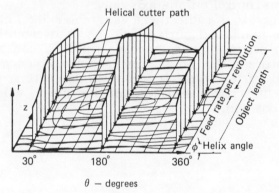

(c) r, θ, z at points on helical cutter path

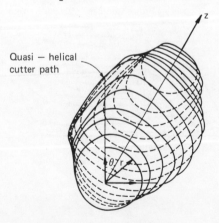

(d) r, θ, z final shape

Figure 7.30 Data transformations to define a quasi-helical cutter path

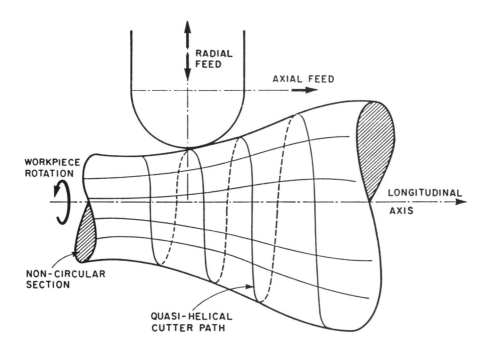

Figure 7.31 Tool and workpiece configuration for quasi-helical mill-turning

7.4.3 Head Shapes - Laser Scanning Data

A laser scanning beam can be used to define objects, as shown in Figure 7.32. The laser beam is scanned across the surface of an object by a rotating galvanometer-driven mirror. The position of the reflected laser spot is recorded and a triangulation scheme employed to calculate surface dimensions. Two synchronized scanning mirrors, or a single two-sided mirror, can be employed to give a wider effective baseline in the triangulation calculation.

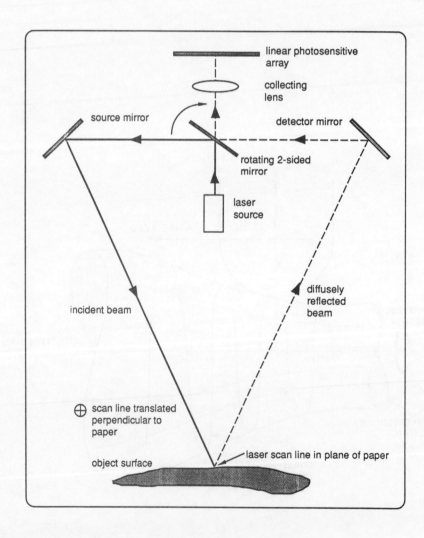

Figure 7.32 Basic features of a 3-D laser scanning digitizer

During scanning, laser digitizers require some controlled translational motion across the object surface. This is usually accomplished with a coordinate measuring machine. However, the laser digitizers can be integrated directly with CNC machine tools and a feedback on position can be provided by the servo motors on the CNC machine tools. This integration saves the expense of coordinate measuring equipment and enables the laser scanner to be used for in-process gauging. A typical scheme for integrating CNC machine tools with laser scanners is given in Figure 7.33.

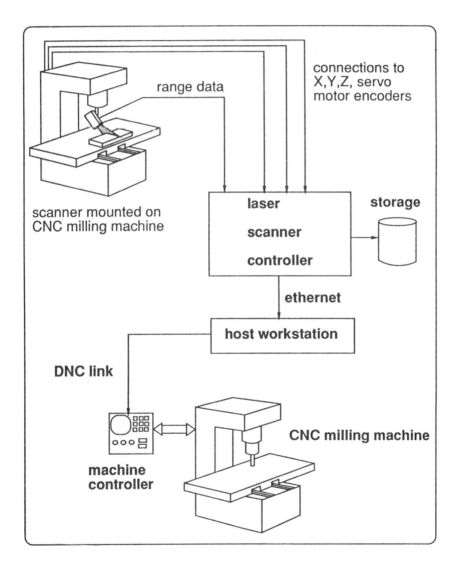

Figure 7.33 Integration of a CNC machine tool and a laser scanner

Laser scanners produce a finely-spaced mass of unstructured digital data, called cloud data. The random nature of the data, together with inherent noise inconsistencies, requires some data smoothing and refinement. A scan of a human head taken with a Hymarc laser scanner is shown in Figure 7.34. The recorded digital data was filtered and reformatted to give 180 longitudinal line scans with 186 points per line.

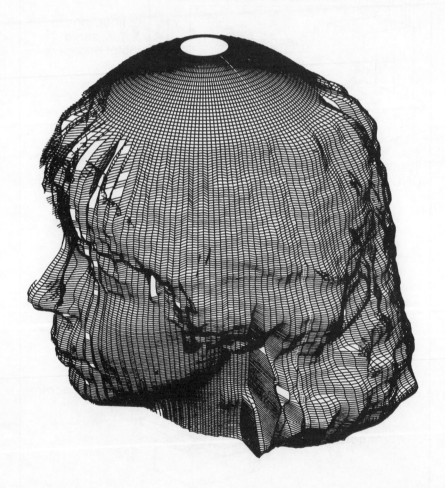

Figure 7.34 Reformatted laser scan data of a human head

Sec.7.4] Biomedical Applications

For CNC machining a block of material was located on the rotary table with the axis of rotation oriented on the machine table in the horizontal X-axis direction. A tailstock with a moving centre was used to support the free end. Cutter offsets can be calculated for a ball-mill cutter as given in Chapter 6. Machining can be carried out with longitudinal passes at incremented rotational positions, i.e. motion in the X and Z axes with incremented rotation after each pass. The machining could be achieved with quasi-helical mill turning, as described for shoe lasts in the previous section. Roughing can be accomplished with an end-mill or a large diameter ball-mill. A CNC machined model of the human head is shown in Figures 7.35 and 7.36. A view of a different head shape machined with a small diameter ball-mill cutter is shown in Figure 7.37.

Figure 7.35 CNC machined model of laser scan data

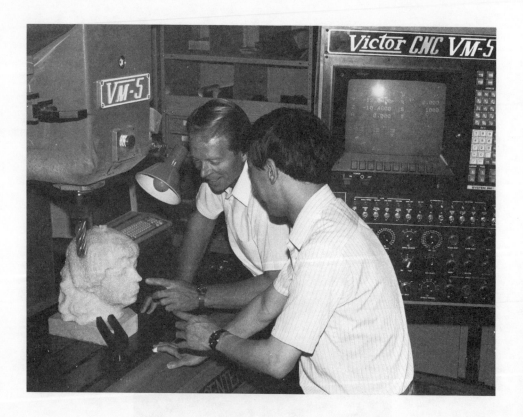

Figure 7.36 CNC machined model of laser scan data

Sec.7.4]   **Biomedical Applications**   221

Figure 7.37   CNC machined model using a small diameter end-mill

7.4.4 Heart Valve Shapes - Surface Definition

The surface of a heart valve mould shape was defined from cross-sectional drawing data; the boundary curve and one section line curve was given. From this a B-spline surface was developed. The machined model using a ball-mill cutter is shown in Figure 7.38. A thin-sheet of biological material is subsequently fitted to the three facets of the mould and sewn at the seams to form a heart valve. The benefits of CNC machining in this case are the accuracy and the repeatability of model shape.

Figure 7.38 CNC machined model of a heart valve mould

Chapter 8
Integrated CAD/CAM for Shipyards: Case Study

An example of integrated computer aided design and manufacture of ship hulls is given by a program called ShipCAM. The description given below illustrates the capabilities of integrated design and adjustment (lofting and fairing of a hull form until it is suitable for manufacture), and subsequent definition of the manufacturing processes including the programming of CNC flame-cutting machines. The flat sheet plates which are cut on the CNC flame-cutting machines are subsequently bent (but not stretched or formed to generate compound curvature) and welded on the frame structure to form the hull shape. The assembly process is like constructing a large three-dimensional jig-saw and it is critical that the bent sheet plates fit together accurately.

ShipCAM, which is based on the approaches and programs given in Chapter 6, is installed in a number of shipyards in Canada and the USA and has been used to manufacture many production vessels. An example of a 12-m tugboat, lofted and manufactured entirely using ShipCAM, is given in Figures 8.1, 8.2 and 8.3.

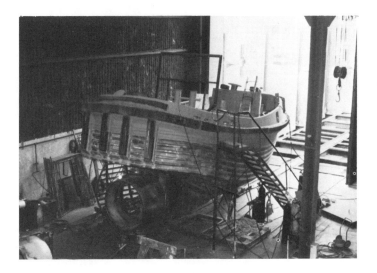

Figure 8.1 View of a 12-m tugboat under construction

Figure 8.2 Front view of a 12-m tugboat

Figure 8.3 Rear view of a 12-m tugboat

Integrated CAD/CAM for Shipyards: Case Study

At the heart of the ShipCAM program are the parametric B-spline routines outlined in Section 6.2. The G-code generation is handled as shown in Section 6.6. Other routines, such as the developable surface module, for example, are handled as sets of suitable density vertex data as shown in Sections 6.3 and 6.4. A flowchart of ShipCAM is given in Figure 8.4.

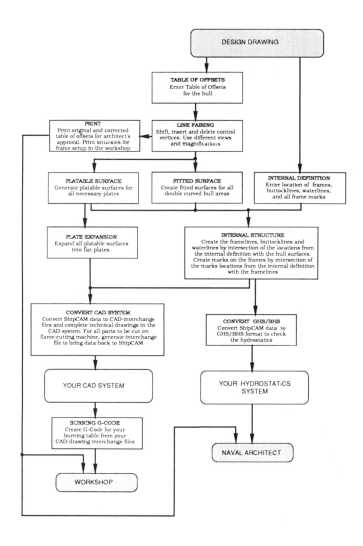

Figure 8.4 Flowchart of ShipCAM

226 Integrated CAD/CAM for Shipyards: Case Study [Ch.8]

Although ShipCAM has been designed as an easy-to-use, Macintosh-style, program for ship manufacture, it nevertheless can be used to rapidly design and adjust any kind of three-dimensional surface shape and establish and orient the vertex grid ready for CNC milling. The initial menu of ShipCAM is shown in Figure 8.5 and a description of the subsequent menu options given in the following sections.

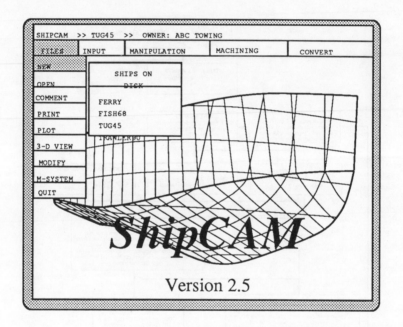

Figure 8.5 ShipCAM start-up screen

8.1 Data Input

The shipbuilder usually starts with a table of offsets that represent the sparse, and irregular, design data from the naval architect. This tabular data and associated small-scale orthogonal cross-section drawings (known as waterlines, stationlines and buttocklines) do not define the ship in sufficient detail for construction. It is the shipbuilder's task to establish the hull shape for manufacture. A spreadsheet-type data input is implemented within ShipCAM, as shown in Figure 8.6. The input module, and all other modules, supports four different data formats widely used in ship construction, namely meters, decimal feet, feet inches eighths and feet inches sixteenths. The input module is capable of automatic data recalculation for station locations.

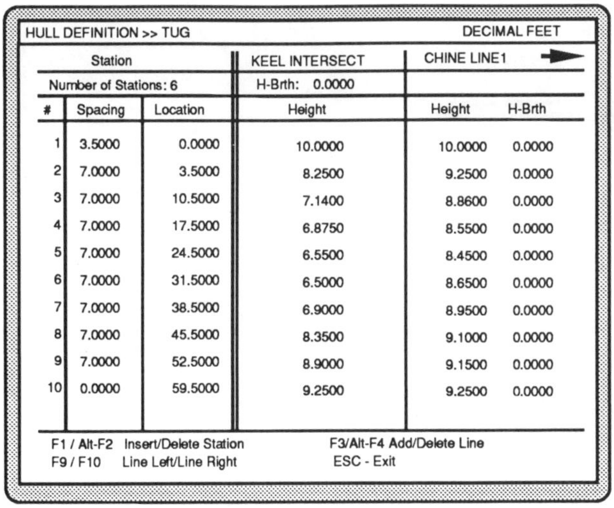

Sec 8.2]   Line Fairing   227

```
HULL DEFINITION >> TUG                              DECIMAL FEET
```

Station		KEEL INTERSECT	CHINE LINE1	
Number of Stations: 6		H-Brth: 0.0000		
#	Spacing	Location	Height	Height	H-Brth
1	3.5000	0.0000	10.0000	10.0000	0.0000
2	7.0000	3.5000	8.2500	9.2500	0.0000
3	7.0000	10.5000	7.1400	8.8600	0.0000
4	7.0000	17.5000	6.8750	8.5500	0.0000
5	7.0000	24.5000	6.5500	8.4500	0.0000
6	7.0000	31.5000	6.5000	8.6500	0.0000
7	7.0000	38.5000	6.9000	8.9500	0.0000
8	7.0000	45.5000	8.3500	9.1000	0.0000
9	7.0000	52.5000	8.9000	9.1500	0.0000
10	0.0000	59.5000	9.2500	9.2500	0.0000

```
F1 / Alt-F2  Insert/Delete Station      F3/Alt-F4 Add/Delete Line
F9 / F10     Line Left/Line Right       ESC - Exit
```

Figure 8.6 Spreadsheet-like data input screen

8.2 Line Fairing

Input data is faired a line at a time, as shown in Figure 8.7. The plan and profile views of a parametric B-spline approximation to any line, or lines, in the data base are displayed on the screen together with line slope or curvature. Using keyboard keys, any line vertex can be readily addressed and interactively adjusted in three-dimensional space. The effect of adjustments on line curvature can be observed during this adjustment process and acts as a means of magnifying the smallest irregularities in the line smoothness. B-spline curve control vertices can be added or deleted. The information window at the lower part of the screen is used to display the magnitude of differences between the input vertices and the adjusted line. Reference lines of different types (straight lines, circles and B-splines) can also be inserted on the screen to provide guides for adjustment. A zoom function allows more accurate adjustment of critical regions. When the adjusted B-splines lines are sufficiently smooth, they are saved as a set of data vertices (of selectable resolution). The resolution depends very much on the size and complexity of the ship hull.

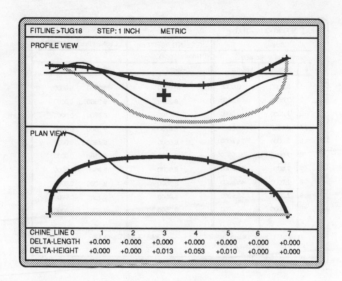

Figure 8.7 Fairing using fourth order B-spline curves

8.3 Surface Generation

Two types of surfaces may be defined within ShipCAM, namely B-spline surfaces and developable surfaces. A B-spline surface is normally fit through a series of previously smoothed lines. Any surface adjustment is accomplished through additional adjustment of the individual lines and a refitting of the full surface; for practical reasons, no direct adjustment of the surface is undertaken. Developable surfaces can be produced, to some predefined accuracy, between any two identified boundary lines. Portions of the surface that are not developable are identified on the screen. In this case the boundary lines may be refaired and a new developable surface fitted. The hull defined through the developable surface module may be subsequently opened out to define flat sheet plate shapes; i.e. developable surfaces allow simpler and more cost-effective manufacture of the hull but do not always provide the vessel with the desired performance and economical operation. Developable surfaces are used, for example, for tug boats which need pulling strength but not speed, while compound curved surfaces may be found on larger ocean-going vessels and are especially visible on bulbous bows of supertankers.

8.3.1 Compound Curve Surface Fitting

The screen display of a fitted surface showing three orthogonal views is given in Figure 8.8. The user interactively selects the modified data lines that are required to define the surface. Parametric B-spline approximation and B-spline interpolation surfaces may be selected. B-spline approximations, which fall within the hull of the enclosing polygon, have a smoothing effect (however the surface does not pass through the input vertices) while B-spline interpolations force the surface through all the data vertices but may introduce unwanted surface oscillations. The surface is eventually stored as a matrix of data vertices as discussed in Chapter 6. The vertex resolution of the final surface may be interactively selected to meet the needs of the manufacturing process.

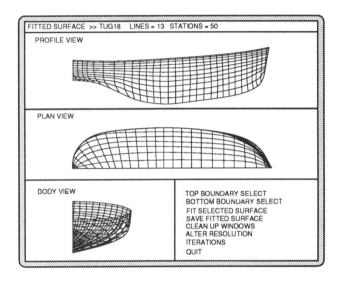

Figure 8.8 Fourth order B-spline surface fitting module

8.3.2 Developable Surface Fitting

The screen display of a developable surface, showing three orthogonal views is given in Figure 8.9. The user interactively selects two faired boundary lines from the data base. A developable surface, if it exists, is then fitted and displayed. A developable surface is one in which vertices on the two boundary lines may be connected by straight lines in three-dimensional space with the surface normal vectors of the boundary vertices lying in the same planes - a cone formed from bent paper is an example of a developable surface. These straight lines connectors are called generators.

The program iteratively checks, to a definable resolution, for the condition of developability along the boundary lines. In general, two fair B-spline boundary lines may contain developable and non-developable regions (areas of compound curvature). Within the module, generator lines are defined for the developable regions and ruled surfaces are defined for the non-developable regions. The use of ruled surface regions make the surface continuous and are important in later stages of the lofting process when the geometry of the internal structure of the ship hull has to be calculated. They have also proved to be useful as approximate templates for plating the compound curvature areas of the hull.

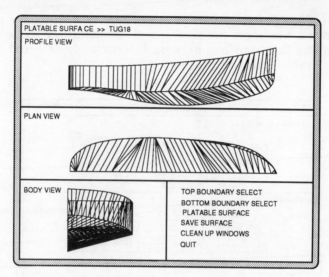

Figure 8.9 Developable surface module

8.4 Plate Expansion

As shown in Figure 8.10, the plate expansion module is used to expand (or unwrap) three-dimensional developable surfaces into flat plate shapes by applying triangulation methods. The defined surface may contain any number of nondevelopable patches. For ease of plating, generator lines may be marked and available as plate roll-lines and frame position marks can be added as plate location points. The expanded vertex polygon is also available data for development of G-code files for the CNC flame cutting machines.

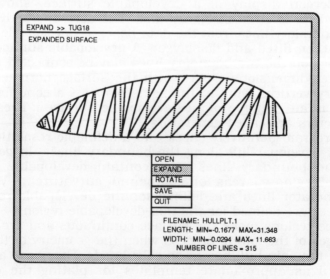

Figure 8.10 Plate expansion module

8.5 Internal Structure

A ship structure that supports outer hull plating consists of a series of frames, longitudinals, bulkheads, etc. The geometrical shapes of these support structures are determined from the final developable and/or compound surface hull patches. The hull/structure intersection lines can be exported to standard CAD package for general two-dimensional and three-dimensional operations on non-curved surface parts.

As shown in Figure 8.11, the module displays the frame shapes, and also intersection marks, which assist in accurate frame assembly. Marking can also be provided for the intersection of longitudinals with frames. Waterlines and buttocklines are created for hydrodynamic purposes.

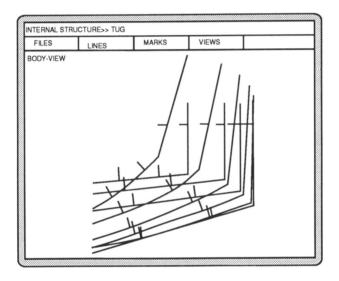

Figure 8.11 Internal structure module

Waterlines, buttocklines and stationlines for a 10-m tugboat, derived from developable surfaces.patches are shown in Figure 8.12.

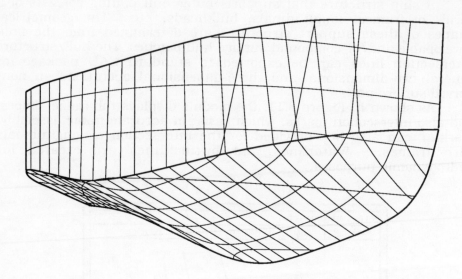

Figure 8.12 Waterlines, buttocklines and frame lines

8.6 Automatic Cutout Insertion

Where frames and longitudinal stiffeners intersect, cutouts or notches have to be provided so that one of the structural members can pass through the other. Since a single ship hull can easily have many hundreds of these cutouts, an automatic cutout insertion module is provided, as shown in Figure 8.13. The cutout shape is designed within a standard CAD program, transported into ShipCAM and inserted at all of the previously automatically generated intersection markings (see Figure 8.11).

Sec 8.7]   **Data Export to CAD Programs**   233

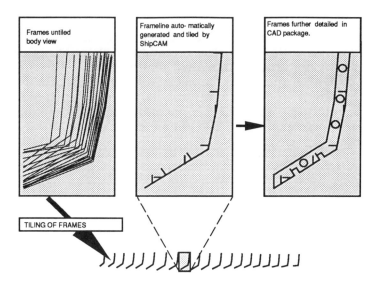

Figure 8.13 Automatic insertion of cutout shapes

8.7 Data Export to CAD Programs

Standard CAD packages, such as AutoCAD, are readily available and have excellent means of defining and generating geometric entities, such as lines, arcs and circles. It is therefore advantageous to be able to interchange data between ShipCAM and these general purpose CAD packages. As shown in Figure 8.14, once the geometry of the ship hull, with its internal structure, has been created, the data format can be converted to formats that standard CAD programs understand. Within ShipCAM two-dimensional and three-dimensional data transfer is possible.

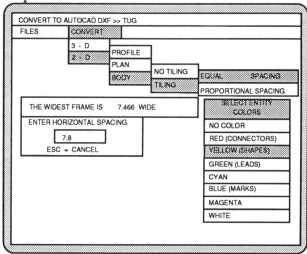

Figure 8.14 CAD interface module

8.8 G-code Generation

Expanded plate shapes are exported to AutoCAD, positioned (called nesting) on a standard size, rectangular sheet of material, and a logical flame-cutting order established for the parts. The drawings are then saved in DXF format and transferred back to ShipCAM for G-code generation. The G-code generation module of ShipCAM, shown in Figure 8.15, interprets the DXF file and automatically creates accurate and error-free instructions for CNC flame-cutting machines.

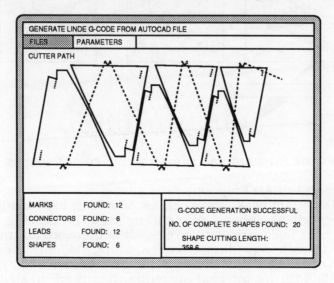

Figure 8.15 G-code generator

Chapter 9
Simulation Testing of CNC Part Programs

Testing of CNC part programs can be expensive and hazardous. Usually testing is done directly on a CNC machine tool which, if there are errors in the G-code program, can be dangerous to the operator as well as to the machine tool. This approach is also an inefficient use of the machine tool when it could be otherwise producing parts.

A computer simulation program can be used to test part programs and to produce a graphic display of the tool path as illustrated in Figure 9.1. The development of such a program is described below. Clearly the program can be modified to suit a variety of needs, and this program is intended as an example of a basic simulation program.

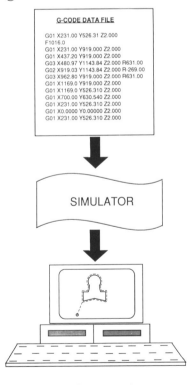

Figure 9.1 Illustration of a simulation testing program

9.1 SIMULATION PROGRAM

The objective of a CNC simulation program is to provide an error check on the accuracy of all G-code statements. A Turbo-C, version-2, program listing, of a fairly basic program is given in Appendix 16. This program should be run on a PC computer with VGA graphics card and mouse. The graphics instructions within the program can be modified to suit the graphic interface of other compilers or computers.

A flowchart of the simulation testing program is given in Figure 9.2. It is divided into three main parts, called the parser, the G-code processor, and the display processor.

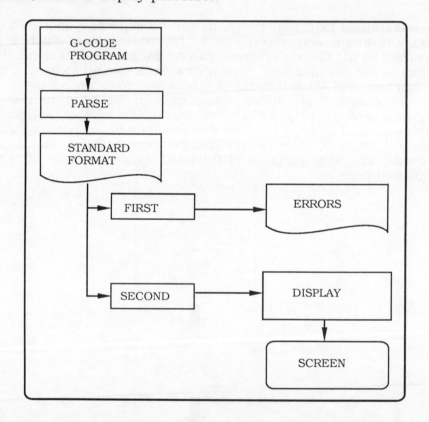

Figure 9.2 Flowchart of the simulation testing program

The program allows the user to read, check and correct a G-code file, and to display the tool and tool path on the screen (in multiple views as well as in different colours for rapid traverse and cutting motions). The program may be processed a block of data at a time or processed continuously. A view of the main screen menu is shown in Figure 9.3. Helpful error messages are generated and displayed as required on the screen.

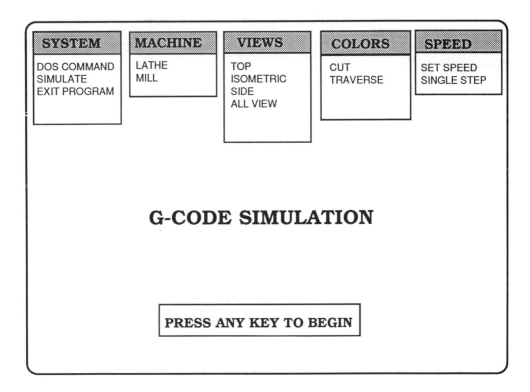

Figure 9.3 Main menu of a simulation testing program

9.1.1 Parser

The parser is used to read the G-code input file (specified by the user) and rewrite it in a single line format. Since G-code line syntax is quite flexible, a variety of different commands can be made on one line. It is more convenient to process the file so that the main G-, F-, S-, T-, and M-codes are placed on separate lines so that the spacing of entries is consistent. An example of the first part of a standard G-code file and the same file after processing through the parser program is given in Table 9.1.

Table 9.1
Standard and Parsed G-code Files

G-code file:	Parsed G-code file
O23;	
N10 G91 G28 Z0.0;	
N20 G28 X0.0 Y0.0;	
N30 G21;	G 21
N40 G92 X500.0 Y250.0 Z400.0;	G 92 X 500.0 Y 250.0 Z 400.0 Y250.0 Z400.0;
N50 T14;	T 14
N60 M06;	M 06
N70 S1400 M03;	S 1400
	M 03
N80 G90 G43 G00 Z100.0 H14;	G 00 Z 100.0
N90 G00 X-170.0 Y-70.0;	G 00 X -170.0 Y -70.0
N100 G01 Z0.0 F500.0;	G 01 Z 0.0
	F 500.0
N110 G01 X170.0 F1000.0;	G 01 X 170.0
	F 1000.0
N120 Y-20.0;	G 01 Y -20.0
N130 X-170.0;	G 01 X -170.0
N140 Y80.0;	G 01 Y 80.0
N150 X170.0;	G 01 X 170.0
N160 Y80.0;	G 01 Y 80.0
N170 X-170.0;	G 01 X -170.0
N180 G91 G28 Z0.0;	
N190 G28 X0.0 Y0.0;	
N1170 G40;	
N1180 G49;	
N1180 M30;	M 30
%	

9.1.2 Error Processor

G-code processing is undertaken in the second operation, in which the parsed G-code file is read and error checks are performed. The processor reads one block of code (consisting of a number of commands) at a time from the parsed G-code file and stores it as a string in memory. All commands consist of a letter followed by a series of numbers followed by a blank (in the parsed version). The initial command letter is compared to a standard list (G, X, Y, Z, I, J, K, R, M, T, S) and, once identified, the associated numbers are recorded. In the error checking part of the processing program, the specified numbers are examined to determine if they fall within a valid size range and also if the code is functionally correct. The maximum and minimum dimensions of the object are determined at this time for input to the display routines later in the program. Examples of valid numbers for the 'G' character are 00, 01, 02, 03, 17, 18, 19, 20, 21, 50, 90, and 91 (or any other combination of required G-codes), while valid numbers for the 'M' character are 01, 03, 04, 05, and 30. Emphasis in the error routines is placed on functionality rather than the G-code syntax. The order of important functions, such as spindle start and stop, tool selection, and feed rate selection, is also checked. An array is created containing both a flag and an integer value for each error condition. Default or modal conditions that are detected are:

- spindle speed, tool, and feedrate selection
- spindle start and stop
- location of G92 and G00 commands
- location of the start of cutting
- location of the program end.

When the above conditions are detected, the corresponding flag is set to TRUE and an integer value is recorded that specifies at what point in time the condition became true. After the complete program is scanned, the error array is interpreted by determining the logical sequence of G-code operations; error messages are produced as needed. An example of error interpretation is, "Has the spindle been started before cutting is performed?" A list of typical error messages is given in Table 9.2.

Table 9.2
G-code Error Messages

1) G-code out of range "G-code number"
2) X Coordinate out of range "x coordinate"
3) Y Coordinate out of range "y coordinate"
4) Z Coordinate out of range "z coordinate"
5) S-code out of range "S-code number"
6) I Coordinate out of range "i value"
7) J Coordinate out of range "j value"
8) K Coordinate out of range "k value"
9) Radius out of range "r value"
10) Feedrate out of range "feedrate value"
11) T-code out of range "T-code number"
12) M-code out of range "M-code number"
13) Spindle speed has not been set
14) Tool has not been selected
15) Feedrate has not been specified
16) Spindle has not been started
17) Spindle started before spindle speed was set
18) Spindle started before tool selected
19) G92/50 Absolute Zero Point not found
20) G code other than G00 encountered before spindle was started
21) End of program encountered before spindle was started
22) No G codes other than G00 encountered, cannot process
23) Spindle has not been stopped
24) End of program not encountered
25) Tape Rewind,%, not found
26) Program set up for Lathe, Y is "y coordinate" must be 0
27) G-code 0/1/2/3 encountered before G92/50

9.1.3 Display Processor

Once the G-code file is error free, the tool path is displayed on the screen through the display processor. The display screen is divided into four equal-sized windows, three of which are used to display the top, side and isometric views. The fourth window is used to display the current values of tool number, spindle speed, feedrate, and X,Y,Z coordinates, as shown in Figure 9.4. The current line of G-code is also displayed and scrolled as the path is generated. The window views may be controlled from the **views** menu option. The top and side views are scaled automatically to their individual windows. The tool path itself is displayed in two different colours representing rapid traverse (G00) and linear interpolation (G01) moves. These colours may also be interpreted as representing cutting and non-cutting moves.

Sec.9.1]   **Simulation Program**   241

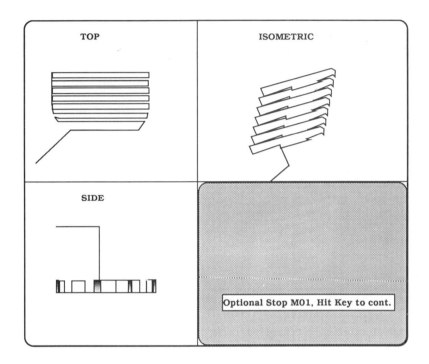

Figure 9.4 Screen tool path display

The user has the option of displaying the tool path in a continuous (automatic) or in a step-by-step (single-step) mode. In automatic mode the speed of G-code block processing is controlled by the SPEED menu (0=fast, 999=slow). The single-step mode is activated and deactivated by the <ESC> key or through the SPEED menu. Once the single-step mode is active, the user must press any key except <ESC> or <SPACE> to execute a G-code step along the tool path. This feature allows the user to observe the tool path more closely.

In order for the dimensions defined within the G-code program (called world coordinates) to be displayed on the screen, they must be normalized and converted to pixel coordinates in the viewport of the screen. A **normalize** function in the display processor converts the maximum and minimum X and Y values and X and Z values of the cutter path to the respective top and side viewport coordinate pixels in order to optimize the use of the viewports. Care is taken to preserve the different proportions of the path. A **lines** function is called to draw lines on the screen between points defined by the G-code coordinates and is also used to calculate the X and Y coordinates for the isometric view. The **monitor** and **d_gcode** functions are responsible for displaying data in the monitor window and lines of Gcode in the G-code window. These two windows are updated with each new line of G-code.

Once the cutter path has been drawn, the **iso_view**, **top_view**, and **side_view** functions call the **scale_path** function in order to scale the cutter path to a full-screen representation. **Scale_path** is also used to scale down the cutter path when it is to be displayed in all views. Whenever the main screen is cleared, it is regenerated using the **set_screen** and **d_path** functions. **Set_screen** draws all of the viewports and titles and **d_path** draws the cutter paths.

9.1.4 Enhancements

The tool path simulator developed for a milling machine is a fairly basic version although it has proved effective in allowing CNC programs to be checked in a safe and reliable manner. The techniques involved could clearly be applied to any other type of CNC machine, such as lathes or flame-cutting machines. Many features could be added to improve the program functionality although there are trade-offs with speed of operation. Some possible enhancements are given below:

- the cutting tool could be drawn (with the appropriate diameter and length offsets) as the tool path is generated and the remaining part material displayed
- error checking could include syntax checks so that errors such as specifying an X without a value, or specifying an arc in the XY plane when the YZ plane is selected, could be detected
- automatic zooming to full screen of a specified portion of the object
- automatic scaling of the isometric view within its window independent of the other orthogonal views
- rotation of the side view so that it reflects the actual orientation of the part on the machine.

In fact, the program could be further developed to include part definition and then one approaches the computer assist programs such as FAPT described in Chapter 3.

Chapter 10
Retrofitting Machine Tools

Manual machine tools can be converted to CNC operation by the suitable addition of motors and controllers (called retrofitting). Good quality servo and stepping motors with integrated controllers are readily available together with a selection of lead screw, ball-nut bushings and encoders or other feedback devices. An example is given of the automation of a cabinet door manufacturing process by retrofitting a standard industrial woodshaper and integrating this into a system that permits the design and manufacture of door components.

10.1 CABINET DOOR MANUFACTURE

10.1.1 Introduction

Cabinet doors, of the type which are fitted into kitchen and bathroom cabinets, are frequently made on copy router machines in small batch manufacturing runs. Each door consists of a curved top rail, two tennoned side rails, a tennoned base rail, and a raised central panel, as shown in Figure 10.1. The copy router machines consist of a high-speed, vertical spindle, cutter head mounted within a fixed worktable. A movable worktable is mounted above the fixed worktable and adjacent to a copy fixture attachment. Workpieces are clamped to the movable table and can be moved manually past the cutter head at a distance determined by a copy template.

In this application the automation process is undertaken using a personal computer as the central controlling component, both for design and manufacture. The door components are defined using a standard, readily available CAD package such as AutoCAD. An automation software package is used to calculate the required cutter offset location files, and this data is used to generate the information necessary to direct the motion control hardware (which consists of three stepping motors and associated drivers together with a stepping motor controller board inserted into an expansion slot on the PC).

All cutter motion is controlled by the menu-driven automation package. The set of commands for producing a specific component (called an application program) is simple instructions that avoid the control codes (G-codes) associated with normal CNC machine tools.

Excellent graphics capabilities for door design and cutter path images are provided by the PC as well as a large off-line data storage capability.

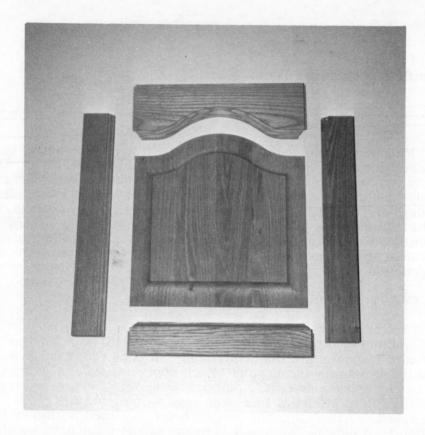

Figure 10.1 Exploded view of a cabinet door

10.1.2 Automated Woodshaper

A photograph of the modified woodshaper and computer controller unit is shown in Figure 10.2. The woodshaper has been retrofitted with an additional movable table, which is mounted on top of the existing table by means of two linear raceways. The additional table contains two pneumatic clamping rams that secure the workpiece to the table top. Both movable tables are fitted with leadscrews and ballnuts. Stepping motors are attached to the leadscrews via flexible couplings. Control of the rotation of the stepping motors can thus be used to position the workpiece relative to the cutter head. This arrangement allows positioning anywhere in a workplane size of 100 cm by 46 cm. The cutter head crank mechanism is also retrofitted with a stepping motor drive so that the cutter height can be automatically adjusted relative to the workpiece.

Sec. 10.1] **Cabinet Door Maufacture** 245

Figure 10.2 Automated woodshaper and computer

Stepping motors were selected in preference to DC servo motors on the basis of price and torque requirements. They were also selected because of their simpler interfacing and software needs; motor shaft position encoders and a feedback loop to the position control system are not required. Each stepping motor has a driver, which provides voltage pulses. The total number of pulses determines the distance travelled, while the frequency of pulses determines the speed of movement.

Interfacing each stepping motor, plus driver, to the PC is via a stepping motor controller board, as shown in Figure 10.3. The controller board fits into an expansion slot in the PC and replaces the dedicated controller found on CNC machining centres. Dual port, random access memory (RAM) allows two-way data transmission between controller board and PC; commands that specify form, position, and speed of motor movement flow from PC to board, whereas, motor position updates flow in the opposite direction. Commands transmitted to the board are in a CNC G-code format and are downloaded line-by-line to the controller board.

The stepping motor controller board provides direct, synchronized control of two axes of motor movement. The stepping motor for cutter height adjustment cannot, therefore, be synchronised to the other two motor movements. Control of the cutter height is effected by writing to its driver through a latching circuit on an additional input/output port on the controller board. This allows simple up/down positioning of the cutter height stepping motor, provided the other two motors are paused.

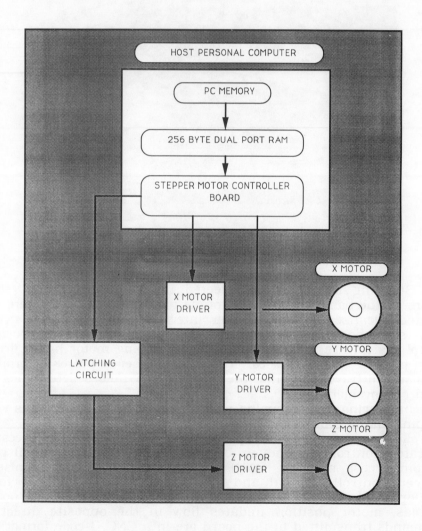

Figure 10.3 Motion control system interconnection

Sec. 10.1]   **Cabinet Door Maufacture**   247

10.1.3 Automation Software

A flow diagram showing an overview of the automation software is given in Figure 10.4. The AutoCad drawing output, in HPGL format, is converted into a cutter offset location file and then into a G-code machine file. The user's application program, which contains all the machining instructions to produce one component (i.e., the curve to be machined, feedrate, cutter height, and position in workplane), is merged to form one final G-code file. A library of application programs, as well as a library of curve profiles, can be assembled as the integrated system is used.

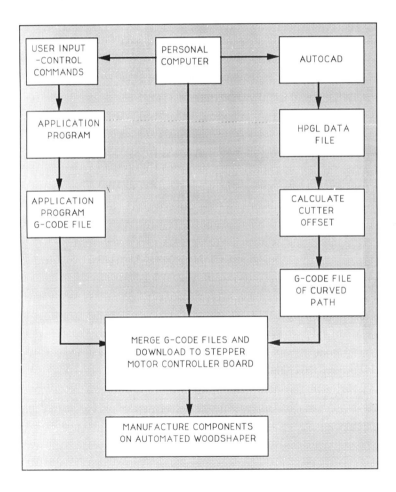

Figure 10.4 Overview of the automation software

Structure of the HPGL and G-code commands is given in Table 10.1. HPGL commands are of a two-letter mnemonic type, where each command is followed by a parameter field (specifying location in the cartesian plane) and a terminator (separating consecutive commands). The mnemonic 'PA' in Table 10.1 stands for 'pen absolute' and is used for positioning the plotter pen at defined cartesian coordinates.

**Table 10.1
Format of HPGL and G-code Commands**

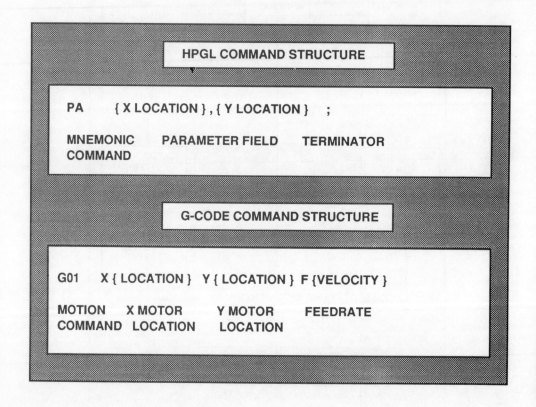

The automation software parses the HPGL commands into correctly structured lines of G-code. The G-code commands, recognized by the controller board, are a subset of the commands found on larger CNC systems. The parsing process involves calculating the offset of the cutter and transforming the positional data from cartesian coordinates to angular coordinates for the motor drivers. Feedrate of the wood past the cutter head is appended to the G-code command line by the F (feedrate) statement.

Curve shape is represented in the HPGL file by a series of points (xn,yn) joined by short line segments. The vertex points of the polygon are arranged in decreasing x magnitude through a sorting exchange algorithm. Cutter offset points (Xn,Yn) are calculated, as illustrated in Figure 10.5, from the unit normal polygon vectors. The number of points on the original and offset curves can be controlled through interactive menu selection. A graphical display of the original curve and the offset curve tool path is automatically generated. The utilization of menus and graphics windows for displaying cutter path is aesthetically pleasing and less prone to incorrect data entry in an industrial environment. Error handling is done by the software and will inform the user if nonstandard commands in the application program are entered or if characters are entered instead of numerical values in the tables or default menu.

Position data in the offset location cutter file are used to form the G-code command lines. The G-code file is normally created in linear interpolation mode, such that motion is in a point-to-point form along short, linear segments. Circular interpolation can be introduced in the automation package to reduce the number of data points required and to give smoother table motion. In this approach two linear interpolation moves are replaced by one circular interpolation move.

Linear interpolation G-code format is shown as:

G01X⟨end point⟩Y⟨end point⟩F⟨feedrate⟩

Movement in response to this command is from the current location to the specified end points at the given feedrate. Position and feedrate are in units of motor pulses and motor pulses per second, respectively.

Circular interpolation G-code format for clockwise (CW) and counter-clockwise (CCW) commands are:

$$\left.\begin{array}{l} \text{CW} - \text{G02} \\ \text{CCW} - \text{G03} \end{array}\right\} X\langle\text{end point}\rangle Y\langle\text{end point}\rangle I\langle\text{centre}\rangle J\langle\text{centre}\rangle F\langle\text{feedrate}\rangle$$

The centre point (I,J) of each arc segment is calculated as follows. Three points are taken from the offset cutter location file and are inserted into the equation for a circle of radius R and centre (I,J) to give:

$$(X_n - I)^2 + (Y_n - J)^2 = R^2 \qquad (1)$$

By rearranging (1), a set of three equations in the variables (I,J,R) can be solved by Gaussian elimination. The centre point of the arc (I,J) is embedded with the end point into either the G02 or G03 command line.

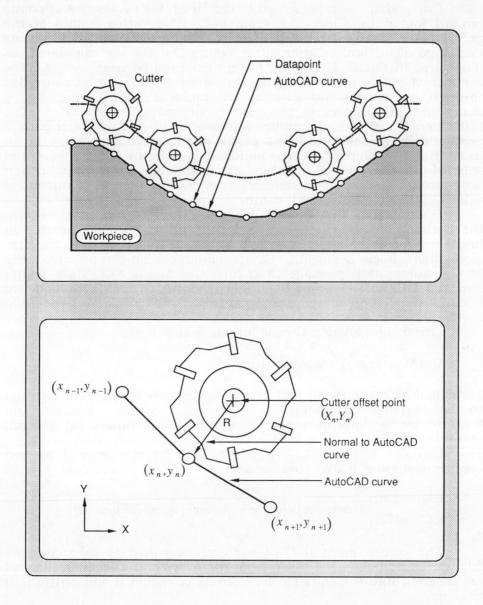

Figure 10.5 Cutter offset position in relation to defined curves

10.1.4 Performance Evaluation

Initial testing of the automated woodshaper revealed good performance characteristics. One undesirable feature was rough table motion at low speeds. This is a common characteristic of stepping motors when the motor torque and motor load are very different. At low shaft angular velocity, there is a 2:1 ratio in torque produced when the motor windings are connected in series rather than in parallel. Changing to parallel motor winding connection gave a closer torque-load match and solved the problem. Another difficulty was that the motors were observed to skip steps, and hence lose position, when operated close to their natural frequency of vibration. A change of driver from 400 pulses per shaft revolution to 2000 pulses per shaft revolution resulted in smoother table motion and no loss of pulses.

The automated woodshaper proved to have good control over the workpiece in terms of positional accuracy and maintaining a steady feedrate. A comparison was made between the AutoCAD curves and the machined workpiece shapes, as shown in Figure 10.6. The curves can be seen to be very close in Figure 10.6(a) and to be within 0.15 mm for the more arched shape in Figure 10.6(b). The predicted positioning accuracy of each motor axis, based on leadscrew pitch and non-cumulative motor error, is approximately 0.15 mm.

The automation package has been used extensively by a cabinet door manufacturer who has shown production time savings (to design, setup, and manufacture a cabinet door) of seventy-five percent.

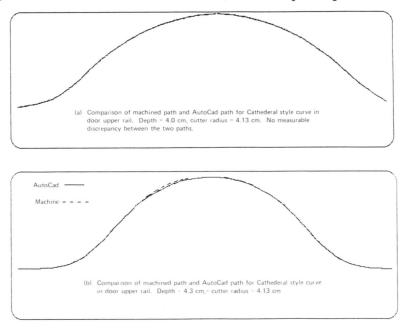

Figure 10.6 Comparison of machined and designed path

Appendix 1
C-Program: DNCLINK.C - Transmit a G-code File to a DNC Machine

```c
/********************************************************************/
/*                       D N C L I N K . C                          */
/********************************************************************/

#include<stdio.h>

#define   CR      0x8d
#define   LF      0x0a

/********************************************************************/
main(argc,argv)
int argc;
char *argv[];
/********************************************************************/
{
   FILE    *fgc,
           *fport,
           *fopen();

   char    c,
           fname[80];

   /*================================================================*/
   /* DISPLAY THE INFORMATION ABOUT THE PROGRAM                      */
   /*================================================================*/
   printf("            PROGRAM *** DNCLINK ***\n");
   printf("      TRANSMIT A G-CODE FILE TO A DNC MACHINE\n\n");
   printf("      INPUT THE GCODE FILE NAME =");
   scanf("%s",fname);
   fgc = fopen(fname,"r");
   if (fgc == NULL)
   {
      printf ("error in opening G-code file");
      exit(1);
   }
   fport = fopen(argv[1],"w");
   if (fport == NULL )
   {
      printf("error in opening communication port");
      exit(1);
   }
```

DNCLINK.C - Transmit a G-code File to a DNC Machine

```
/*================================================================*/
/* INITIALIZE THE PROGRAM WITH '%' 'LF' 'CR' 'CR'                 */
/* TAPE START CHARACTER %, PROGRAM START CHARACTER 00001010 LF,   */
/* AND TWO 10001101 CR CHARACTERS                                 */
/*================================================================*/
putc('%',fport);
putc(LF,fport);
putc(CR,fport);
putc(CR,fport);

/*================================================================*/
/* LOOP THROUGH EVERY CHARACTER UNTIL END OF FILE WITH A '%'      */
/*================================================================*/
while ( ( c = getc(fgc)) != '%')
{

   /*================================================================*/
   /* WRITE THE ISO CHARACTER TO PORT,                               */
   /* AND IGNORE ALL SPACES, END OF BLOCKS AND NEWLINES,             */
   /*================================================================*/
      if ( c != ' ' && c != ';' && c != '\n')
      {
         putc(c,fport);
         putc(c,stdout);
      }


   /*================================================================*/
   /* REPLACE END OF BLOCK CHARACTER ';' with LF and CR,CR           */
   /*================================================================*/
      if ( c == ';')
      {
         putc(LF,fport);
         putc(CR,fport);
         putc(CR,fport);
      }
}

/*================================================================*/
/* WRITE END OF TAPE CHARACTER AND CLEAR UP THE OUTPUT BUFFER,    */
/* CLOSE ALL FILES                                                */
/*================================================================*/
putc('%',fport);
fflush(fport);
fclose(fgc);
fclose(fport);
}
```

Appendix 2
C-Program: POSTPRO.C - Convert an APT-IV File to G-code Format

```c
/**********************************************************************/
/*                        P O S T P R O . C                           */
/**********************************************************************/

#include<stdio.h>
#include<math.h>

/**********************************************************************/
main()
/**********************************************************************/
{
    /*------------------------------------------------------------------*/
    /* DEFINE VARIABLES USED                                            */
    /*------------------------------------------------------------------*/

    int direct,i,j;
    int cir_fg,fedrate_fg,from_fg;
    int intol_fg,outtol_fg,spind_fg,unit_fg;

    float x,y,z;
    float x_prev,y_prev,z_prev;
    float xprev,yprev,zprev;
    float xdumc,ydumc,zdumc;
    float from_x,from_y,from_z;
    float x_cir_ctr,y_cir_ctr,z_cir_ctr;
    float x_cir_start,y_cir_start ;
    float theta_start,cir_rad;
    float feed_rate,spind_rpm,in_tol,out_tol,tool_no;
    float px,py,xprod,sx,sy;

    double atan(),sqrt(),fabs();
    double theta1,theta2,rad1,radlow,radhigh;
    double xtest,ytest,ztest;

    char c,gcodefn[80],aptfn[80];
    char pname[10];
    char unit_type[6];
    char spind_direct[3];
    char cool_state[3];
    char dum[4];
    char cmd[4];
    char previous_cmd[80];
    char dummy[80];
```

POSTPRO.C - Convert an APT-IV File to G-code Format

```
FILE *fopen(),*fp1,*fp2;

/*------------------------------------------------------------*/
/* INITIALIZE APT COMMANDS                                    */
/*------------------------------------------------------------*/
static char circle[] = {"SURF"};
static char coolant[] = {"COOL"};
static char spindle[] = {"SPIN"};
static char feedrate[] = {"FEDR"};
static char fini[] = {"FINI"};
static char from[] = {"FROM"};
static char gto[] = {"GOTO"};
static char intol[] = {"INTO"};
static char outtol[] = {"OUTT"};
static char rapid[] = {"RAPI"};
static char stop[] = {"STOP"};
static char toolno[] = {"LOAD"};
static char units[] = {"UNIT"};

/*------------------------------------------------------------*/
/* INITIALIZE CONTROL FLAGS                                   */
/*------------------------------------------------------------*/
intol_fg=outtol_fg=spind_fg=from_fg=unit_fg = 0;
cir_fg = fedrate_fg = 0;
direct = 4;

/*============================================================*/
/* DISPLAY THE INFORMATION ABOUT THE PROGRAM                  */
/*============================================================*/
printf("     PROGRAM POSTPRO ***\n");
printf("     CONVERT AN APT-IV FILE TO G-CODE FROMAT\n");
printf("\nINPUT APT IV FILE NAME = ");
scanf("%s",aptfn);
fp1 = fopen(aptfn,"r");
if (fp1 == NULL)
{
   printf("ERROR IN OPENING APT-IV FILE\n");
   exit(1);
}
printf("\n OUTPUT G-CODE FILE NAME = ");
scanf("%s",gcodefn);
fp2 = fopen(gcodefn,"w");

/*============================================================*/
/* READ PARTNO AND PARTNAME                                   */
/*============================================================*/
fscanf(fp1,"%6s %s",previous_cmd,pname);
i = compstr(previous_cmd,"PARTNO");
if (i==4)
{
   fprintf(fp2,"O111;\n");
}

/*============================================================*/
/* READ IN THE FIRST COMMAND                                  */
/*============================================================*/
fscanf(fp1,"\n%4s",cmd);
```

Appendix 2

```c
/*==============================================================*/
/* CHECK IF THE COMMAND "CMD" IS "FINI" THROUGH A 'DO-WHILE' LOOP */
/*==============================================================*/

j = 0;
do
    {
    /*==============================================================*/
    /* CHECK IF THE COMMAND "CMD" IS "GOTO"                         */
    /* AND READ THE FULL DATA LINE                                  */
    /*==============================================================*/
    j = compstr(cmd,gto);
    if (j == 4)
    {
        fscanf(fp1,"%1s%f,%f,%f",&c,&x,&y,&z);

        /*==============================================================*/
        /* IF THE COMMAND IS "GOTO", CHECK THE PREVIOUS COMMAND         */
        /*==============================================================*/
        i = compstr(previous_cmd,"RAPI");

        /*==============================================================*/
        /* IF PREVIOUS COMMAND != RAPI                                  */
        /*==============================================================*/
        if (i != 4)
        {

        /*==============================================================*/
        /* IF PREVIOUS COMMAND = CIRCLE,                                */
        /* CHECK IF THE POINT IS ON THE CIRCLE                          */
        /*==============================================================*/
            if(cir_fg == 1)
            {
                if(intol_fg==0) in_tol =0.2;
                if(outtol_fg==0) out_tol = 0.2;
                xtest = x_cir_ctr - x;
                ytest = y_cir_ctr - y;
                ztest = z_cir_ctr - z;
                rad1 = sqrt(xtest*xtest + ytest*ytest + ztest*ztest);
                radhigh = fabs(cir_rad) + in_tol + out_tol;
                radlow = fabs(cir_rad) - in_tol - out_tol;

        /*==============================================================*/
        /* IF THE POINT IS ON THE CIRCLE,                               */
        /* CHECK IF THE CIRCLE IS CW OR CCW                             */
        /*==============================================================*/
                if((rad1 >= radlow) && (rad1 <= radhigh) &&
                ((fabs(100.0*z)) == (fabs(100.0*z_prev))) )
                {

        /*==============================================================*/
        /* AVIOD DIVIDING BY ZERO OR  HAVING ONLY ONE POINT             */
        /*==============================================================*/
                    if(direct == 4 || (x == x_cir_ctr) ||
                    (xprev==x_cir_ctr))
                    {
                        if(direct == 4 ) direct = 3 ;
```

POSTPRO.C - Convert an APT-IV File to G-code Format

```c
         xprev = x;
         yprev = y;
         zprev = z;
      }

      /*===================================================*/
      /* CALCULATE THETA1, THETA2 AND                      */
      /* CHECK IF THE CIRCLE IS CW OR CCW                  */
      /*===================================================*/
      else
      {
         theta1 = atan((y-y_cir_ctr)/(x-x_cir_ctr));
         theta2 = atan((yprev-y_cir_ctr)/(xprev-x_cir_ctr));
         yprev = y;
         xprev = x;
         zprev = z;
         if(theta1>theta2 && direct == 3 )     direct = 1;
         if(theta1<theta2 && direct == 3 )     direct = 0;
         if(theta1 == theta2 && direct == 3)   direct = 3;
      }
   }
   else

   /*======================================================*/
   /* OTHERWISE THE POINT IS NOT ON THE CIRCLE,            */
   /* THUS THE 'CIRC'IS COMPLETE                           */
   /*======================================================*/
   {
      cir_fg = 0;
      if(direct == 0)

         /*===================================================*/
         /* WRITE THE CW CIRCLE TO THE GCODEFN                */
         /*===================================================*/
         {
            sx = x_cir_start - x_cir_ctr;
            sy = y_cir_start - y_cir_ctr;
            px = xprev - x_cir_ctr ;
            py = yprev - y_cir_ctr ;

            /*===============================================*/
            /* CROSS PRODUCT OF START AND END VECTOR         */
            /* TO DETERMINE IF 0 < ARC < 180                 */
            /*===============================================*/
            xprod = (py*sx - px*sy);
            if(xprod > 0 ) cir_rad = - cir_rad;
            fprintf(fp2,"G02 X%8.3f Y%8.3f Z%8.3f R%8.3f;\n",
            xprev,yprev,zprev,cir_rad);
            fprintf(fp2,"G01 X%8.3f Y%8.3f Z%8.3f;\n",x,y,z);
            x_prev = x;
            y_prev = y;
            z_prev = z;
            direct=4;
         }

         /*===================================================*/
         /* WRITE THE CCW CIRCLE TO GCODEFN                   */
         /*===================================================*/
```

Appendix 2

```
                    else
                    {
                        sx = x_cir_start - x_cir_ctr ;
                        sy = y_cir_start - y_cir_ctr ;
                        px = x - x_cir_ctr ;
                        py = y - y_cir_ctr ;
                        xprod = py*sx - px*sy ;
                        if(xprod < 0 ) cir_rad = - cir_rad;
                        fprintf(fp2,"G03 X%8.3f Y%8.3f Z%8.3f R%8.3f;\n",
                        xprev,yprev,zprev,cir_rad);
                        fprintf(fp2,"G01 X%8.3f Y%8.3f Z%8.3f;\n",x,y,z);
                        x_prev = x;
                        y_prev = y;
                        z_prev = z;
                        direct = 4;
                    }
                }
            }
            else

            /*============================================================*/
            /* IF PREVIOUS COMMAND IS NOT A 'CIRCLE'                      */
            /* WRITE G01 TO GCODEFN FILE                                  */
            /*============================================================*/
            {
                fprintf(fp2,"G01 X%8.3f Y%8.3f Z%8.3f;\n",x,y,z);
                x_prev = x;
                y_prev = y;
                z_prev = z;
            }
        }
        else

        /*============================================================*/
        /* IF PREVIOUS COMMAND IS A 'RAPID'                           */
        /* WRITE G00 TO   GCODEFN FILE                                */
        /*============================================================*/
        {
            fprintf(fp2,"G00 X%8.3f Y%8.3f Z%8.3f;\n",x,y,z);
            x_prev = x;
            y_prev = y;
            z_prev = z;
        }
    }

    /*============================================================*/
    /* CHECK IF THE COMMAND 'CMD' IS 'RAPID'                      */
    /*============================================================*/
    j = 0;
    j = compstr(cmd,rapid);
    if ( j == 4)
    {
       fscanf(fp1,"%s",dummy);
       strcpy(previous_cmd,"RAPI");
    }

    /*============================================================*/
    /* CHECK IF THE COMMAND 'CMD' IS 'FEEDRATE'                   */
```

```
/*==============================================================*/
j = 0;
j = compstr(cmd,feedrate);
if ( j == 4)
{
   fedrate_fg = 1;
   fscanf(fp1,"%3s%f",dummy,&feed_rate);
   fprintf(fp2,"F%8.3f;\n",feed_rate);
   strcpy(previous_cmd,"FEDR");
}

/*==============================================================*/
/* CHECK IF THE COMMAND 'CMD' IS 'SPINDLE SPEED'                */
/*==============================================================*/
j = 0;
j = compstr(cmd,spindle);
if ( j == 4)
{
   spind_fg = 1;
   fscanf(fp1,"%3s%f,%s",dummy,&spind_rpm,spind_direct);
   i = compstr(spind_direct,"CLW");
   if ( i == 4)
   {
      fprintf(fp2,"M03 S%8.3f;\n",spind_rpm);
   }
   else
   {
      if ( (compstr(spind_direct,"OFF")) != 4)
      {
         fprintf(fp2,"M04 S%8.3f;\n",spind_rpm);
      }
   }
}

/*==============================================================*/
/* CHECK IF THE COMMAND 'CMD' IS 'CIRCLE'                       */
/*==============================================================*/
j = 0;
j = compstr(cmd,circle);
if ( j == 4)
{
   if (cir_fg == 1)
   {
      cir_fg = 0;
      if (direct == 1)
      {
         sx = x_cir_start - x_cir_ctr ;
         sy = y_cir_start - y_cir_ctr ;
         px = x - x_cir_ctr ;
         py = y - y_cir_ctr ;
         xprod = py*sx - px*sy ;
         if(xprod < 0 ) cir_rad = - cir_rad;
         fprintf(fp2,"G03 X%8.3f Y%8.3f Z%8.3f R%8.3f;\n",
         xprev,yprev,zprev,cir_rad);
         x_prev = x;
         y_prev = y;
         z_prev = z;
         direct =4;
```

```
                    }
                else
                    {
                      sx = x_cir_start - x_cir_ctr ;
                      sy = y_cir_start - y_cir_ctr ;
                      px = x - x_cir_ctr ;
                      py = y - y_cir_ctr ;
                      xprod = py*sx - px*sy ;
                      if(xprod > 0 ) cir_rad = - cir_rad;
                      fprintf(fp2,"G02 X%8.3f Y%8.3f Z%8.3f R%8.3f;\n",
                      xprev,yprev,zprev,cir_rad);
                      x_prev = x;
                      y_prev = y;
                      z_prev = z;
                      direct =4;
                    }
              }
        cir_fg = 1;
        fscanf(fp1,"%4s%f,%f,%f,%f,%f,%f",dummy,&x,&y,&z,
        &xdumc,&ydumc,&zdumc,&cir_rad);
        x_cir_ctr = x;
        y_cir_ctr = y;
        z_cir_ctr = z;
        x_cir_start = x_prev ;
        y_cir_start = y_prev ;
        if (x_cir_start == x_cir_ctr)
           theta_start = 11.0/7.0 ;
        else
           theta_start = atan((y_cir_start-y_cir_ctr)/
           (x_cir_start-x_cir_ctr));
        if ( (theta_start < 0.0) && (y_cir_start < y_cir_ctr) )
           theta_start = theta_start ;
        if ( (theta_start < 0.0) && (y_cir_start > y_cir_ctr) )
           theta_start = theta_start + (22.0/7.0) ;
        if ( (theta_start > 0.0) && (y_cir_start < y_cir_ctr) )
           theta_start = theta_start + (22.0/7.0) ;
}

/*==============================================================*/
/* CHECK IF THE COMMAND 'CMD' IS 'STOP'                         */
/*==============================================================*/
j = 0;
j = compstr(cmd,stop);
if ( j == 4)
{
   if(spind_fg == 1) spind_fg = 0;
   fprintf(fp2,"M05;\n");
}


/*==============================================================*/
/* CHECK IF THE COMMAND 'CMD' IS 'INTOL'                        */
/*==============================================================*/
j = 0;
j = compstr(cmd,intol);
if ( j == 4)
{
   intol_fg = 1;
   fscanf(fp1,"%2s%f",dummy,&in_tol);
```

```c
}

/*================================================================*/
/* CHECK IF THE COMMAND 'CMD' IS 'OUTTOL'                         */
/*================================================================*/
j = 0;
j = compstr(cmd,outtol);
if ( j == 4)
{
   outtol_fg = 1;
   fscanf(fp1,"%3s%f",dummy,&out_tol);
}

/*================================================================*/
/* CHECK IF THE COMMAND 'CMD' IS 'UNITS'                          */
/*================================================================*/
j = 0;
j = compstr(cmd,units);
if ( j == 4)
{
   unit_fg = 1;
   fscanf(fp1,"%2s%s",dummy,unit_type);
   i = compstr(unit_type,"MM");
if (i == 4)
   fprintf(fp2,"G21;\n");
else
   fprintf(fp2,"G20;\n");
}

/*================================================================*/
/* CHECK IF THE COMMAND 'CMD' IS 'COOLANT'                        */
/*================================================================*/
j = 0;
j = compstr(cmd,coolant);
if ( j == 4)
   fscanf(fp1,"%3s%3s",dummy,cool_state);

/*================================================================*/
/* CHECK IF THE COMMAND 'CMD' IS 'FROM'                           */
/*================================================================*/
j = 0;
j = compstr(cmd,from);
if ( j == 4)
{
   from_fg = 1;
   fscanf(fp1,"%1s%f,%f,%f",dummy,&from_x,&from_y,&from_z);
   fprintf(fp2,"G92 X%8.3f Y%8.3f Z%8.3f ;\n"
   ,from_x,from_y,from_z);
   fprintf(fp2,"G90;\n");
}

/*================================================================*/
/* CHECK IF THE COMMAND 'CMD' IS 'TOOLNO'                         */
/*================================================================*/
j = 0;
j = compstr(cmd,toolno);
if ( j == 4)
   fscanf(fp1,"%3s%s",dummy,tool_no);
```

Appendix 2

```c
    /*================================================================*/
    /* CHECK IF THE COMMAND 'CMD' IS 'FINI'                           */
    /*================================================================*/
    j = 0;
    j = compstr(cmd,fini);
    if ( j == 4 ) break;

    /*================================================================*/
    /* READ THE NEXT LINE OF THE APTFN FILE                           */
    /*================================================================*/
    fscanf(fp1,"\n%4s",cmd);
    j = compstr(cmd,fini);
  }
    /*================================================================*/
    /* END OF THE DO-WHILE LOOP WHEN THE COMMAND 'CMD' IS 'FINI'      */
    /*================================================================*/
  while ( j != 4) ;
  fprintf(fp2,"M30;\n%%\n");
}



/**********************************************************************/
/*                  FUNCTION COMPSTR ( COMPARE STRING )               */
/*         RETURN '4' IF TWO STRINGS 'S' AND 'T' ARE EQUAL            */
/**********************************************************************/

compstr(s,t)
char s[] , t[];
{
   int i ;

   i = 0;
   while (s[i] == t[i])
   {
      if ( s[i++] == '\0') return(4);
      i = i+1;
   }
   return(i);
}
```

Appendix 3
Library Routines for Frequently Used Functions

```
/****************************************************************/
/*                L I B R A R Y . H                             */
/****************************************************************/

#define TRUE            1
#define FALSE           0

FILE * open_input_file
    (
        char *Name
    );

FILE * open_output_file
    (
        char *Name
    );

void  beep
    (
        void
    );

double get_one_double_value
    (
        char *Question
    );

void check_parameters_two
    (
        int    NumberOfParameters,
        char **FileNames
    );

void check_parameters_three
    (
        int    NumberOfParameters,
        char **FileNames
    );

void  allocate_memory_for_one_line
    (
        float    **X,
        float    **Y,
        float    **Z,
```

```
        int      NVertices
    );

void read_file_allocate
    (
        float ***X,
        float ***Y,
        float ***Z,
        int     *NLines,
        int     *NVertices,
        char    *FileName
    );

int read_one_line_allocate
    (
        float **X,
        float **Y,
        float **Z,
        FILE   *fin
    );

float **allocate_two_d_float_array
    (
        int NRows,
        int NCols
    );

int write_file
    (
        float **X,
        float **Y,
        float **Z,
        int    NLines,
        int    NVertices,
        char   FileName[]
    );

void write_one_line
    (
        float X[],
        float Y[],
        float Z[],
        int   NVertices,
        FILE *fout
    );

void free_two_d_float_array
    (
        int     NRows,
        float **A
    );


/******************************************************************/
/*                    L I B R A R Y                               */
/******************************************************************/
```

```c
#include <stdio.h>
#include <math.h>
#include <stdlib.h>
#include <ctype.h>
#include <conio.h>
#include <dos.h>

#include "library.h"


/*******************************************************************/
FILE *open_input_file
    (
        char *Name
    )
/*******************************************************************/
{
    FILE    *fin;

    /*================================================================*/
    /* OPEN THE FILE FOR READ ACCESS                                 */
    /*================================================================*/
    fin = fopen( Name, "r" );

    /*================================================================*/
    /* IF OPEN UNSUCCESSFUL ABORT PROGRAM                            */
    /*================================================================*/
    if( fin == NULL )
    {
        printf(" UNABLE TO OPEN FILE: %s\n", Name);
        exit(1);
    }

    /*================================================================*/
    /* RETURN THE FILE POINTER                                       */
    /*================================================================*/
    return( fin );
}

/*******************************************************************/
FILE *open_output_file
    (
        char *Name
    )
/*******************************************************************/
{
    FILE    *fout;

    /*================================================================*/
    /* OPEN THE FILE FOR WRITE ACCESS                                */
    /*================================================================*/
    fout = fopen( Name, "w" );

    /*================================================================*/
    /* IF OPEN UNSUCCESSFUL ABORT PROGRAM                            */
    /*================================================================*/
    if(fout == NULL)
```

```
        {
            printf(" UNABLE TO OPEN FILE: %s\n", Name);
            exit(1);
        }

        /*================================================================*/
        /* RETURN THE FILE POINTER                                        */
        /*================================================================*/
        return( fout );
    }
/********************************************************************/
void  beep
        (
            void
        )
/********************************************************************/
    {
        /*================================================================*/
        /* SEND A BEEP LIKE SOUND TO THE SPEAKER FOR ONE TENTH OF A SECOND */
        /*================================================================*/
        sound( 1000 );
        delay( 100 );
        nosound( );
    }


/********************************************************************/
double get_one_double_value
        (
            char *Question
        )
/********************************************************************/
    {
        double Answer;
        int    NFieldsRead;


        /*================================================================*/
        /* DISPLAY THE QUESTION ON THE SCREEN                             */
        /*================================================================*/
        printf( "%s", Question );

        /*================================================================*/
        /* READ THE FLOAT NUMBER UNTIL INPUT IS VALID                     */
        /*================================================================*/
        do
        {
            NFieldsRead = scanf( "%lf", &Answer );
            /*============================================================*/
            /* CHECK IF INPUT WAS VALID                                   */
            /*============================================================*/
            if( NFieldsRead != 1 )
            {
                beep();
            }
        }while( NFieldsRead != 1 );
```

Library Routines for Frequently Used Functions   267

```
    /*==============================================================*/
    /* ADVANCE CURSOR TO A NEW LINE                                 */
    /*==============================================================*/
    printf( "\n" );

    /*==============================================================*/
    /* RETURN THE READ VALUE                                        */
    /*==============================================================*/
    return( Answer );
}
/********************************************************************/
void check_parameters_two
      (
          int    NumberOfParameters,
          char **FileNames
      )
/********************************************************************/
{
    if( NumberOfParameters != 3 )
    {
        /*==============================================================*/
        /* CLEAR THE SCREEN                                             */
        /*==============================================================*/
        clrscr( );

        printf( "ERROR IN PROGRAM %s\n", FileNames[0] );
        printf( "YOU HAVE TO PASS THE NAMES OF THE INPUT AND OUTPUT FILE\n" );
        printf( "USE! >>> %s INPUTFILE OUTPUTFILE\n", FileNames[0] );
        exit( 1 );
    }

}


/********************************************************************/
void check_parameters_three
      (
          int    NumberOfParameters,
          char **FileNames
      )
/********************************************************************/
{
    if( NumberOfParameters != 4 )
    {
        /*==============================================================*/
        /* CLEAR THE SCREEN                                             */
        /*==============================================================*/
        clrscr( );

        printf( "ERROR IN PROGRAM %s\n", FileNames[0] );
        printf( "YOU HAVE TO PASS THE NAMES OF TWO INPUT FILES \n" );
        printf( "AND ONE OUTPUT FILE\n" );
        printf( "USE!>>> %s INPUTFILE1 INPUTFILE2 OUTPUTFILE\n", FileNames[0] );
        exit( 1 );
    }
```

Appendix 3

```c
}
/************************************************************************/
void allocate_memory_for_one_line
    (
        float    **X,
        float    **Y,
        float    **Z,
        int      NVertices
    )
/************************************************************************/
{
   /*====================================================================*/
   /* ALLOCATE THE MEMORY                                                */
   /*====================================================================*/

   *X = (float* ) malloc( (unsigned) (NVertices * sizeof(float) ) );
   *Y = (float* ) malloc( (unsigned) (NVertices * sizeof(float) ) );
   *Z = (float* ) malloc( (unsigned) (NVertices * sizeof(float) ) );

   /*====================================================================*/
   /* CHECK IF ALLOACTION WAS SUCCESSFUL - IF NOT ABORT THE PROGRAM      */
   /*====================================================================*/
   if( (*X==NULL)  || (*Y==NULL)  || (*Z==NULL)   )
   {
      printf( "ERROR IN FUNCTION allocate_memory_for_one_line\n" );
      printf( "TRYING TO ALLOCATE MEMORY %3d VERTICES\n", NVertices );
      printf( "OUT OF MEMORY\n" );
      exit( 1 );
   }
}

/************************************************************************/
void read_file_allocate
    (
        float   ***X,
        float   ***Y,
        float   ***Z,
        int     *NLines,
        int     *NVertices,
        char    *FileName
    )
/************************************************************************/
{
   register int Line;
   FILE         *fin;


   /*====================================================================*/
   /* OPEN THE FILE                                                      */
   /*====================================================================*/
   fin = open_input_file( FileName );

   gotoxy( 5, 10 );
   printf( "READING %s", FileName );

   /*====================================================================*/
   /* READ HOW MANY LINES IN THIS FILE                                   */
```

```c
/*===============================================================*/
   fscanf( fin, "%d", NLines );

/*===============================================================*/
/* ALLOCATE THE MEMORY FOR THE POINTER ARRAY FOR THE COORDINATES */
/*===============================================================*/
   *X = (float **) malloc( (unsigned) ( *NLines * sizeof(float *) ) );
   *Y = (float **) malloc( (unsigned) ( *NLines * sizeof(float *) ) );
   *Z = (float **) malloc( (unsigned) ( *NLines * sizeof(float *) ) );
   if( !*X  ||  !*Y  ||  !*Z )
   {
      printf( "allocation error 1 in n_read_ship_file_allocate\n" );
      exit( 1 );
   }

/*===============================================================*/
/* READ THE FILE INTO MEMORY                                     */
/*===============================================================*/
   for( Line=0 ; Line<*NLines ; Line++ )
   {
      *NVertices = read_one_line_allocate( &((*X)[Line]),
                                           &((*Y)[Line]),
                                           &((*Z)[Line]),
                                           fin );
   }

/*===============================================================*/
/* CLOSE THE FILE AND RETURN 'SUCCESS'                           */
/*===============================================================*/
   fclose( fin );
}


/******************************************************************/
int read_one_line_allocate
     (
        float  **X,
        float  **Y,
        float  **Z,
        FILE    *fin
     )
/******************************************************************/
{
   register int  n;
   int           NVertices;

/*===============================================================*/
/* READ THE NUMBER OF VERTICES IN THIS LINE                      */
/*===============================================================*/
   fscanf(fin, "%d\n", &NVertices);

   if( NVertices == 0 )
   {
      /*===============================================================*/
      /* ALLOCATE THE MINIMUM OF 1 VERTEX                              */
      /*===============================================================*/
      allocate_memory_for_one_line( X, Y, Z, 1 );
   }
```

```c
      else
      {
         /*================================================================*/
         /* ALLOCATE THE MEMORY FOR THE LINE                               */
         /*================================================================*/
         allocate_memory_for_one_line( X, Y, Z, NVertices );
      }


      /*===================================================================*/
      /* READ THE LINE COORDINATES                                         */
      /*===================================================================*/
      for( n = 0 ; n < NVertices ; n++ )
      {
         fscanf(fin, " %f      %f      %f  ",
                     &((*X)[n]), &((*Y)[n]), &((*Z)[n]) );
      }

      /*===================================================================*/
      /* RETURN HOW MANY VERTICES IN THIS LINE                             */
      /*===================================================================*/
      return( NVertices );
}
/*************************************************************************/
float **allocate_two_d_float_array
      (
            int   NRows,
            int   NCols
      )
/*************************************************************************/
{
   float   **A;
   int       n;


      /*===================================================================*/
      /* ALLOCATE THE MEMORY FOR THE POINTER ARRAY                         */
      /*===================================================================*/
      A = (float **) malloc( (unsigned) ( NRows * sizeof(float *) )  );
      if( !A )
      {
         printf( "allocation error 1 in alloc_two_d_float_array\n" );
         return( NULL );
      }

      /*===================================================================*/
      /* ALLOCATE THE MEMORY FOR EACH COLUMN                               */
      /*===================================================================*/
      for( n=0 ; n<NRows ; n++ )
      {
         A[n] = (float *) malloc( (unsigned) ( NCols * sizeof(float) )  );
         if( !A[n] )
         {
            printf( "allocation error 1 in alloc_two_d_float_array\n" );
            return( NULL );
         }
```

```
    }
    /*================================================================*/
    /* RETURN THE DOUBLE INDIRECTED POINTER TO THE CALLING FUNCTION   */
    /*================================================================*/
    return( A );
}
/******************************************************************/
int write_file
    (
        float   **X,
        float   **Y,
        float   **Z,
        int     NLines,
        int     NVertices,
        char    FileName[]
    )
/******************************************************************/
{
    register int Line;
    FILE *fout;


    /*================================================================*/
    /* OPEN THE FILE                                                  */
    /*================================================================*/
    fout = open_output_file( FileName );

    gotoxy( 5, 20 );
    printf( "WRITING %s", FileName );

    /*================================================================*/
    /* WRITE THE FILE TO THE DISK                                     */
    /*================================================================*/
    fprintf( fout, "%d\n", NLines );

    for( Line=0 ; Line<NLines ; Line++ )
    {
        write_one_line( X[Line], Y[Line], Z[Line], NVertices, fout );
    }

    /*================================================================*/
    /* CLOSE THE FILE AND RETURN 'SUCCESS'                            */
    /*================================================================*/
    fclose( fout );
}
/******************************************************************/
void write_one_line
    (
        float   *X,
        float   *Y,
        float   *Z,
        int     NVertices,
        FILE    *fout
    )
/******************************************************************/
```

```c
{
    register int Vertex;

    /*====================================================================*/
    /* WRITE THE NUMBER OF VERTICES TO THE OUTPUT FILE                    */
    /*====================================================================*/
    fprintf( fout, "%d\n", NVertices );

    /*====================================================================*/
    /* WRITE THE VALUES FOR EACH VERTEX TO THE OUTPUT FILE                */
    /*====================================================================*/
    for( Vertex=0 ; Vertex<NVertices ; Vertex++ )
    {
        fprintf( fout, "%10.6f %10.6f %10.6f\n",
                    X[Vertex], Y[Vertex], Z[Vertex] );
    }

}

/**********************************************************************/
void free_two_d_float_array
    (
        int     NRows,
        float   **A
    )
/**********************************************************************/
{
    register int n;

    /*====================================================================*/
    /* FREE THE MEMORY LINKED TO THE POINTER ARRAY                        */
    /*====================================================================*/
    for( n=0 ; n<NRows ; n++ )
    {
        free( A[n] );
    }

    /*====================================================================*/
    /* FREE THE POINTER ARRAY LINKED TO THE DOUBLE INDIRECTED POINTER     */
    /*====================================================================*/
    free( A );

}
```

Appendix 4
C- Program: FITLINE - Fit a B-spline Line

```
/****************************************************************/
/*                   F I T L I N E 2 . H                        */
/****************************************************************/

int bspline_3d_curve
    (
        float   XOrg[],
        float   YOrg[],
        float   ZOrg[],
        float   SplnX[],
        float   SplnY[],
        float   SplnZ[],
        int     NVrtcs,
        double  DeltaT
    );

int add_vertices_in_end_intervals
    (
        float X[],
        float Y[],
        float Z[],
        int   NVertices
    );

int knot_vector
    (
        float   KnotVector[],
        int     Order,
        int     NVrtcs
    );

int spline_3d
    (
        float   *X,
        float   *Y,
        float   *Z,
        float   SplnX[],
        float   SplnY[],
        float   SplnZ[],
        float   KnotVector[],
        int     MaxOrdr,
        int     NVrtcs,
        int     NKnots,
        double  DeltaT
```

```c
    );

void basis_order_1
    (
        float   *Basis1,
        int     NKnots,
        int     KnotNumber
    );

void basis_order_n
    (
        float   **Basis,
        float   KnotVector[],
        double  T,
        int     Ordr,
        int     NKnots
    );


/**********************************************************************/
/*      F I T L I N E 1 . C                                           */
/**********************************************************************/

#include <stdio.h>
#include <stdlib.h>
#include <conio.h>

#include "library.h"

#include "fitline2.h"


/**********************************************************************/
/* PROTOTYPED FUNCTIONS                                               */
/**********************************************************************/
void information
    (
        char **FileNames
    );

void fit_3d_curves
    (
        char    *InputFileName,
        char    *OutputFileName
    );

/**********************************************************************/
main
    (
        int     NParameters,
        char    **ParameterStrings
    )
/**********************************************************************/
{

    /*==============================================================*/
```

FITLINE - Fit a B-spline Line

```c
   /* CHECK IF RIGHT NUMBER OF PARAMETERS HAVE BEEN PASSED     */
   /*==========================================================*/
   check_parameters_two( NParameters, ParameterStrings );

   /*==========================================================*/
   /* DISPLAY INFORMATION ABOUT THE PROGRAM                    */
   /*==========================================================*/
   information( ParameterStrings );

   /*==========================================================*/
   /* FIT A 3 DIMENSIONAL CURVE FOR EVERY LINE ON THE SPARSE DATA SET */
   /*==========================================================*/
   fit_3d_curves( ParameterStrings[1], ParameterStrings[2] );

   gotoxy( 10, 22 );
   printf( "PROGRAM SUCCESSFUL\n" );
}
/****************************************************************/
void information
     (
         char **FileNames
     )
/****************************************************************/
{
   /*==========================================================*/
   /* CLEAR THE SCREEN                                         */
   /*==========================================================*/
   clrscr( );

   /*==========================================================*/
   /* DISPLAY THE INFORMATION ABOUT THE PROGRAM                */
   /*==========================================================*/
   printf( "          PROGRAM *** F I T L I N E ***\n");
   printf( " FIT A B-SPLINE LINE (OR LINES) TO A SET OF SPARSE DATA
           \n\n");

   printf( "  SPARSE DATA FILE NAME  = %s\n", FileNames[1] );
   printf( "  FITTED LINE FILE NAME  = %s\n", FileNames[2] );
}


/****************************************************************/
void fit_3d_curves
     (
         char   *InputFileName,
         char   *OutputFileName
     )
/****************************************************************/
{
   FILE     *FinSprs;
   float    *XSprs,
            *YSprs,
            *ZSprs;
   int       NVrtcsSprs;
   int       NLinesSprs;

   FILE     *FoutFit;
```

Appendix 4

```
   float     *XFit,
             *YFit,
             *ZFit;
   int       NVrtcsFit;

   double    DeltaT;

   int       Line,
             Vrtx;

/*=============================================================*/
/* OPEN THE SPARSE DATA FILE AND THE FILE FOR THE FITTED SPLINES */
/*=============================================================*/
FinSprs = open_input_file( InputFileName );
FoutFit = open_output_file( OutputFileName );

/*=============================================================*/
/* READ THE NUMBER OF LINES AND WRITE TO THE OUTPUT FILE       */
/*=============================================================*/
fscanf( FinSprs, "%d", &NLinesSprs );
fprintf( FoutFit, "%d\n", NLinesSprs );

/*=============================================================*/
/* DISPLAY HOW MANY LINES TO PROCESS                           */
/*=============================================================*/
gotoxy( 10, 15 );
printf( "THE SURFACE HAS %3d LINES", NLinesSprs );
gotoxy( 10, 16 );
printf( "PROCESSING LINE" );

/*=============================================================*/
/* HANDLE THE FIRST LINE ON THE SURFACE                        */
/*=============================================================*/
Line = 0;

/*=============================================================*/
/* UPDATE SCREEN, SHOW WHICH LINE IN PROCESS                   */
/*=============================================================*/
gotoxy( 30, 16 );
printf( "%4d", Line+1 );

/*=============================================================*/
/* READ HOW MANY VERTICES ARE IN THE SPARSE DATA LINE          */
/*=============================================================*/
fscanf( FinSprs, "%d", &NVrtcsSprs );

/*=============================================================*/
/* ALLOCATE THE MEMORY FOR ONE LINE OF THE SPARSE DATA SET SURFACE */
/*=============================================================*/
XSprs = (float*) malloc( (unsigned)( NVrtcsSprs * sizeof(float) ) );
YSprs = (float*) malloc( (unsigned)( NVrtcsSprs * sizeof(float) ) );
ZSprs = (float*) malloc( (unsigned)( NVrtcsSprs * sizeof(float) ) );
if( (XSprs==NULL) || (YSprs==NULL) || (ZSprs==NULL) )
{
   printf( "OUT OF MEMORY\n" );
   exit( 1 );
}
```

FITLINE - Fit a B-spline Line

```c
/*================================================================*/
/* GET THE DELTA-T FROM THE USER                                  */
/*================================================================*/
gotoxy( 5,18 );
DeltaT = get_one_double_value( "ENTER DELTA-T" );

/*================================================================*/
/* CALCULATE THE NUMBER OF VERTICES ON THE SPLINE                 */
/*================================================================*/
NVrtcsFit   =   (int)( ( (NVrtcsSprs - 1) / DeltaT ) + 1.5 );

/*================================================================*/
/* ALLOCATE THE MEMORY FOR THE FITTED SURFACE                     */
/*================================================================*/
XFit = (float*) malloc( (unsigned)( NVrtcsFit * sizeof(float) )  );
YFit = (float*) malloc( (unsigned)( NVrtcsFit * sizeof(float) )  );
ZFit = (float*) malloc( (unsigned)( NVrtcsFit * sizeof(float) )  );
if( (XFit==NULL) || (YFit==NULL) || (ZFit==NULL) )
{
   printf( "OUT OF MEMORY\n" );
   exit( 1 );
}

/*================================================================*/
/* READ ONE LINE FROM THE SPARSE DATA FILE                        */
/*================================================================*/
for( Vrtx=0 ; Vrtx<NVrtcsSprs ; Vrtx++ )
{
   fscanf( FinSprs, "%f %f %f",
                    &(XSprs[Vrtx]), &(YSprs[Vrtx]), &(ZSprs[Vrtx]) );
}

/*================================================================*/
/* FIT B-SPLINE CURVE TO THE SPARSE DATA SET                      */
/*================================================================*/
bspline_3d_curve( XSprs, YSprs, ZSprs,
                  XFit,  YFit,  ZFit,
                  NVrtcsSprs, DeltaT    );

/*================================================================*/
/* WRITE ONE LINE TO THE OUTPUT FILE                              */
/*================================================================*/
write_one_line( XFit, YFit, ZFit, NVrtcsFit, FoutFit );


for( Line=1; Line<NLinesSprs ; Line++ )
{
   /*=============================================================*/
   /* UPDATE SCREEN, SHOW WHICH LINE IN PROCESS                   */
   /*=============================================================*/
   gotoxy( 30, 16 );
   printf( "%4d", Line+1 );

   /*=============================================================*/
   /* READ HOW MANY VERTICES ARE IN THE SPARSE DATA LINE          */
   /*=============================================================*/
   fscanf( FinSprs, "%d", &NVrtcsSprs );
```

```
            /*=============================================================*/
            /* READ ONE LINE FROM THE SPARSE DATA FILE                     */
            /*=============================================================*/
            for( Vrtx=0 ; Vrtx<NVrtcsSprs ; Vrtx++ )
            {
               fscanf( FinSprs, "%f %f %f",
                          &(XSprs[Vrtx]), &(YSprs[Vrtx]), &(ZSprs[Vrtx]) );
            }

            /*=============================================================*/
            /* FIT B-SPLINE CURVE TO THE SPARSE DATA SET                   */
            /*=============================================================*/
            bspline_3d_curve( XSprs, YSprs, ZSprs,
                              XFit,  YFit,  ZFit,
                              NVrtcsSprs, DeltaT      );

            /*=============================================================*/
            /* WRITE ONE LINE TO THE OUTPUT FILE                           */
            /*=============================================================*/
            write_one_line( XFit, YFit, ZFit, NVrtcsFit, FoutFit );

         }


}



/**********************************************************************/
/*          F I T L I N E 2 . C                                       */
/**********************************************************************/

#include <stdio.h>
#include <alloc.h>
#include <stdlib.h>

#include "fitline2.h"
#include "library.h"

#define  MAX_ORDER      4


/**********************************************************************/
int bspline_3d_curve
     (
        float     XSprs[],
        float     YSprs[],
        float     ZSprs[],
        float     SplnX[],
        float     SplnY[],
        float     SplnZ[],
        int       NVrtcs,
        double    DeltaT
     )
/**********************************************************************/
{
   register int   Vrtx,
```

FITLINE - Fit a B-spline Line

```
                NKnots;
float       *KnotVector;
int         NSplinedVrtcs;
float       *X,
            *Y,
            *Z;


/*===============================================================*/
/* ALLOCATE ALL THE MEMORY                                       */
/*===============================================================*/
X = (float*) malloc( (unsigned) ( (NVrtcs+2) * sizeof(float) ) );
Y = (float*) malloc( (unsigned) ( (NVrtcs+2) * sizeof(float) ) );
Z = (float*) malloc( (unsigned) ( (NVrtcs+2) * sizeof(float) ) );
if( (X==NULL) || (Y==NULL) || (Z==NULL)  )
{
   printf( "ERROR IN bspline_3d_curve: NOT ENOUGH MEMORY\n" );
   exit( 1 );
}

KnotVector = (float*)
        malloc( (unsigned) ( NVrtcs + MAX_ORDER*2) *sizeof(double) );
if( KnotVector == NULL )
{
   printf( "ERROR IN bspline_3d_curve: NOT ENOUGH MEMORY\n" );
   exit( 1 );
}


/*===============================================================*/
/* MAKE A COPY OF THE SPARSE DATA POINTS                         */
/*===============================================================*/
for( Vrtx=0 ; Vrtx<NVrtcs ; Vrtx++ )
{
   X[Vrtx] = XSprs[Vrtx];
   Y[Vrtx] = YSprs[Vrtx];
   Z[Vrtx] = ZSprs[Vrtx];
}

/*===============================================================*/
/* ADD VERTICES IN END INTERVALES                                */
/*===============================================================*/
NVrtcs = add_vertices_in_end_intervals( X, Y, Z, NVrtcs );

/*===============================================================*/
/* CALCULATE THE KNOTVECTOR                                      */
/*===============================================================*/
NKnots = knot_vector( KnotVector, MAX_ORDER, NVrtcs );

/*===============================================================*/
/* SPLINE THE CURVE                                              */
/*===============================================================*/
NSplinedVrtcs = spline_3d( X, Y, Z, SplnX, SplnY, SplnZ, KnotVector,
                       MAX_ORDER, NVrtcs, NKnots, DeltaT );
```

```c
/*===================================================================*/
/* FREE THE USED MEMORY                                              */
/*===================================================================*/
free( X );
free( Y );
free( Z );
free( KnotVector );

/*===================================================================*/
/* RETURN HOW MANY VERTICES ARE IN THE FITTED CURVE                  */
/*===================================================================*/
return( NSplinedVrtcs );

}

/**********************************************************************/
int add_vertices_in_end_intervals
    (
        float   X[],
        float   Y[],
        float   Z[],
        int     NVrtcs
    )
/**********************************************************************/
{
    register int Vrtx;

    double          XFirst,
                    YFirst,
                    ZFirst,
                    XLast,
                    YLast,
                    ZLast;

    double          One3rd = 1.0/3.0;
    double          Two3rd = 2.0/3.0;

    /*===============================================================*/
    /* CALCULATE THE NEW VERTEX IN THE FIRST INTERVALL               */
    /*===============================================================*/
    XFirst = X[0] + One3rd * ( X[1] - X[0] );
    YFirst = Y[0] + One3rd * ( Y[1] - Y[0] );
    ZFirst = Z[0] + One3rd * ( Z[1] - Z[0] );

    /*===============================================================*/
    /* CALCULATE THE NEW VERTEX IN THE LAST INTERVALL                */
    /*===============================================================*/
    XLast = X[NVrtcs-2] + Two3rd * ( X[NVrtcs-1] - X[NVrtcs-2] );
    YLast = Y[NVrtcs-2] + Two3rd * ( Y[NVrtcs-1] - Y[NVrtcs-2] );
    ZLast = Z[NVrtcs-2] + Two3rd * ( Z[NVrtcs-1] - Z[NVrtcs-2] );

    /*===============================================================*/
    /* SHIFT LAST VERTEX FOR TWO POSITIONS                           */
    /*===============================================================*/
    X[NVrtcs+1] = X[NVrtcs-1];
    Y[NVrtcs+1] = Y[NVrtcs-1];
    Z[NVrtcs+1] = Z[NVrtcs-1];
```

```c
/*====================================================================*/
/* SHIFT ALL VERTICES EXCEPT FIRST AND LAST FOR ONE POSITION          */
/*====================================================================*/
    for( Vrtx=NVrtcs-1 ; Vrtx>1 ; Vrtx-- )
    {
        X[Vrtx] = X[Vrtx-1];
        Y[Vrtx] = Y[Vrtx-1];
        Z[Vrtx] = Z[Vrtx-1];
    }

/*====================================================================*/
/* COPY THE VERTEX IN THE FIRST INTERVAL INTO THE ARRAY               */
/*====================================================================*/
    X[1] = XFirst;
    Y[1] = YFirst;
    Z[1] = ZFirst;

/*====================================================================*/
/* COPY THE VERTEX IN THE LAST INTERVAL INTO THE ARRAY                */
/*====================================================================*/
    X[NVrtcs] = XLast;
    Y[NVrtcs] = YLast;
    Z[NVrtcs] = ZLast;

/*====================================================================*/
/* THERE ARE TWO MORE VERTICES NOW IN THE ARRAY                       */
/*====================================================================*/
    return( NVrtcs + 2 );

}
/**********************************************************************/
int knot_vector
    (
        float   KnotVector[],
        int     Ordr,
        int     NVrtcs
    )
/**********************************************************************/
{
    register int Knot;

    int         Vrtx;
    int         NKnots;

    double      LastPoint;

/*====================================================================*/
/* CALC. THE NEW NUMBER OF KNOTS ADDING ORDER-1 POINTS AT EACH END    */
/*====================================================================*/
    NKnots = NVrtcs + (Ordr-1) * 2 ;

/*====================================================================*/
/* THE FIRST VALUES OF THE KNOTVECTOR ARE 0 UP TO ORDER TIMES         */
/*====================================================================*/
    for( Knot=0 ; Knot<(Ordr-1) ; Knot++ )
    {
        KnotVector[Knot] = 0.0;
```

Appendix 4

```c
    }

    /*==================================================================*/
    /* NOW THE VALUE TAKES THE VALUE OF THE POINT NUMBER                */
    /*==================================================================*/
    for( Vrtx=0 ; Vrtx<(NVrtcs-2) ; Vrtx++, Knot++ )
    {
        KnotVector[Knot] = (double) Vrtx;
    }

    LastPoint = (double) Vrtx-1 ;

    /*==================================================================*/
    /* AT THE END ALL THE KNOTVECTOR VALUES TAKE THE VALUE OF LASTPOINT */
    /*==================================================================*/
    for( ; Knot<NKnots ; Knot++ )
    {
        KnotVector[Knot] = LastPoint;
    }

    /*==================================================================*/
    /* RETURN HOW MANY KNOTS THERE ARE IN THE KNOTVECTOR                */
    /*==================================================================*/
    return( NKnots );
}
/*********************************************************************/
int    spline_3d
       (
           float    *X,
           float    *Y,
           float    *Z,
           float    SplnX[],
           float    SplnY[],
           float    SplnZ[],
           float    KnotVector[],
           int      MaxOrdr,
           int      NVrtcs,
           int      NKnots,
           double   DeltaT
       )
/*********************************************************************/
{
    register int  Knot,
                  SplnVrtx;

    int           Vrtx;
    int           Ordr;

    float         **Basis;

    double        TStart,
                  TEnd,
                  T;

    int           LastKnot;
    int           MaxVrtcsSpln;
```

FITLINE - Fit a B-spline Line

```
   MaxVrtcsSpln = (int) ( (NVrtcs-3) / DeltaT) + 1.5 );

   /*===============================================================*/
   /* ALLOCATE THE MEMORY                                           */
   /*===============================================================*/
   Basis = allocate_two_d_float_array( MAX_ORDER, NKnots );
   if( Basis == NULL )
   {
      printf( "ERROR IN bspline_3d_curve: NOT ENOUGH MEMORY\n" );
      exit( 1 );
   }

   /*===============================================================*/
   /* INITIALIZE VARIABLES                                          */
   /*===============================================================*/
   SplnVrtx = 0;
   SplnX[SplnVrtx] = 0.0;
   SplnY[SplnVrtx] = 0.0;
   SplnZ[SplnVrtx] = 0.0;

   /*===============================================================*/
   /* CALCULATE LAST KNOT                                           */
   /*===============================================================*/
   LastKnot = (NVrtcs - 2) + (MAX_ORDER - 2) ;

   /*===============================================================*/
   /* LOOP THROUGH THE KNOTS                                        */
   /*===============================================================*/
   for( Knot=3 ; Knot<=LastKnot ; Knot++ )
   {
      /*===============================================================*/
      /* CALCULATE THE VALUES FOR 1.ORDER BASIS FUNCTION FOR THIS KNOT*/
      /*===============================================================*/
      if( Knot < LastKnot)
      {
         basis_order_1( Basis[0], NKnots, Knot );
      }

      /*===============================================================*/
      /* COMPUTE BOUNDARIES OF THE INTERVAL TO CALCULATE BSPLINE IN    */
      /*===============================================================*/
      TStart = KnotVector[Knot];
      if( Knot == NVrtcs+MaxOrdr-2 )
      {
         /*===============================================================*/
         /* IN THE LAST INTERVAL THE BOUNDARY IS INCLUDED IN THE          */
         /* INTERVAL                                                      */
         /*===============================================================*/
         TEnd = KnotVector[Knot+1] + 0.000001;
      }
      else
      {
         /*===============================================================*/
         /* IN ALL OTHER INTERVALS THE BOUNDARY IS EXCLUDED FROM THE      */
         /* INTERVAL                                                     .*/
         /*===============================================================*/
         TEnd = KnotVector[Knot+1] - DeltaT + 0.000001;
```

```c
      }

      /*==============================================================*/
      /* SET T TO THE START OF THE INTERVAL                           */
      /*==============================================================*/
      T = TStart;

      /*==============================================================*/
      /* CALCULATE VALUES FOR EACH VERTEX ON THE SPLINE AT EACH       */
      /*  T WITHIN THE INTERVAL STEPPING THROUGH IT WITH DELTA-T      */
      /*==============================================================*/
      do
      {
         /*===========================================================*/
         /* CALCULATE VALUES OF BASIS FUNCTION  IN THIS INTERVAL      */
         /* FOR THIS T FOR ORDERS OF ONE TO MAXORDER                  */
         /*===========================================================*/
         for( Ordr=2 ; Ordr<=MAX_ORDER ; Ordr++ )

         {
            basis_order_n( Basis[Ordr-1], Basis[Ordr-2],
                           KnotVector, T, Ordr, NKnots );
         }


         for( Vrtx=0 ; Vrtx < NVrtcs ; Vrtx++ )
         {
            /*========================================================*/
            /* ADD UP THE SPLINED VALUES FOR THIS SPLINED VERTEX      */
            /*========================================================*/
            SplnX[SplnVrtx] += ( X[Vrtx] * Basis[3][Vrtx] );
            SplnY[SplnVrtx] += ( Y[Vrtx] * Basis[3][Vrtx] );
            SplnZ[SplnVrtx] += ( Z[Vrtx] * Basis[3][Vrtx] );
         }

         /*===========================================================*/
         /* GO TO THE NEXT VERTEX ON SPLINE AND INITIALIZE ITS        */
         /* VALUE TO 0.0                                              */
         /*===========================================================*/
         SplnVrtx++;
         if( SplnVrtx<MaxVrtcsSpln )
         {
            SplnX[SplnVrtx] = 0.0;
            SplnY[SplnVrtx] = 0.0;
            SplnZ[SplnVrtx] = 0.0;
         }

         /*===========================================================*/
         /* NEXT T IN THE INTERVAL                                    */
         /*===========================================================*/
         T = T + DeltaT;

      }while( T < TEnd );

}
```

```c
    /*===============================================================*/
    /* FREE THE TWO DIMENSIONAL POINTER ARRAY FOR THE BASIS FUNCTION */
    /* VALUES                                                        */
    /*===============================================================*/
    free_two_d_float_array( MAX_ORDER, Basis );;

    /*===============================================================*/
    /* RETURN HOW MANY VERTICES THERE ARE ON THE SPLINE              */
    /*===============================================================*/
    return( SplnVrtx );
}
/*******************************************************************/
void basis_order_1
    (
        float   *Basis1,
        int     NKnots,
        int     KnotNumber
    )
/*******************************************************************
/
{
    register int Knot;

    /*===============================================================*/
    /* LOOP THROUGH THE KNOTS                                        */
    /*===============================================================*/
    for( Knot=0 ; Knot<NKnots ; Knot++ )
    {
        /*===========================================================*/
        /* INITIALIZE KNOT VALUES                                    */
        /*===========================================================*/
        if( (Knot == KnotNumber) )
        {
            Basis1[Knot] = 1.0;
        }
        else
        {
            Basis1[Knot] = 0.0;
        }
    }
}
/*******************************************************************/
void basis_order_n
    (
        float   **Basis,
        float   KnotVector[],
        double  T,
        int     Ordr,
        int     NKnots
    )
/*******************************************************************/
{   double      C,
                CDiv,
                D,
                DDiv;

    register int BOrdr;
```

Appendix 4

```c
   int        Knot;


/*====================================================================*/
/* BECAUSE A C-ARRAY GOES FROM 0 TO N-1                               */
/*====================================================================*/
   BOrdr = Ordr - 1;

/*====================================================================*/
/* LOOP THROUGH THE LAST ORDER BASIS FUNCTION VALUES                  */
/*====================================================================*/
   for( Knot=0 ; Knot <= NKnots-BOrdr-1 ; Knot++)
   {
      /*====================================================================*/
      /* FIRST TERM OF BASIS FUNCTION                                       */
      /*====================================================================*/
      if( Basis[BOrdr-1][Knot] == 0.0 )
      {
         C = 0.0;
      }
      else
      {
         CDiv = KnotVector[Knot+Ordr-1] - KnotVector[Knot];
         if( CDiv == 0.0 )
         {
            C = 1.0;
         }
         else
         {
            C = Basis[BOrdr-1][Knot] * ( T - KnotVector[Knot] ) / CDiv;
         }
      }

      /*====================================================================*/
      /* SECOND TERM OF BASIS FUNCTION                                      */
      /*====================================================================*/
      if( Basis[BOrdr-1][Knot+1] == 0.0)
      {
         D = 0.0;
      }
      else
      {
         DDiv = KnotVector[Knot+Ordr] - KnotVector[Knot+1];
         if( DDiv == 0.0)
            D = 1.0;
         else
            D = Basis[BOrdr-1][Knot+1] * ( KnotVector[Knot+Ordr] - T ) /
                DDiv;
      }

      /*====================================================================*/
      /* VALUE OF BASIS FUNCTION AT KNOT AND FOR ORDER                      */
      /*====================================================================*/
      Basis[BOrdr][Knot] = C + D;
   }
}
```

Appendix 5
C- Program: FITSURF - Fit a B-spline Surface

```
/*******************************************************************/
/*                    F I T S U R F 2 . H                          */
/*******************************************************************/

void bspline_surface
    (
        float   **XOrg,
        float   **YOrg,
        float   **ZOrg,
        float   **SplnX,
        float   **SplnY,
        float   **SplnZ,
        int       NLines,
        int       NVrtcs,
        double    DeltaTLine,
        double    DeltaTVrtx,
        int       NSplnLines,
        int       NSplnVrtcs
    );

int   add_lines_in_endintervals
    (
        float   **X,
        float   **Y,
        float   **Z,
        int       NLines,
        int  NVrtcs
    );

int   add_vertices_in_endintervals
    (
        float   **X,
        float   **Y,
        float   **Z,
        int       NLines,
        int       NVrtcs
    );

void   knot_vector
    (
        float  KnotVector[],
        int     NVrtcs,
        int     NKnots
    );
```

Appendix 5

```c
void spline_surface
    (
        float   **X,
        float   **Y,
        float   **Z,
        float   **SplnX,
        float   **SplnY,
        float   **SplnZ,
        float   KnotVectorLine[],
        float   KnotVectorVrtx[],
        int     NLines,
        int     NVrtcs,
        int     NKnotsLine,
        int     NKnotsVrtx,
        double  DeltaTLine,
        double  DeltaTVrtx,
        int     NSplnVrtcs,
        int     NSplnLines
    );

void basis_order_1
    (
        float   **Basis,
        int     NKnots,
        int     KnotNumber
    );

void basis_order_n
    (
        float   **Basis,
        float   KnotVector[],
        double  T,
        int     Ordr,
        int     NKnots
    );


/**********************************************************************/
/*      F I T S U R F 1 . C                                           */
/**********************************************************************/

#include <stdio.h>
#include <stdlib.h>
#include <string.h>
#include <conio.h>
#include <process.h>
#include <graphics.h>
#include <alloc.h>
#include <ctype.h>

#include "library.h"

#include "fitsurf2.h"
```

FITSURF - Fit a B-spline Surface

```c
/**********************************************************************/
/* PROTOTYPED FUNCTIONS                                               */
/**********************************************************************/
void information
    (
        char **FileNames
    );

/**********************************************************************/
main
    (
        int    NParameters,
        char **ParameterStrings
    )
/**********************************************************************/
{
    float    **XSrf,
             **YSrf,
             **ZSrf;
    int       NVrtcsSrf;
    int       NLinesSrf;

    float    **FitX,
             **FitY,
             **FitZ;

    int       NLinesFit;
    int       NVrtcsFit;

    double    DTLine,
              DTVrtx;


    /*==================================================================*/
    /* CHECK IF RIGHT NUMBER OF PARAMETERS HAVE BEEN PASSED             */
    /*==================================================================*/
    check_parameters_two( NParameters, ParameterStrings );

    /*==================================================================*/
    /* DISPLAY INFORMATION ABOUT THE PROGRAM                            */
    /*==================================================================*/
    information( ParameterStrings );

    /*==================================================================*/
    /* READ THE SPARSE DATA FILE INTO MEMORY                            */
    /*==================================================================*/
    read_file_allocate( &XSrf, &YSrf, &ZSrf,
                        &NLinesSrf, &NVrtcsSrf,
                        ParameterStrings[1] );

    /*==================================================================*/
    /* DISPLAY SIZE OF SPARSE DATA MATRIX FILE                          */
    /*==================================================================*/
    gotoxy( 5, 12 );
    printf( "THE SURFACE HAS %3d LINES AND %3d VERTICES", NLinesSrf,
                                                          NVrtcsSrf );
```

Appendix 5

```c
   /*================================================================*/
   /* GET THE DELTA-T FOR THE LINE DIRECTION FROM THE USER           */
   /*================================================================*/
   gotoxy( 5,14 );
   DTLine = get_one_double_value( "ENTER DELTA-T  FOR THE LINES " );

   /*================================================================*/
   /* GET THE DELTA-T FOR THE VERTEX DIRECTION FROM THE USER         */
   /*================================================================*/
   gotoxy( 5, 15);
   DTVrtx = get_one_double_value( "ENTER DELTA-T  FOR THE VERTICES " );

   /*================================================================*/
   /* CALCULATE THE NUMBER OF SPLINED LINES AND SPLINED VERTICES     */
   /*================================================================*/
   NLinesFit   =   (int)( ( (NLinesSrf - 1) / DTLine ) + 1.5 );
   NVrtcsFit   =   (int)( ( (NVrtcsSrf - 1) / DTVrtx ) + 1.5 );

   /*================================================================*/
   /* ALLOCATE THE MEMORY FOR THE FITTED SURFACE                     */
   /*================================================================*/
   FitX = allocate_two_d_float_array( NLinesFit, NVrtcsFit );
   FitY = allocate_two_d_float_array( NLinesFit, NVrtcsFit );
   FitZ = allocate_two_d_float_array( NLinesFit, NVrtcsFit );
   if( (FitX==NULL) || (FitY==NULL) || (FitZ==NULL) )
   {
      printf( "OUT OF MEMORY\n" );
      exit( 1 );
   }

   /*================================================================*/
   /* FIT A B-SPLINE OVER THE PATCH                                  */
   /*================================================================*/
   bspline_surface( XSrf, YSrf, ZSrf, FitX, FitY, FitZ,
                    NLinesSrf, NVrtcsSrf,
                    DTLine, DTVrtx,
                    NLinesFit, NVrtcsFit );

   /*================================================================*/
   /* WRITE THE FITTED SURFACE TO THE OUTPUT FILE                    */
   /*================================================================*/
   write_file( FitX, FitY, FitZ, NLinesFit, NVrtcsFit,
ParameterStrings[2] );

   gotoxy( 10, 22 );
   printf( "PROGRAM SUCCESSFUL\n" );
}
/**********************************************************************/
void information
     (
         char **FileNames
     )
/**********************************************************************/
{
   /*================================================================*/
   /* CLEAR THE SCREEN                                               */
   /*================================================================*/
```

FITSURF - Fit a B-spline Surface

```c
    clrscr( );

    /*====================================================================*/
    /* DISPLAY THE INFORMATION ABOUT THE PROGRAM                          */
    /*====================================================================*/
    printf( "           PROGRAM *** B S P L I N E  ***\n");
    printf( " FITS A 4TH ORDER B-SPLINE SURFACE\n\n");

    printf( "  SURFACE FILE NAME        = %s\n", FileNames[1] );
    printf( "  FITTED SURFACE FILE NAME = %s\n", FileNames[2] );
}


/**********************************************************************/
/*       F I T S U R F 2 . C                                          */
/**********************************************************************/
#include <stdio.h>
#include <process.h>
#include <alloc.h>
#include <conio.h>
#include <graphics.h>

#include "library.h"
#include "fitsurf2.h"

/*====================================================================*/
/* DEFINES                                                            */
/*====================================================================*/
#define MAXORDER    4

/*====================================================================*/
/* GLOBAL VARIABLES.                                                  */
/*====================================================================*/
double OneThird         = 1.0 / 3.0;
double OneSixths        = 1.0 / 6.0;
double TwoThird         = 2.0 / 3.0;
double TwoThirdSquared  = 4.0 / ( 3.0 * 3.0 );
double OneNinths        = 1.0 / 9.0;
double OneSixthsSquared = 1.0 / ( 6.0 * 6.0 );

/**********************************************************************/
void bspline_surface
    (
        float **XOrg,
        float **YOrg,
        float **ZOrg,
        float **SplnX,
        float **SplnY,
        float **SplnZ,
        int    NLines,
        int    NVrtcs,
        double DeltaTLine,
        double DeltaTVrtx,
        int    NSplnLines,
        int    NSplnVrtcs
    )
/**********************************************************************/
```

```
{
   register int  Vrtx;

   int      Line;
   int      NKnotsLine,
            NKnotsVrtx;

   float    **XCopy,
            **YCopy,
            **ZCopy;

   float    *KnotVectorLine,
            *KnotVectorVrtx;

   int      NCopyLines,
            NCopyVrtcs;

/*==============================================================*/
/* INITIALIZE PARAMETERS                                        */
/*==============================================================*/
NCopyLines  = NLines + 2;
NCopyVrtcs  = NVrtcs + 2;
NKnotsLine = NCopyLines + (MAXORDER-1) * 2;
NKnotsVrtx = NCopyVrtcs + (MAXORDER-1) * 2;


/*==============================================================*/
/* ALLOCATE MEMORY FOR COPY OF SPARSE DATA PLUS VERTICES IN     */
/* ENDINTERVALS                                                 */
/*==============================================================*/
XCopy = allocate_two_d_float_array( NCopyLines, NCopyVrtcs );
YCopy = allocate_two_d_float_array( NCopyLines, NCopyVrtcs );
ZCopy = allocate_two_d_float_array( NCopyLines, NCopyVrtcs );

if( XCopy==NULL || YCopy==NULL || ZCopy==NULL )
{
   printf( "ERROR IN : >bspline_surface< NOT ENOUGH MEMORY\n" );
   exit( 0 );
}


/*==============================================================*/
/* ALLOCATE THE MEMORY FOR THE KNOTVECTOR                       */
/*==============================================================*/
KnotVectorLine = (float*) malloc( (unsigned) (NKnotsLine *
                  sizeof(float)) );
KnotVectorVrtx = (float*) malloc( (unsigned) (NKnotsVrtx *
                  sizeof(float)) );

if( KnotVectorLine==0 || KnotVectorVrtx==0 )
{
   printf( "ERROR IN : >bspline_surface< NOT ENOUGH MEMORY\n" );
   exit( 0 );
}


/*==============================================================*/
/* MAKE A COPY OF THE ORIGINAL DATA VERTICES                    */
/*==============================================================*/
for( Line=0 ; Line<NLines ; Line++)
```

FITSURF - Fit a B-spline Surface

```
      {
         for( Vrtx=0 ; Vrtx<NVrtcs ; Vrtx++ )
         {
            XCopy[Line][Vrtx] = XOrg[Line][Vrtx];
            YCopy[Line][Vrtx] = YOrg[Line][Vrtx];
            ZCopy[Line][Vrtx] = ZOrg[Line][Vrtx];
         }
      }

      /*================================================================*/
      /* ADD LINEAR INTERPOLATION VERTICES IN THE END INTERVALS         */
      /*================================================================*/
      NCopyLines = add_lines_in_endintervals( XCopy, YCopy, ZCopy, NLines,
                                              NVrtcs );
      NCopyVrtcs = add_vertices_in_endintervals( XCopy, YCopy, ZCopy,
                                                 NCopyLines, NVrtcs );

      /*================================================================*/
      /* CALCULATE THE KNOTVECTOR FOR LINE AND VERTEX DIRECTION         */
      /*================================================================*/
      knot_vector( KnotVectorLine, NCopyLines, NKnotsLine );
      knot_vector( KnotVectorVrtx, NCopyVrtcs, NKnotsVrtx);

      /*================================================================*/
      /* BSPLINE THE SURFACE                                            */
      /*================================================================*/
      spline_surface( XCopy, YCopy, ZCopy,
                      SplnX, SplnY, SplnZ,
                      KnotVectorLine, KnotVectorVrtx,
                      NCopyLines, NCopyVrtcs,
                      NKnotsLine, NKnotsVrtx,
                      DeltaTLine, DeltaTVrtx,
                      NSplnVrtcs, NSplnLines );

      /*================================================================*/
      /* FREE THE MEMORY                                                */
      /*================================================================*/
      free_two_d_float_array( NCopyLines, XCopy );
      free_two_d_float_array( NCopyLines, YCopy );
      free_two_d_float_array( NCopyLines, ZCopy );
      free( KnotVectorVrtx );
      free( KnotVectorLine );
}
/*******************************************************************/
int  add_lines_in_endintervals
        (
           float **X,
           float **Y,
           float **Z,
           int NLines,
           int NVrtcs
        )
/*******************************************************************/
{
   register int Vrtx,
                Line;
```

Appendix 5

```
/*================================================================*/
/* THERE WILL BE TWO MORE LINES                                   */
/*================================================================*/
NLines = NLines + 2;

/*================================================================*/
/* COPY VERTICES ON LAST LINE TWO LINES UP                        */
/*================================================================*/
Line = NLines - 1;

for( Vrtx=0 ; Vrtx<NVrtcs ; Vrtx++ )
{
   X[Line][Vrtx] = X[Line-2][Vrtx];
   Y[Line][Vrtx] = Y[Line-2][Vrtx];
   Z[Line][Vrtx] = Z[Line-2][Vrtx];
}

/*================================================================*/
/* ADD LINEAR INTERPOLATION VERTEX IN LAST INTERVAL               */
/*================================================================*/
Line = NLines - 2;

for( Vrtx=0 ; Vrtx<NVrtcs ; Vrtx++ )
{
   X[Line][Vrtx] = X[Line-2][Vrtx] + TwoThird * ( X[Line+1][Vrtx] -
                                                  X[Line-2][Vrtx] );
   Y[Line][Vrtx] = Y[Line-2][Vrtx] + TwoThird * ( Y[Line+1][Vrtx] -
                                                  Y[Line-2][Vrtx] );
   Z[Line][Vrtx] = Z[Line-2][Vrtx] + TwoThird * ( Z[Line+1][Vrtx] -
                                                  Z[Line-2][Vrtx] );
}

/*================================================================*/
/* COPY ALL LINES EXCEPT FOR THE FIRST ONE UP BY ONE POSITION     */
/*================================================================*/
for( Line=NLines-3 ; Line>0 ; Line-- )
{
   for( Vrtx=0 ; Vrtx<NVrtcs ; Vrtx++ )
   {
      X[Line][Vrtx] = X[Line-1][Vrtx];
      Y[Line][Vrtx] = Y[Line-1][Vrtx];
      Z[Line][Vrtx] = Z[Line-1][Vrtx];
   }
}

/*================================================================*/
/* ADD LINER INTERPOLATION VERTICES IN SECOND LINE                */
/*================================================================*/
Line = 1;

for( Vrtx=0 ; Vrtx<NVrtcs ; Vrtx++ )
{
   X[Line][Vrtx] = X[Line-1][Vrtx] + OneThird * ( X[Line+1][Vrtx] -
                                                  X[Line-1][Vrtx] );
   Y[Line][Vrtx] = Y[Line-1][Vrtx] + OneThird * ( Y[Line+1][Vrtx] -
                                                  Y[Line-1][Vrtx] );
   Z[Line][Vrtx] = Z[Line-1][Vrtx] + OneThird * ( Z[Line+1][Vrtx] -
```

FITSURF - Fit a B-spline Surface

```
                                                    Z[Line-1][Vrtx] );
   }

   return( NLines );

}
/**********************************************************************/
int  add_vertices_in_endintervals
     (
        float   **X,
        float   **Y,
        float   **Z,
        int     NLines,
        int     NVrtcs
     )
/**********************************************************************/
{
   register int Vrtx,
                Line;

   /*================================================================*/
   /* THERE WILL BE TWO MORE VERTICES                                */
   /*================================================================*/
   NVrtcs = NVrtcs + 2 ;

   /*================================================================*/
   /* COPY VERTICES ON LAST VERTEX TWO VERTICES UP                   */
   /*================================================================*/
   Vrtx = NVrtcs - 1;

   for( Line=0 ; Line<NLines ; Line++ )
   {
      X[Line][Vrtx] = X[Line][Vrtx-2];
      Y[Line][Vrtx] = Y[Line][Vrtx-2];
      Z[Line][Vrtx] = Z[Line][Vrtx-2];
   }

   /*================================================================*/
   /* ADD LINEAR INTERPOLATION VERTEX IN LAST INTERVAL               */
   /*================================================================*/
   Vrtx = NVrtcs - 2;

   for( Line=0 ; Line<NLines ; Line++ )
   {
      X[Line][Vrtx] = X[Line][Vrtx-2] + TwoThird * ( X[Line][Vrtx+1] -
                                                    X[Line][Vrtx-2] );
      Y[Line][Vrtx] = Y[Line][Vrtx-2] + TwoThird * ( Y[Line][Vrtx+1] -
                                                    Y[Line][Vrtx-2] );
      Z[Line][Vrtx] = Z[Line][Vrtx-2] + TwoThird * ( Z[Line][Vrtx+1] -
                                                    Z[Line][Vrtx-2] );
   }

   /*================================================================*/
   /* COPY ALL LINES EXCEPT FOR THE FIRST ONE UP BY ONE POSITION     */
   /*================================================================*/
   for( Vrtx=NVrtcs-3 ; Vrtx>0 ; Vrtx-- )
   {
```

```c
        for( Line=0 ; Line<NLines ; Line++ )
        {
            X[Line][Vrtx] = X[Line][Vrtx-1];
            Y[Line][Vrtx] = Y[Line][Vrtx-1];
            Z[Line][Vrtx] = Z[Line][Vrtx-1];
        }
    }

    /*====================================================================*/
    /* ADD LINER INTERPOLATION VERTICES IN SECOND LINE                    */
    /*====================================================================*/
    Vrtx = 1;

    for( Line=0 ; Line<NLines ; Line++ )
    {
        X[Line][Vrtx] = X[Line][Vrtx-1] + OneThird * (  X[Line][Vrtx+1] -
                                                        X[Line][Vrtx-1]   );
        Y[Line][Vrtx] = Y[Line][Vrtx-1] + OneThird * (  Y[Line][Vrtx+1] -
                                                        Y[Line][Vrtx-1]   );
        Z[Line][Vrtx] = Z[Line][Vrtx-1] + OneThird * (  Z[Line][Vrtx+1] -
                                                        Z[Line][Vrtx-1]   );
    }

    return( NVrtcs );
}

/**********************************************************************/
void knot_vector
        (
            float   KnotVector[],
            int     NVrtcs,
            int     NKnots
        )
/**********************************************************************/
{
    int            Vrtx;
    register int   Knot;
    double         LastVrtx;

    /*====================================================================*/
    /* MAKE KNOTVECTOR VALUE 0.0 FOR UP TO MAXORDER-1 KNOTS               */
    /*====================================================================*/
    for( Knot=0 ; Knot<MAXORDER-1 ; Knot++ )
    {
        KnotVector[Knot] = 0.0;
    }

    /*====================================================================*/
    /* NOW THE THE KNOTVECTOR TAKES THE VALUE OF THE VERTEX NUMBER.       */
    /*====================================================================*/
    for( Vrtx=0; Vrtx<(NVrtcs-2) ; Vrtx++, Knot++ )
    {
        KnotVector[Knot] = (float) (Vrtx);
    }

    LastVrtx = (double) Vrtx-1;

    /*====================================================================*/
```

```
    /* MAKE THE VALUE OF THE REST TO VALUE OF VERTEX NUMBER         */
    /*================================================================*/
    for( ; Knot<NKnots ; Knot++ )
    {
       KnotVector[Knot] = LastVrtx;
    }

}
/******************************************************************/
void spline_surface
      (
         float  **X,
         float  **Y,
         float  **Z,
         float  **SplnX,
         float  **SplnY,
         float  **SplnZ,
         float    KnotVectorLine[],
         float    KnotVectorVrtx[],
         int      NLines,
         int      NVrtcs,
         int      NKnotsLine,
         int      NKnotsVrtx,
         double   DeltaTLine,
         double   DeltaTVrtx,
         int      NSplnVrtcs,
         int      NSplnLines
      )
/******************************************************************/
{
    register int Vrtx,
                 Line;

    float  **BasisLine,
           **BasisVrtx;

    float  **Basis4Line,
           **Basis4Vrtx;

    int    KnotV,
           KnotL;

    int    SplnV,
           SplnL;

    int      Ordr;

    double T,
           TStart,
           TEnd;

    int    LastKnotLine,
           LastKnotVrtx;

    /*================================================================*/
    /* ALLOCATE MEMORY FOR BASIS FUNCTION                           */
    /*================================================================*/
```

Appendix 5

```
   BasisLine = allocate_two_d_float_array( MAXORDER, NKnotsLine );
   BasisVrtx = allocate_two_d_float_array( MAXORDER, NKnotsVrtx );
   if( BasisLine==NULL || BasisVrtx==NULL )
   {
      printf( " ERROR IN BSPLINE_SURFACE: NOT ENOUGH MEMORY\n" );
      exit( 1 );
   }

   Basis4Line = allocate_two_d_float_array( NSplnLines, NLines );
   Basis4Vrtx = allocate_two_d_float_array( NSplnVrtcs, NVrtcs );
   if( Basis4Line==NULL || Basis4Vrtx==NULL )
   {
      printf( " ERROR IN BSPLINE_SURFACE: NOT ENOUGH MEMORY\n" );
      exit( 1 );
   }


   /*================================================================*/
   /* MAKE THE FOURTH ORDER BASIS FUNCTION VALUES FOR THE VERTEX     */
   /* DIRECTION                                                      */
   /*================================================================*/
   /* INITIALIZE VARIABLES                                           */
   /*================================================================*/
   LastKnotVrtx = NVrtcs - 2 + MAXORDER - 2;
   SplnV = 0;

   /*================================================================*/
   /* STEP 3 GO THROUGH THE KNOTS AND CALCULATE THE VALUES OF        */
   /* THE BASIS FUN.                                                 */
   /*================================================================*/
   for( KnotV=3 ; KnotV<=LastKnotVrtx ; KnotV++ )
   {
      if( KnotV < LastKnotVrtx )
      {
         basis_order_1( BasisVrtx, NKnotsVrtx, KnotV );
      }

      /*=============================================================*/
      /* CALCULATE LENGTH OF INTERVAL                                */
      /*=============================================================*/
      TStart = KnotVectorVrtx[KnotV];

      if( KnotV==NVrtcs+MAXORDER-2)
      {
         TEnd = KnotVectorVrtx[KnotV+1] + 0.001;
      }
      else
      {
         TEnd = KnotVectorVrtx[KnotV+1] - DeltaTVrtx + 0.001;
      }

      T = TStart;

      /*=============================================================*/
      /* CALCULATE BASIS FUNCTION IN THIS INTERVAL                   */
      /*=============================================================*/
      do
      {  /*==========================================================*/
```

FITSURF - Fit a B-spline Surface

```c
        /* CALCULATE VALUES FOR BASIS FUNCTION FOR ORDERS > 1   */
        /*=====================================================*/
        for( Ordr=2 ; Ordr<=MAXORDER ; Ordr++ )
        {
            basis_order_n( BasisVrtx, KnotVectorVrtx, T, Ordr,
                          NKnotsVrtx );
        }

        /*=====================================================*/
        /* COPY BASIS FUNCTION OF MAXORDER INTO BASIS4LINE      */
        /*=====================================================*/
        for( Vrtx=0; Vrtx<NVrtcs ; Vrtx++ )
        {
            Basis4Vrtx[SplnV][Vrtx] = BasisVrtx[MAXORDER-1][Vrtx];
        }

        T = T + DeltaTVrtx;
        SplnV++;

    }while( T <= TEnd );

}

/*=========================================================*/
/* MAKE THE FOURTH ORDER BASIS FUNCTION VALUES FOR          */
/* THE LINE DIRECTION                                       */
/*=========================================================*/
/* INITIALIZE VARIABLES                                     */
/*=========================================================*/
LastKnotLine = NLines    - 2 + MAXORDER - 2;
SplnL = 0;

/*=========================================================*/
/* STEP 3 GO THROUGH THE KNOTS AND CALCULATE THE VALUES OF  */
/* THE BASIS FUN.                                           */
/*=========================================================*/
for( KnotL=3 ; KnotL<=LastKnotLine ; KnotL++ )
{
    if( KnotL < LastKnotLine )
    {
        basis_order_1( BasisLine, NKnotsLine, KnotL );
    }

        /*=====================================================*/
        /* CALCULATE LENGTH OF INTERVAL                         */
        /*=====================================================*/
        TStart = KnotVectorLine[KnotL];

        if( KnotL==NLines+MAXORDER-2)
        {
            TEnd = KnotVectorLine[KnotL+1] + 0.001;
        }
        else
        {
            TEnd = KnotVectorLine[KnotL+1] - DeltaTLine + 0.001;
        }

        T = TStart;
```

Appendix 5

```c
/*==============================================================*/
/* CALCULATE BASIS FUNCTION IN THIS INTERVAL                    */
/*==============================================================*/
do
{
    /*==========================================================*/
    /* CALCULATE VALUES FOR BASIS FUNCTION FOR ORDERS > 1       */
    /*==========================================================*/
    for( Ordr=2 ; Ordr<=MAXORDER ; Ordr++ )
    {
        basis_order_n( BasisLine, KnotVectorLine, T, Ordr,
                       NKnotsLine );
    }

    /*==========================================================*/
    /* COPY BASIS FUNCTION OF MAXORDER INTO BASIS4LINE          */
    /*==========================================================*/
    for( Line=0 ; Line<NLines ; Line++ )
    {
        Basis4Line[SplnL][Line] = BasisLine[MAXORDER-1][Line];
    }

    T = T + DeltaTLine;
    SplnL++;

}while( T <= TEnd );

}

/*==============================================================*/
/* B-SPLINE THE SURFACE                                         */
/*==============================================================*/
gotoxy( 5, 17 );
printf( "SPLINING THE SURFACE - %d LINES BY %d VERTICES TO
        CALCULATE",NSplnLines, NSplnVrtcs );
gotoxy( 5, 18 );
printf( "PROCESSING LINE #" );

/*==============================================================*/
/* LOOP THROUGH ALL SPLINE LINES                                */
/*==============================================================*/
for( SplnL=0 ; SplnL<NSplnLines ; SplnL++)
{
    gotoxy( 22, 18 );
    printf( "%4d", SplnL+1 );

    /*==========================================================*/
    /* LOOP THROUGH ALL SPLINE VERTICES                         */
    /*==========================================================*/
    for( SplnV=0 ; SplnV<NSplnVrtcs ; SplnV++)
    {
        /*======================================================*/
        /* INITIALIZE EACH COMPONENT VALUE AT THIS VERTEX TO ZERO*/
        /*======================================================*/
        SplnX[SplnL][SplnV] = 0.0;
        SplnY[SplnL][SplnV] = 0.0;
```

FITSURF - Fit a B-spline Surface

```c
            SplnZ[SplnL][SplnV] = 0.0;

            /*===============================================================*/
            /* LOOP THROUGH ALL SPARSE DATA SET LINE                         */
            /*===============================================================*/
            for( Line=0 ; Line<NLines ; Line++ )
            {

               /*===========================================================*/
               /* LOOP THROUGH ALL SPARSE DATA SET VERTICES                 */
               /*===========================================================*/
               for( Vrtx=0 ; Vrtx<NVrtcs ; Vrtx++ )
               {
                  /*========================================================*/
                  /* CALCULATE PORTION FOR EACH THE COMPONENT AT THIS       */
                  /* LINE AND VERTEX LOCATION AND ADD TO SPLINE COMPONENT*/
                  /*========================================================*/
                  SplnX[SplnL][SplnV]  += X[Line][Vrtx] *
                                          Basis4Vrtx[SplnV][Vrtx] *
                                          Basis4Line[SplnL][Line];

                  SplnY[SplnL][SplnV]  += Y[Line][Vrtx] *
                                          Basis4Vrtx[SplnV][Vrtx] *
                                          Basis4Line[SplnL][Line]   ;

                  SplnZ[SplnL][SplnV]  += Z[Line][Vrtx] *
                                          Basis4Vrtx[SplnV][Vrtx] *
                                          Basis4Line[SplnL][Line]   ;
               }
            }
         }
      }


      /*=================================================================*/
      /* FREE THE MEMORY                                                 */
      /*=================================================================*/
      free_two_d_float_array( MAXORDER, BasisLine );
      free_two_d_float_array( MAXORDER, BasisVrtx );
      free_two_d_float_array( NSplnLines, Basis4Line );
      free_two_d_float_array( NSplnVrtcs, Basis4Vrtx );

}
/*******************************************************************/
void basis_order_1
      (
         float    **Basis,
         int      NKnots,
         int      KnotNumber
      )
/*******************************************************************/
{
   register int Knot;

   /*=================================================================*/
   /* LOOP THROUGH ALL KNOTS AND SET BASIS FUNCTION VALUE             */
   /* LIKE DE BOOR SPECIFIED                                          */
```

```c
   /*===============================================================*/
   for( Knot=0 ; Knot<NKnots ; Knot++ )
   {
      if( Knot == KnotNumber )
      {
         Basis[0][Knot] = 1.0;
      }
      else
      {
         Basis[0][Knot] = 0.0;
      }

   }
}
/***********************************************************************/
void basis_order_n
     (
        float   **Basis,
        float   KnotVector[],
        double  T,
        int     Ordr,
        int     NKnots
     )
/***********************************************************************/
{  double      C,
               CDiv,
               D,
               DDiv;

   register int BOrdr;

   int          Knot;


   /*===============================================================*/
   /* BECAUSE A C-ARRAY GOES FROM 0 TO N-1                           */
   /*===============================================================*/
   BOrdr = Ordr - 1;

   /*===============================================================*/
   /* LOOP THROUGH THE LAST ORDER BASIS FUNCTION VALUES              */
   /*===============================================================*/
   for( Knot=0 ; Knot <= NKnots-BOrdr-1 ; Knot++)
   {
      /*===============================================================*/
      /* FIRST TERM OF BASIS FUNCTION                                   */
      /*===============================================================*/
      if( Basis[BOrdr-1][Knot] == 0.0 )
      {
         C = 0.0;
      }
      else
      {
         CDiv = KnotVector[Knot+Ordr-1] - KnotVector[Knot];
         if( CDiv == 0.0 )
         {
            C = 1.0;
```

```
        }
        else
        {
            C = Basis[BOrdr-1][Knot] * ( T - KnotVector[Knot] ) / CDiv;
        }
    }

    /*================================================================*/
    /* SECOND TERM OF BASIS FUNCTION                                  */
    /*================================================================*/
    if( Basis[BOrdr-1][Knot+1] == 0.0)
    {
        D = 0.0;
    }
    else
    {
        DDiv = KnotVector[Knot+Ordr] - KnotVector[Knot+1];
        if( DDiv == 0.0)
        {
            D = 1.0;
        }
        else
        {
            D = Basis[BOrdr-1][Knot+1] * ( KnotVector[Knot+Ordr] - T ) /
                DDiv;
        }
    }

    /*================================================================*/
    /* VALUE OF BASIS FUNCTION AT KNOT AND FOR ORDER                  */
    /*================================================================*/
    Basis[BOrdr][Knot] = C + D;
    }
}
```

Appendix 6
C-Program: ROTATE.C - Rotate a Surface around Principal Axes

```c
/**********************************************************************/
/*                  R O T A T E . C                                   */
/**********************************************************************/


#include <stdio.h>
#include <math.h>
#include <stdlib.h>
#include <ctype.h>
#include <conio.h>
#include <dos.h>


/*====================================================================*/
/* INCLUDE FILE FOR OUR OWN LIBRARY FUNCTIONS                         */
/*====================================================================*/
#include <library.h>


/**********************************************************************/
/* PROTOYPES OF FUNCTIONS                                             */
/**********************************************************************/
void information
    (
        char **FileNames
    );

void rotate_surface_file
    (
        char    *InputFile,
        char    *OutputFile,
        double  Alpha,
        double  Beta,
        double  Gamma
    );


/**********************************************************************/
main
    (
        int NumberOfParameters,
        char **ParameterStrings
    )
/**********************************************************************/
```

ROTATE - Rotate a Surface Around Principal Axes

```c
{
    double   Alpha,
             Beta,
             Gamma;

    /*==================================================================*/
    /* CHECK FOR RIGHT NUMBER OF PARAMETERS BEING PASSED                */
    /*==================================================================*/
    check_parameters_two( NumberOfParameters, ParameterStrings );

    /*==================================================================*/
    /* DISPLAY INFORMATION ABOUT THE PROGRAM                            */
    /*==================================================================*/
    information( ParameterStrings );

    /*==================================================================*/
    /* READ THE ROTATION ANGLES FROM THE USER                           */
    /*==================================================================*/
    gotoxy( 10, 10 );
    Alpha = get_one_double_value( "ROTATION ABOUT X-ACHIS?" );
    Alpha *= (M_PI/180.0);

    gotoxy( 10, 11 );
    Beta = get_one_double_value( "ROTATION ABOUT Y-ACHIS?" );
    Beta *= (M_PI/180.0);

    gotoxy( 10, 12 );
    Gamma = get_one_double_value( "ROTATION ABOUT Z-ACHIS?" );
    Gamma *= (M_PI/180.0);

    /*==================================================================*/
    /* ROTATE A SURFACE FILE                                            */
    /*==================================================================*/
    rotate_surface_file( ParameterStrings[1], ParameterStrings[2],
                         Alpha, Beta, Gamma );

    printf( "\n\nPROGRAM SUCCESSFULLY COMPLETED\n" );
}
/******************************************************************/
void information
     (
         char    **FileNames
     )
/******************************************************************/
{
    /*==================================================================*/
    /* CLEAR THE SCREEN                                                 */
    /*==================================================================*/
    clrscr( );

    /*==================================================================*/
    /* DISPLAY THE INFORMATION ABOUT THE PROGRAM                        */
    /*==================================================================*/
    printf( "         PROGRAM *** X F O R Y ***\n");
    printf( "      EXCHANGES THE X AND Y COMPONENTS \n\n");
```

```
    printf( "  INPUT FILE NAME = %s\n", FileNames[1] );
    printf( " OUTPUT FILE NAME = %s\n", FileNames[2] );
}
/**********************************************************************/
void rotate_surface_file
    (
        char    *InputFile,
        char    *OutputFile,
        double  Alpha,
        double  Beta,
        double  Gamma
    )
/**********************************************************************/
{
    int     Vertex,
            Line;

    FILE    *fIn,
            *fOut;

    double  X,
            Y,
            Z;

    double  RotX,
            RotY,
            RotZ;

    int     NLines,
            NVertices;


    /*================================================================*/
    /* OPEN THE INPUT AND THE OUTPUT FILE                             */
    /*================================================================*/
    fIn  = open_input_file(  InputFile  );
    fOut = open_output_file( OutputFile );

    /*================================================================*/
    /* READ AND WRITE THE NUMBER OF LINES IN THIS FILE                */
    /*================================================================*/
    fscanf(  fIn,  "%d",   &NLines );
    fprintf( fOut, "%d\n",  NLines );

    /*================================================================*/
    /* DISPLAY HOW MANY LINES TO PROCESS                              */
    /*================================================================*/
    gotoxy( 10, 13 );
    printf("%d LINES TO PROCESS\n", NLines);
    gotoxy( 10, 14 );
    printf( "PROCESSING LINE # " );

    /*================================================================*/
    /* LOOP THROUGH EVERY LINE                                        */
    /*================================================================*/
    for( Line=0 ; Line<NLines ; Line++ )
```

ROTATE - Rotate a Surface Around Principal Axes

```
{
   /*====================================================================*/
   /* UPDATE THE SCREEN - SHOW WHICH LINE NUMBER IN PROCESS              */
   /*====================================================================*/
   gotoxy( 27, 14 );
   printf("%4d", Line+1);

   /*====================================================================*/
   /* READ AND WRITE THE NUMBER OF VERTICES IN THIS LINE                 */
   /*====================================================================*/
   fscanf( fIn,  "%d",    &NVertices );
   fprintf( fOut,     "%d\n",   NVertices );

   /*====================================================================*/
   /* LOOP THROUGH ALL VERTICES ON THIS LINE                             */
   /*====================================================================*/
   for( Vertex=0 ; Vertex<NVertices ; Vertex++ )
   {
      /*=================================================================*/
      /* READ THE COMPONENTS FOR ONE VERTEX                              */
      /*=================================================================*/
      fscanf( fIn, "%lf %lf %lf", &X, &Y, &Z );

      RotX =
        ( X * ( sin(Gamma)*cos(Beta) ) +
          Y * ( sin(Gamma)*sin(Beta)*sin(Alpha) +
                cos(Gamma)*cos(Alpha) ) +
          Z * ( sin(Gamma)*sin(Beta)*cos(Alpha) -
                cos(Gamma)*sin(Alpha) )
        );

      RotY =
        ( X * ( cos(Gamma)*cos(Beta) ) +
          Y * ( cos(Gamma)*sin(Beta)*sin(Alpha) -
                sin(Gamma)*cos(Alpha) ) +
          Z * ( cos(Gamma)*sin(Beta)*cos(Alpha) -
                sin(Gamma)*sin(Alpha) )
        );

      RotZ =
        (-X *   sin(Beta)   +
          Y * ( cos(Beta)*sin(Alpha) ) +
          Z * ( cos(Beta)*cos(Alpha) )
        );




      /*=================================================================*/
      /* WRITE THE ROTATED OUTPUT COORDINATES                            */
      /*=================================================================*/
      fprintf( fOut, "%10.6lf %10.6lf %10.6lf\n", RotX, RotY, RotZ );
   }

 }

}
```

Appendix 7
C-Program: TRANSLAT.C - Translate a Surface along Principal Axes

```c
/**********************************************************************/
/*                T R A N S L A T . C                                 */
/**********************************************************************/


#include <stdio.h>
#include <math.h>
#include <stdlib.h>
#include <ctype.h>
#include <conio.h>
#include <dos.h>

#include <library.h>

#define X_COMPONENT    0
#define Y_COMPONENT    1
#define Z_COMPONENT    2


/**********************************************************************/
/* PROTOYPES OF FUNCTIONS                                             */
/**********************************************************************/
void  information
    (
        char **FileNames
    );

void  translate_surface_file
    (
        char    *SurfaceFile,
        char    *OffsetsFileFile,
        double  TransX,
        double  TransY,
        double  TransZ
    );


/**********************************************************************/
main
    (
        int NumberOfParameters,
        char **ParameterStrings
    )
/**********************************************************************/
```

```
{
   double  TransX,
           TransY,
           TransZ;

   /*===================================================================*/
   /* CHECK FOR RIGHT NUMBER OF PARAMETERS BEING PASSED                 */
   /*===================================================================*/
   check_parameters_two(   NumberOfParameters, ParameterStrings );

   /*===================================================================*/
   /* DISPLAY INFORMATION ABOUT THE PROGRAM                             */
   /*===================================================================*/
   information( ParameterStrings );

   /*===================================================================*/
   /* GET THE TRANSLATION VALUES IN ALL THREE ORTHOGONAL DIRECTIONS     */
   /*===================================================================*/
   TransX = get_one_double_value( "TRANSLATION IN X-DIRECTION?:" );
   TransY = get_one_double_value( "TRANSLATION IN Y-DIRECTION?:" );
   TransZ = get_one_double_value( "TRANSLATION IN Z-DIRECTION?:" );

   /*===================================================================*/
   /* ADD THE VALUE TO THE COORDIATE COMPONENTS                         */
   /*===================================================================*/
   translate_surface_file( ParameterStrings[1],
                           ParameterStrings[2],
                           TransX, TransY, TransZ
                         );

   printf( "\n\nPROGRAM SUCCESSFULLY COMPLETED\n" );

}
/*********************************************************************/
void information
     (
         char **FileNames
     )
/*********************************************************************/
{
   /*===================================================================*/
   /* CLEAR THE SCREEN                                                  */
   /*===================================================================*/
   clrscr( );

   /*===================================================================*/
   /* DISPLAY THE INFORMATION ABOUT THE PROGRAM                         */
   /*===================================================================*/
   printf( "        PROGRAM *** T R A N S L A T E ***\n");
   printf( "        READS THE SURFCACE FILE\n\n");
   printf( "        GETS VALUE AND COMPONENT FROM THE USER\n" );
   printf( "        ADDS THE BOTH TOGETHER\n\n");
   printf( "        WRITES THE RESULTS TO THE OUTPUT FILE\n\n");

   printf( " SURFACE FILE NAME      = %s\n", FileNames[1] );
```

```c
    printf( "   TRANSLATED FILE NAME   = %s\n", FileNames[2] );
    printf( "\n" );
}
/***********************************************************************/
void  translate_surface_file
      (
          char    *SurfaceFile,
          char    *TrnsltedFile,
          double  TransX,
          double  TransY,
          double  TransZ
      )
/***********************************************************************/
{
    /*---------------------------------------------------------------*/
    /* VARIABLES FOR SURFACE FILE                                    */
    /*---------------------------------------------------------------*/
    FILE     *fSurfaceFile;
    double   SurfaceX,
             SurfaceY,
             SurfaceZ;
    int      NLinesSurface,
             NVerticesSurface;

    /*---------------------------------------------------------------*/
    /* VARIABLES FOR OUTPUT FILE                                     */
    /*---------------------------------------------------------------*/
    FILE     *fTrnsltedFile;
    double   TrnsltedX,
             TrnsltedY,
             TrnsltedZ;

    /*---------------------------------------------------------------*/
    /* OTHER VARIABLES                                               */
    /*---------------------------------------------------------------*/
    register int    Line,
                    Vertex;

    /*===============================================================*/
    /* OPEN THE INPUT AND THE OUTPUT FILE                            */
    /*===============================================================*/
    fSurfaceFile      = open_input_file(  SurfaceFile  );
    fTrnsltedFile     = open_output_file( TrnsltedFile );

    /*===============================================================*/
    /* READ AND WRITE THE NUMBER OF LINES IN THIS FILE               */
    /*===============================================================*/
    fscanf(  fSurfaceFile,  "%d",   &NLinesSurface );
    fprintf( fTrnsltedFile, "%d\n",  NLinesSurface );

    /*===============================================================*/
    /* DISPLAY HOW MANY LINES TO PROCESS                             */
    /*===============================================================*/
    gotoxy( 10, 17 );
    printf("%d LINES TO PROCESS\n", NLinesSurface );
    gotoxy( 10, 18 );
    printf( "PROCESSING LINE # " );
```

TRANSLATE - Translate a surface along a principal axes

```
/*==============================================================*/
/* LOOP THROUGH EVERY LINE                                      */
/*==============================================================*/
for( Line=0 ; Line<NLinesSurface ; Line++ )
{
    /*==========================================================*/
    /* UPDATE THE SCREEN - SHOW WHICH LINE NUMBER IN PROCESS    */
    /*==========================================================*/
    gotoxy( 27, 18 );
    printf("%4d", Line+1);

    /*==========================================================*/
    /*    READ AND WRITE THE NUMBER OF VERTICES IN THIS LINE    */
    /*==========================================================*/
    fscanf(  fSurfaceFile, "%d",    &NVerticesSurface );
    fprintf( fTrnsltedFile, "%d\n",  NVerticesSurface );

    /*==========================================================*/
    /* LOOP THROUGH ALL VERTICES ON THIS LINE                   */
    /*==========================================================*/
    for( Vertex=0 ; Vertex<NVerticesSurface ; Vertex++ )
    {
        /*======================================================*/
        /* READ THE SURFACE NORMAL VECTOR FOR THIS VERTEX       */
        /*======================================================*/
        fscanf( fSurfaceFile, "%lf %lf %lf",
                         &SurfaceX, &SurfaceY, &SurfaceZ );

        /*======================================================*/
        /* TRANSLATE THE SURFACE VERTEX                         */
        /*======================================================*/
        TrnsltedX = SurfaceX + TransX;
        TrnsltedY = SurfaceY + TransY;
        TrnsltedZ = SurfaceZ + TransZ;

        /*======================================================*/
        /* WRITE OFFSETS FOR THIS VERTEX TO OUTPUT FILE         */
        /*======================================================*/
        fprintf( fTrnsltedFile, " %9.6f %9.6f %9.6f\n",
                                         TrnsltedX,
                                         TrnsltedY,
                                         TrnsltedZ );
    }

}

}
```

Appendix 8
C- Program: SURFNORM.C - Calculate Surface Normal Vectors

```c
/**********************************************************************/
/*                      S U R F N O R M . C                           */
/**********************************************************************/

#include <stdio.h>
#include <math.h>
#include <stdlib.h>
#include <ctype.h>
#include <conio.h>
#include <dos.h>


/*====================================================================*/
/* INCLUDE FILE FOR OUR OWN LIBRARY FUNCTIONS                         */
/*====================================================================*/
#include <library.h>


/**********************************************************************/
/* PROTOYPES OF FUNCTIONS                                             */
/**********************************************************************/
void information
    (
        char **FileNames
    );

void make_normal_vectors_on_surface
    (
        char *InputFile,
        char *OutputFile
    );

void make_normal_vector_on_one_line
    (
        float X[][1000],
        float Y[][1000],
        float Z[][1000],
        float NX[],
        float NY[],
        float NZ[],
        int   NVertices
    );
```

```
void read_one_line
    (
        float   *X,
        float   *Y,
        float   *Z,
        int     *NVertices,
        FILE    *fin
    );
void write_one_line
    (
        float *X,
        float *Y,
        float *Z,
        int    NVertices,
        FILE  *fout
    );
void copy_line
    (
        float   XFrom[],
        float   YFrom[],
        float   ZFrom[],
        float   XTo[],
        float   YTo[],
        float   ZTo[],
        int     NVertices
    );


/*******************************************************************/
main
    (
        int NumberOfParameters,
        char **ParameterStrings
    )
/*******************************************************************/
{
    /*=================================================================*/
    /* CHECK FOR RIGHT NUMER OF PARAMETERS BEING PASSED                */
    /*=================================================================*/
    check_parameters_two( NumberOfParameters, ParameterStrings );

    /*=================================================================*/
    /* DISPLAY INFORMATION ABOUT THE PROGRAM                           */
    /*=================================================================*/
    information( ParameterStrings );

    /*=================================================================*/
    /* MAKE THE NORMAL VECTORS ON THE SURFACE                          */
    /*=================================================================*/
    make_normal_vectors_on_surface( ParameterStrings[1],
                                    ParameterStrings[2] );

    printf( "\n\nPROGRAM SUCCESSFULLY COMPLETED\n" );
```

```
}
/**********************************************************************/
void information
    (
        char **FileNames
    )
/**********************************************************************/
{
    /*====================================================================*/
    /* CLEAR THE SCREEN                                                   */
    /*====================================================================*/
    clrscr( );

    /*====================================================================*/
    /* DISPLAY THE INFORMATION ABOUT THE PROGRAM                          */
    /*====================================================================*/
    printf( "         PROGRAM *** S U R F N O R M ***\n");
    printf( "         CALCULATE SURFACE NORMAL VECTORS ON A GRID
                      SURFACE\n\n");

    printf( " INPUT FILE NAME  = %s\n", FileNames[1] );
    printf( " OUTPUT FILE NAME = %s\n", FileNames[2] );
}


/**********************************************************************/
void make_normal_vectors_on_surface
    (
        char *InputFile,
        char *OutputFile
    )
/**********************************************************************/
{
    /*--------------------------------------------------------------------*/
    /* LINE AND VERTEX COUNTERS                                           */
    /*--------------------------------------------------------------------*/
    register int  Line;

    int           NLines,
                  NVertices;

    /*--------------------------------------------------------------------*/
    /* FILE ACCESS STRUCTURES FOR INPUT AND OUTPUT FILE                   */
    /*--------------------------------------------------------------------*/
    FILE    *fSurfaceFile,
            *fNormalVectorFile;

    /*--------------------------------------------------------------------*/
    /* ARRAYS FOR THREE LINES FROM THE SURFACE                            */
    /*--------------------------------------------------------------------*/
    static float  X[3][1000],
                  Y[3][1000],
                  Z[3][1000];

    /*--------------------------------------------------------------------*/
    /* ARRAYS FOR ONE LINE OF NORMAL VECTORS                              */
    /*--------------------------------------------------------------------*/
```

SURFNORM - Calculate Surface Normal Vectors

```
static float  NX[1000],
              NY[1000],
              NZ[1000];

/*================================================================*/
/* OPEN THE INPUT AND THE OUTPUT FILE                             */
/*================================================================*/
fSurfaceFile      = open_input_file( InputFile );
fNormalVectorFile = open_output_file( OutputFile );

/*================================================================*/
/* READ THE NUMBER OF LINES THAT ARE IN THE NORMAL VECTOR FILE    */
/* AND WRITE TO OFFSET FILE                                       */
/*================================================================*/
fscanf(  fSurfaceFile, "%d", &NLines );
fprintf( fNormalVectorFile, "%d\n", NLines);

/*================================================================*/
/* DISPLAY HOW MANY LINES TO PROCESS                              */
/*================================================================*/
gotoxy( 10, 10 );
printf("%d LINES TO PROCESS\n", NLines);
gotoxy( 10, 11 );
printf( "PROCESSING LINE # " );

/*================================================================*/
/* UPDATE THE SCREEN - SHOW WHAT LINE NUMBER IN PROCESS           */
/*================================================================*/
gotoxy( 27, 11 );
printf("%4d", 1);

/*================================================================*/
/* READ FIRST TWO LINES FROM THE SURFACE FILE                     */
/*================================================================*/
read_one_line( X[1], Y[1], Z[1], &NVertices, fSurfaceFile );
read_one_line( X[2], Y[2], Z[2], &NVertices, fSurfaceFile );

/*================================================================*/
/* COPY VALUES FROM LINE 0 INTO ARRAY ZERO                        */
/*================================================================*/
copy_line( X[1], Y[1], Z[1], X[0], Y[0], Z[0], NVertices );

/*================================================================*/
/* MAKE SURFACE NORMAL VECTOR FOR ONE LINE                        */
/*================================================================*/
make_normal_vector_on_one_line( X, Y, Z, NX, NY, NZ, NVertices );

/*================================================================*/
/* WRITE ONE LINE OF OFFSETS TO THE OUTPUT FILE                   */
/*================================================================*/
write_one_line( NX, NY, NZ, NVertices, fNormalVectorFile );

/*================================================================*/
/* LOOP THROUGH ALL LINES                                         */
/*================================================================*/
for( Line=1 ; Line<NLines-1 ; Line++ )
{
    /*============================================================*/
```

```c
        /* UPDATE THE SCREEN - SHOW WHAT LINE NUMBER IN PROCESS       */
        /*==========================================================*/
        gotoxy( 27, 11 );
        printf("%4d", Line+1);

        /*==========================================================*/
        /* COPY TWO LINES ONCE OVER                                  */
        /*==========================================================*/
        copy_line( X[1], Y[1], Z[1], X[0], Y[0], Z[0], NVertices );
        copy_line( X[2], Y[2], Z[2], X[1], Y[1], Z[1], NVertices );

        /*==========================================================*/
        /* READ NEXT LINE                                            */
        /*==========================================================*/
        read_one_line( X[2], Y[2], Z[2], &NVertices, fSurfaceFile );

        /*==========================================================*/
        /* MAKE SURFACE NORMAL VECTOR FOR ONE LINE                   */
        /*==========================================================*/
        make_normal_vector_on_one_line( X, Y, Z, NX, NY, NZ, NVertices );

        /*==========================================================*/
        /* WRITE ONE LINE OF OFFSETS TO THE OUTPUT FILE              */
        /*==========================================================*/
        write_one_line( NX, NY, NZ, NVertices, fNormalVectorFile );

    }

    /*==============================================================*/
    /* UPDATE THE SCREEN - SHOW WHAT LINE NUMBER IN PROCESS          */
    /*==============================================================*/
    gotoxy( 27, 11 );
    printf("%4d", Line+1);

    /*==============================================================*/
    /* COPY TWO LINES ONCE OVER                                      */
    /*==============================================================*/
    copy_line( X[1], Y[1], Z[1], X[0], Y[0], Z[0], NVertices );
    copy_line( X[2], Y[2], Z[2], X[1], Y[1], Z[1], NVertices );

    /*==============================================================*/
    /* MAKE NORMAL VECTOR FOR END OF SURFACE                         */
    /*==============================================================*/
    make_normal_vector_on_one_line( X, Y, Z, NX, NY, NZ, NVertices );

    /*==============================================================*/
    /* WRITE ONE LINE OF OFFSETS TO THE OUTPUT FILE                  */
    /*==============================================================*/
    write_one_line( NX, NY, NZ, NVertices, fNormalVectorFile );

}

/******************************************************************/
void make_normal_vector_on_one_line
        (
        float X[][1000],
        float Y[][1000],
        float Z[][1000],
```

SURFNORM - Calculate Surface Normal Vectors

```
          float   NX[],
          float   NY[],
          float   NZ[],
          int     NVertices
     )
/**********************************************************************/
{
     /*--------------------------------------------------------------*/
     /* VERTEX COUNTER                                               */
     /*--------------------------------------------------------------*/
     register int Vrtx;

     /*--------------------------------------------------------------*/
     /* DISTANCE BETWEEN VERTICES IN VERTEX DIRECTION                */
     /*--------------------------------------------------------------*/
     double        DVX,
                   DVY,
                   DVZ;

     /*--------------------------------------------------------------*/
     /* DISTANCE BETWEEN VERTICES IN LINE DIRECTION                  */
     /*--------------------------------------------------------------*/
     double        DLX,
                   DLY,
                   DLZ;

     /*--------------------------------------------------------------*/
     /* LENGTH OF THE NORMAL VECTOR BEFORE NORMALIZATION             */
     /*--------------------------------------------------------------*/
     double        LengthNormalVector;

     /*==============================================================*/
     /* LOOP THROUGH ALL VERTICES                                    */
     /*==============================================================*/
     for( Vrtx=1 ; Vrtx<(NVertices+1) ; Vrtx++ )
     {
          /*=========================================================*/
          /* CALCULATE DISTANCES IN VERTEX DIRECTION                 */
          /*=========================================================*/
          DVX = X[1][Vrtx+1] - X[1][Vrtx-1];
          DVY = Y[1][Vrtx+1] - Y[1][Vrtx-1];
          DVZ = Z[1][Vrtx+1] - Z[1][Vrtx-1];

          /*=========================================================*/
          /* CALCULATE DISTANCES IN LINE DIRECTION                   */
          /*=========================================================*/
          DLX = X[2][Vrtx] - X[0][Vrtx];
          DLY = Y[2][Vrtx] - Y[0][Vrtx];
          DLZ = Z[2][Vrtx] - Z[0][Vrtx];

          /*=========================================================*/
          /* CALCULATE THE THREE ORTHOGONAL COMPONENTS OF NORMAL VECTOR */
          /*=========================================================*/
          NX[Vrtx-1] =            DVY * DLZ - DLY * DVZ;
          NY[Vrtx-1] = (-1) *  ( DVX * DLZ - DLX * DVZ );
          NZ[Vrtx-1] =            DVX * DLY - DLX * DVY;

          /*=========================================================*/
```

```c
      /* CALCULATE LENGTH OF NORMAL VECTOR                              */
      /*==============================================================*/
      LengthNormalVector = sqrt( NX[Vrtx-1] * NX[Vrtx-1]  +
                                 NY[Vrtx-1] * NY[Vrtx-1]  +
                                 NZ[Vrtx-1] * NZ[Vrtx-1]
                               );

      /*==============================================================*/
      /* NORMALIZE NORMAL VECTOR                                        */
      /*==============================================================*/
      if( LengthNormalVector > 0.0 )
      {
         NX[Vrtx-1] /= LengthNormalVector;
         NY[Vrtx-1] /= LengthNormalVector;
         NZ[Vrtx-1] /= LengthNormalVector;
      }
      else
      {
         gotoxy( 20, 15 );
         printf( "NORMAL VECTOR LENGTH = 0.0 ON VERTEX %d\n", Vrtx );
         NX[Vrtx-1] = NY[Vrtx-1] = NZ[Vrtx-1] = 1.0;
         getch();
      }
   }

}

/************************************************************************/
void read_one_line
    (
       float  *X,
       float  *Y,
       float  *Z,
       int    *NVertices,
       FILE   *fin
    )
/************************************************************************/
{
   register int  Vertex;

   /*==============================================================*/
   /* READ THE NUMBER OF VERTICES ON THIS LINE                       */
   /*==============================================================*/
   fscanf( fin, "%d", NVertices );

   /*==============================================================*/
   /* READ THE VALUES INTO THE ALLOCATED MEMORY                      */
   /*==============================================================*/
   for( Vertex=1  ; Vertex<(*NVertices+1) ; Vertex++ )
   {
      fscanf( fin, "%f %f %f", &(X[Vertex]),
                               &(Y[Vertex]),
                               &(Z[Vertex]) );
   }

   /*==============================================================*/
   /* COPY VALUES FROM LOCATION ONE INTO LOCATION ZERO               */
   /*==============================================================*/
```

```
        X[0] = X[1];
        Y[0] = Y[1];
        Z[0] = Z[1];

        /*====================================================================*/
        /* COPY VALUES FROM SECOND LAST POSION INTO LAST POSITION             */
        /*====================================================================*/
        X[*NVertices+1] = X[*NVertices];
        Y[*NVertices+1] = Y[*NVertices];
        Z[*NVertices+1] = Z[*NVertices];

}

/*********************************************************************/
void copy_line
    (
        float   XFrom[],
        float   YFrom[],
        float   ZFrom[],
        float   XTo[],
        float   YTo[],
        float   ZTo[],
        int     NVertices
    )
/*********************************************************************/
{
    register int Vertex;

    /*====================================================================*/
    /* LOOP THROUGH ALL VERTICES AND COPY VERTEX VALUES                   */
    /*====================================================================*/
    for( Vertex=0 ; Vertex<(NVertices+2) ; Vertex++ )
    {
        XTo[Vertex] = XFrom[Vertex];
        YTo[Vertex] = YFrom[Vertex];
        ZTo[Vertex] = ZFrom[Vertex];
    }
}
```

Appendix 9
C-Program: OFFGEN.C - Calculate Generalized Cutter Offsets

```c
/**********************************************************************/
/*              O F F G E N . C                                       */
/**********************************************************************/

#include <stdio.h>
#include <math.h>
#include <stdlib.h>
#include <ctype.h>
#include <conio.h>
#include <dos.h>

#include <library.h>


/**********************************************************************/
/* PROTOTYPES OF FUNCTIONS                                            */
/**********************************************************************/
void information
     (
         char **FileNames
     );


void calculate_offsets_general_mill
     (
         double  CutterEdgeRadius,
         double  CutterOuterRadius,
         char    *InputFile,
         char    *OutputFile
     );
/**********************************************************************/
main
    (
        int NumberOfParameters,
        char **ParameterStrings
    )
/**********************************************************************/
{
    double          CutterEdgeRadius,
                    CutterOuterRadius;


    /*================================================================*/
```

OFFGEN - Calculate Generalized Cutter Offsets

```c
    /* CHECK FOR RIGHT NUMER OF PARAMETERS BEING PASSED          */
    /*============================================================*/
    check_parameters_two( NumberOfParameters, ParameterStrings );

    /*============================================================*/
    /* DISPLAY INFORMATION ABOUT THE PROGRAM                     */
    /*============================================================*/
    information( ParameterStrings );

    /*============================================================*/
    /* GET THE RADII FOR THE GENERAL SHAPED CUTTER FROM THE USER */
    /*============================================================*/
    CutterOuterRadius  = get_one_double_value( "SHANK RADIUS OF CUTTER?:
                           " );
    CutterEdgeRadius   = get_one_double_value( "FILLET RADIUS OF CUTTER?:
                           " );

    if( fabs(CutterOuterRadius-CutterEdgeRadius) < 0.000001 )
    {
       printf( "CALCULATING FOR BALL-MILL CUTTER\n" );
    }
    else if( CutterEdgeRadius == 0.0 )
    {
       printf( "CALCULATING FOR END-MILL CUTTER\n" );
    }
    else
    {
       printf( "CALCULATING FOR GENERAL SHAPE CUTTER\n" );
    }

    /*============================================================*/
    /* CALCULATE THE OFFSETS FOR ONE LINE                        */
    /*============================================================*/
    calculate_offsets_general_mill( CutterEdgeRadius,
                                    CutterOuterRadius,
                                    ParameterStrings[1],
                                    ParameterStrings[2]
                                  );

    printf( "\n\nPROGRAM SUCCESSFULLY COMPLETED\n" );

}
/********************************************************************/
void information
     (
         char **FileNames
     )
/********************************************************************/
{
    /*============================================================*/
    /* CLEAR THE SCREEN                                          */
    /*============================================================*/
    clrscr( );

    /*============================================================*/
    /* DISPLAY THE INFORMATION ABOUT THE PROGRAM                 */
```

```c
/*===================================================================*/
    printf( "          PROGRAM *** O F F G E N ***\n");
    printf( "          CALCULATES OFFSETS FOR A GENERAL SHAPE CUTTER\n\n");

    printf( "  INPUT FILE NAME = %s\n", FileNames[1] );
    printf( " OUTPUT FILE NAME = %s\n", FileNames[2] );
    printf( "\n" );
}
/*********************************************************************/
void  calculate_offsets_general_mill
      (
          double  CutterEdgeRadius,
          double  CutterOuterRadius,
          char    *InputFile,
          char    *OutputFile
      )
/*********************************************************************/
{
    /*---------------------------------------------------------------*/
    /* VARIABLES FOR NORMAL VECTOR FILE                              */
    /*---------------------------------------------------------------*/
    FILE    *fNormalVectorFile;
    double  NormalVectorX,
            NormalVectorY,
            NormalVectorZ;
    int     NLines,
            NVertices;

    /*---------------------------------------------------------------*/
    /* NORMAL VECTOR PROJECTED INTO THE XY-PLANE                     */
    /*---------------------------------------------------------------*/
    double  NormalXYPlaneX,
            NormalXYPlaneY,
            NormalXYPlaneZ;

    /*---------------------------------------------------------------*/
    /* VARIABLES FOR OFFSET FILE                                     */
    /*---------------------------------------------------------------*/
    FILE    *fOffsetFile;
    double  OffsetX,
            OffsetY,
            OffsetZ;

    /*---------------------------------------------------------------*/
    /* OTHER VARIABLES                                               */
    /*---------------------------------------------------------------*/
    double          O1X,
                    O1Y,
                    O1Z;

    double          O2X,
                    O2Y,
                    O2Z;

    double          O3X,
                    O3Y,
                    O3Z;
```

OFFGEN - Calculate Generalized Cutter Offsets

```
        double              LengthNormalVectorXYPlane;
        register int        Line,
                            Vertex;

/*================================================================*/
/* OPEN THE INPUT AND THE OUTPUT FILE                             */
/*================================================================*/
fNormalVectorFile  = open_input_file(  InputFile );
fOffsetFile        = open_output_file( OutputFile );

/*================================================================*/
/* READ AND WRITE THE NUMBER OF LINES IN THIS FILE                */
/*================================================================*/
fscanf( fNormalVectorFile, "%d",    &NLines );
fprintf( fOffsetFile,      "%d\n",  NLines );

/*================================================================*/
/* DISPLAY HOW MANY LINES TO PROCESS                              */
/*================================================================*/
gotoxy( 10, 12 );
printf("%d LINES TO PROCESS\n", NLines);
gotoxy( 10, 13 );
printf( "PROCESSING LINE # " );

/*================================================================*/
/* LOOP THROUGH EVERY LINE                                        */
/*================================================================*/
for( Line=0 ; Line<NLines ; Line++ )
{
   /*================================================================*/
   /* UPDATE THE SCREEN - SHOW WHICH LINE NUMBER IN PROCESS          */
   /*================================================================*/
   gotoxy( 27, 13 );
   printf("%4d", Line+1);

   /*================================================================*/
   /*    READ AND WRITE THE NUMBER OF VERTICES IN THIS LINE          */
   /*================================================================*/
   fscanf(  fNormalVectorFile, "%d",    &NVertices );
   fprintf( fOffsetFile,       "%d\n",  NVertices );

   /*================================================================*/
   /* LOOP THROUGH ALL VERTICES ON THIS LINE                         */
   /*================================================================*/
   for( Vertex=0 ; Vertex<NVertices ; Vertex++)
   {
      /*================================================================*/
      /* READ THE SURFACE NORMAL VECTOR FOR THIS VERTEX                 */
      /*================================================================*/
      fscanf( fNormalVectorFile, "%lf %lf %lf", &NormalVectorX,
                                                &NormalVectorY,
                                                &NormalVectorZ
            );

      /*================================================================*/
      /* CALCULATE THE LENGTH OF THE NORMAL VECTOR IN THE X-Y PLANE*/
      /*================================================================*/
```

Appendix 9

```
            LengthNormalVectorXYPlane =
                    sqrt( ( NormalVectorX * NormalVectorX ) +
                          ( NormalVectorY * NormalVectorY )
                        );

            /*=============================================================*/
            /* NORMALIZE THE NORMAL VECTOR IN THE X-Y PLANE                */
            /*=============================================================*/
            NormalXYPlaneX = NormalVectorX / LengthNormalVectorXYPlane;
            NormalXYPlaneY = NormalVectorY / LengthNormalVectorXYPlane;
            NormalXYPlaneZ = 0.0;


            /*=============================================================*/
            /* CALCULATE THE OFFSET FOR THIS VERTEX                        */
            /*=============================================================*/
            O1X = NormalXYPlaneX  * (CutterOuterRadius - CutterEdgeRadius);
            O1Y = NormalXYPlaneY  * (CutterOuterRadius - CutterEdgeRadius);
            O1Z = NormalXYPlaneZ  * (CutterOuterRadius - CutterEdgeRadius);

            O2X = NormalVectorX   * CutterEdgeRadius;
            O2Y = NormalVectorY   * CutterEdgeRadius;
            O2Z = NormalVectorZ   * CutterEdgeRadius;



            O3X = 0;
            O3Y = 0;
            O3Z = CutterEdgeRadius;


            OffsetX = O1X + O2X - O3X;
            OffsetY = O1Y + O2Y - O3Y;
            OffsetZ = O1Z + O2Z - O3Z;

            /*=============================================================*/
            /* WRITE OFFSETS FOR THIS VERTEX TO OUTPUT FILE                */
            /*=============================================================*/
            fprintf( fOffsetFile, "%10.6f %10.6f %10.6f\n", OffsetX,
                                                            OffsetY,
                                                            OffsetZ );
         }

      }

   }
```

Appendix 10
C-Program: ADDOFFST.C - Add Cutter Offsets to Polyhedron

```c
/**********************************************************************/
/*                   A D D O F F S T . C                              */
/**********************************************************************/

#include <stdio.h>
#include <math.h>
#include <stdlib.h>
#include <ctype.h>
#include <conio.h>
#include <dos.h>

#include <library.h>


/**********************************************************************/
/* PROTOYPES OF FUNCTIONS                                             */
/**********************************************************************/
void information
    (
        char **FileNames
    );

void  add_offsets_to_surface
    (
        char    *SurfaceFile,
        char    *OffsetsFileFile,
        char    *CutterLocationFile
    );

void check_for_equal
    (
        int Number1,
        int Number2
    );
/**********************************************************************/
main
   (
      int NumberOfParameters,
      char **ParameterStrings
   )
/**********************************************************************/
{

   /*================================================================*/
```

```c
    /* CHECK FOR RIGHT NUMBER OF PARAMETERS BEING PASSED            */
    /*==============================================================*/
    check_parameters_three( NumberOfParameters, ParameterStrings );

    /*==============================================================*/
    /* DISPLAY INFORMATION ABOUT THE PROGRAM                        */
    /*==============================================================*/
    information( ParameterStrings );

    /*==============================================================*/
    /* ADD OFFSETS TO THE SURFACE AND WRITE TO FILE                 */
    /*==============================================================*/
    add_offsets_to_surface( ParameterStrings[1],
                            ParameterStrings[2],
                            ParameterStrings[3]
                          );

    printf( "\n\nPROGRAM SUCCESSFULLY COMPLETED\n" );

}
/***********************************************************************/
void information
    (
        char **FileNames
    )
/***********************************************************************/
{
    /*==============================================================*/
    /* CLEAR THE SCREEN                                             */
    /*==============================================================*/
    clrscr( );

    /*==============================================================*/
    /* DISPLAY THE INFORMATION ABOUT THE PROGRAM                    */
    /*==============================================================*/
    printf( "         PROGRAM  ***  A D D O F F S T   ***\n");
    printf( "       READS THE SURFACE FILE\n\n");
    printf( "       READS THE OFFSET FILE\n\n");
    printf( "       ADDS BOTH FILES TOGETHER\n\n");
    printf( "  WRITES THE RESULTS TO THE CUTTER LOCATION FILE\n\n");

    printf( "  SURFACE FILE NAME         = %s\n", FileNames[1] );
    printf( "  OFFSET FILE NAME          = %s\n", FileNames[2] );
    printf( "  CUTTER LOCATION FILE NAME = %s\n", FileNames[3] );
}
/***********************************************************************/
void  add_offsets_to_surface
        (
            char    *SurfaceFile,
            char    *OffsetsFile,
            char    *CutterLocationFile
        )
/***********************************************************************/
{
    /*--------------------------------------------------------------*/
```

ADDOFFST - Add Cutter Offsets to Polyhedron

```
/* VARIABLES FOR SURFACE                                              */
/*--------------------------------------------------------------------*/
FILE    *fSurfaceFile;
double  SurfaceX,
        SurfaceY,
        SurfaceZ;
int     NLinesSurface,
        NVerticesSurface;

/*--------------------------------------------------------------------*/
/* VARIABLES FOR OFFSETS                                              */
/*--------------------------------------------------------------------*/
FILE    *fOffsetsFile;
double  OffsetsX,
        OffsetsY,
        OffsetsZ;
int     NLinesOffsets,
        NVerticesOffsets;

/*--------------------------------------------------------------------*/
/* VARIABLES FOR CUTTER LOCATION                                      */
/*--------------------------------------------------------------------*/
FILE    *fCutterLocationFile;
double  CutterLocationX,
        CutterLocationY,
        CutterLocationZ;

/*--------------------------------------------------------------------*/
/* OTHER VARIABLES                                                    */
/*--------------------------------------------------------------------*/
register int    Line,
                Vertex;

/*====================================================================*/
/* OPEN THE INPUT AND THE OUTPUT FILE                                 */
/*====================================================================*/
fSurfaceFile        = open_input_file( SurfaceFile );
fOffsetsFile        = open_input_file( OffsetsFile );
fCutterLocationFile = open_output_file( CutterLocationFile );

/*====================================================================*/
/* READ AND WRITE THE NUMBER OF LINES ON THE SURFACE                  */
/*====================================================================*/
fscanf( fSurfaceFile, "%d", &NLinesSurface );
fscanf( fOffsetsFile, "%d", &NLinesOffsets );

/*====================================================================*/
/* MAKE SURE THAT THE NUMBER OF LINES IN BOTH FILES ARE THE SAME      */
/*====================================================================*/
check_for_equal( NLinesSurface, NLinesOffsets );

/*====================================================================*/
/* WRITE THE NUMBER OF LINES TO THE CUTTER LOCATION FILE              */
/*====================================================================*/
fprintf( fCutterLocationFile, "%d\n", NLinesSurface );

/*====================================================================*/
/* DISPLAY HOW MANY LINES TO PROCESS                                  */
```

Appendix 10

```c
/*================================================================*/
gotoxy( 10, 10 );
printf("%d LINES TO PROCESS\n", NLinesSurface );
gotoxy( 10, 11 );
printf( "PROCESSING LINE # " );

/*================================================================*/
/* LOOP THROUGH EVERY LINE                                        */
/*================================================================*/
for( Line=0 ; Line<NLinesSurface ; Line++ )
{
   /*=============================================================*/
   /* UPDATE THE SCREEN - SHOW WHICH LINE NUMBER IN PROCESS       */
   /*=============================================================*/
   gotoxy( 27, 11 );
   printf("%4d", Line+1);

   /*=============================================================*/
   /* READ AND WRITE THE NUMBER OF VERTICES IN THIS LINE          */
   /*=============================================================*/
   fscanf( fSurfaceFile, "%d",  &NVerticesSurface );
   fscanf( fOffsetsFile, "%d",  &NVerticesOffsets );

   /*=============================================================*/
   /* MAKE SURE THAT THE NUMBER OF VERTICES IN BOTH FILES ARE THE */
   /* SAME                                                        */
   /*=============================================================*/
   check_for_equal( NVerticesSurface, NVerticesOffsets );

   /*=============================================================*/
   /* WRITE THE NUMBER OF VERTICES TO THE CUTTER LOCATION FILE    */
   /*=============================================================*/
   fprintf( fCutterLocationFile, "%d\n",  NVerticesSurface );

   /*=============================================================*/
   /* LOOP THROUGH ALL VERTICES ON THIS LINE                      */
   /*=============================================================*/
   for( Vertex=0 ; Vertex<NVerticesSurface ; Vertex++)
   {
      /*==========================================================*/
      /* READ THE SURFACE NORMAL VECTOR FOR THIS VERTEX           */
      /*==========================================================*/
      fscanf( fSurfaceFile, "%lf %lf %lf",
                        &SurfaceX, &SurfaceY, &SurfaceZ );
      fscanf( fOffsetsFile, "%lf %lf %lf",
                        &OffsetsX, &OffsetsY, &OffsetsZ );

      /*==========================================================*/
      /* CALCULATE THE OFFSET THIS FOR VERTEX                     */
      /*==========================================================*/
      CutterLocationX = SurfaceX + OffsetsX;
      CutterLocationY = SurfaceY + OffsetsY;
      CutterLocationZ = SurfaceZ + OffsetsZ;

      /*==========================================================*/
      /* WRITE CUTTER LOCATION FOR THIS VERTEX TO OUTPUT FILE     */
      /*==========================================================*/
      fprintf( fCutterLocationFile, "%10.6f %10.6f %10.6f\n",
```

ADDOFFST - Add Cutter Offsets to Polyhedron

```
                                    CutterLocationX,
                                    CutterLocationY,
                                    CutterLocationZ );
        }

    }

}
/*******************************************************************/
void check_for_equal
        (
            int Number1,
            int Number2
        )
/*******************************************************************/
{
    if( Number1 != Number2 )
    {
        gotoxy( 20, 20 );
        printf( "ERROR!: SURFACE & OFFSETS FILE ARE NOT THE SAME
                SIZE\n" );
        printf( "PROGRAM ABORTED\n" );
        exit( 1 );
    }
}
```

Appendix 11
C-Program: MAKEGCDC - Generate a G-Code File

```c
/**********************************************************************/
/*                    M A K E G C D                                   */
/**********************************************************************/

#include <stdio.h>
#include <math.h>
#include <stdlib.h>
#include <ctype.h>
#include <conio.h>
#include <dos.h>

#include "library.h"

/**********************************************************************/
/* PROTOYPES OF FUNCTIONS                                             */
/**********************************************************************/
void information
    (
        char **FileNames
    );

int  get_machining_procedure
    (
        void
    );

void make_gcode_for_surface
    (
        char *CutterLocationFile,
        char *GCodeFile
    );

int  make_gcode_forward
    (
        FILE    *fCutterLocationFile,
        FILE    *fGCodeFile,
        float   ZClear,
        float   FeedRate,
        int     CodeLine
    );

int  make_gcode_backward
    (
```

MAKEGCDC - Translate a Cutter Location file into G-Code

```
        FILE    *fCutterLocationFile,
        FILE    *fGCodeFile,
        float   ZClear,
        float   FeedRate,
        int     CodeLine
    );

int make_gcode_alternating
    (
        FILE    *fCutterLocationFile,
        FILE    *fGCodeFile,
        float   ZClear,
        float   FeedRate,
        int     CodeLine
    );

int header
    (
        FILE    *fGCodeFile,
        float   XStart,
        float   YStart,
        float   ZStart,
        int     SpindleSpeed,
        int     MetricImperial

    );


int   make_gcode_forward_one_line
    (
        FILE  *fGCodeFile,
        float *X,
        float *Y,
        float *Z,
        int    NVrtcs,
        float  ZClear,
        int    CodeLine,
        float  FeedRate,
        int    StartCondition
    );

int   make_gcode_backward_one_line
    (
        FILE  *fGCodeFile,
        float *X,
        float *Y,
        float *Z,
        int    NVrtcs,
        float  ZClear,
        int    CodeLine,
        float  FeedRate,
        int    StartCondition
    );

int gcode_begin_line_g00
    (
        FILE    *fGCodeFile,
```

Appendix 11

```c
        float   X,
        float   Y,
        float   Z,
        float   ZClear,
        float   FeedRate,
        int     CodeLine
    );

int ender
    (
        FILE    *fGCodeFile,
        int     CodeLine,
        float   XStart,
        float   YStart,
        float   ZStart,
        float   ZClear
    );

void read_one_line
    (
        float   *X,
        float   *Y,
        float   *Z,
        int     *NVertices,
        FILE    *fin
    );

/*==================================================================*/
/* DEFINES FOR EASIER READING                                       */
/*==================================================================*/
#define FORWARD      0
#define BACKWARD     1
#define ALTERNATING  2

#define METRIC       0
#define IMPERIAL     1

#define MAX_VERTICES 2000

#define G00_START    0
#define G01_START    1


/********************************************************************/
main
    (
        int     NumberOfParameters,
        char    **ParameterStrings
    )
/********************************************************************/
{

    /*==================================================================*/
    /* CHECK FOR RIGHT NUMER OF PARAMETERS BEING PASSED                 */
    /*==================================================================*/
    check_parameters_two( NumberOfParameters, ParameterStrings );

    /*==================================================================*/
```

```c
    /* DISPLAY INFORMATION ABOUT THE PROGRAM                                */
    /*======================================================================*/
    information( ParameterStrings );

    /*======================================================================*/
    /* MAKE THE G-CODE                                                      */
    /*======================================================================*/
    make_gcode_for_surface( ParameterStrings[1], ParameterStrings[2] );

    printf( "\n\nPROGRAM SUCCESSFULLY COMPLETED\n" );
}
/**************************************************************************/
void information
    (
        char **FileNames
    )
/**************************************************************************/
{
    /*======================================================================*/
    /* CLEAR THE SCREEN                                                     */
    /*======================================================================*/
    clrscr( );

    /*======================================================================*/
    /* DISPLAY THE INFORMATION ABOUT THE PROGRAM                            */
    /*======================================================================*/
    printf( "          PROGRAM *** M A K E G C D  ***\n");
    printf( "          GENERATES GCODE FROM A CUTTER LOCATION FILE\n\n");

    printf( "CUTTER LOCATION FILE NAME = %s\n", FileNames[1] );
    printf( "G-CODE FILE NAME          = %s\n", FileNames[2] );
}
/**************************************************************************/
int  get_machining_procedure
    (
        void
    )
/**************************************************************************/
{
    int    Procedure,
           Continue;

    char   Answer;

    gotoxy( 1, 14 );
    printf("    ===============================================\n" );
    printf("    THERE ARE THREE METHODS FOR MACHINING THE PART\n" );
    printf("    ===============================================\n" );
    printf("    1. IN THE ORDER THE VERTICES ARE DEFINED    =
                   (F)ORWARD      \n");
    printf("    2. IN REVERSE ORDER                         =
                   (B)ACKWARD     \n");
    printf("    3. IN ALTERNATING ORDER = FORWARD, BACKWARD =
                   (A)LTERNATING\n\n");
```

Appendix 11

```c
    do
    {
       gotoxy( 10, 21 );
       printf("CHOOSE (F) (B) (A): ? ");
       Answer = toupper( getch() );
       putch( Answer ) ;

       switch( Answer )
       {
          /*==============================================================*/
          case 'F':
          /*==============================================================*/
             Procedure = FORWARD;
             gotoxy( 10, 13 );
             printf( "MACHINING PROCEDURE IS : FORWARD" );
             Continue = FALSE;
          break;

          /*==============================================================*/
             case 'B':
          /*==============================================================*/
             Procedure = BACKWARD;
             gotoxy( 10, 13 );
             printf( "MACHINING PROCEDURE IS : BACKWARD" );
             Continue = FALSE;
          break;

          /*==============================================================*/
             case 'A':
          /*==============================================================*/
             Procedure = ALTERNATING;
             gotoxy( 10, 13 );
             printf( "MACHINING PROCEDURE IS : ALTERNATING" );
             Continue = FALSE;
          break;

          /*==============================================================*/
          /* REJECT ALL OTHER CHARACTERS                                  */
          /*==============================================================*/
             default:
          /*==============================================================*/
             beep();
             Continue = TRUE;
          break;
       }

    }while( Continue == TRUE );

    /*==============================================================*/
    /* CLEAN USED PART OF THE SCREEN                                */
    /*==============================================================*/
    gotoxy( 1, 14 );
    printf( "                                                      \n" );
    printf( "                                                      \n" );
    printf( "                                                      \n" );
    printf( "                                                      \n" );
    printf( "                                                      \n" );
```

```
        printf( "                                                          \n" );
        printf( "                                                          \n" );
        printf( "                                                          \n" );

        return( Procedure );
}
/****************************************************************/
int   get_measurement_system
        (
            void
        )
/****************************************************************/
{
    int    System,
           Continue;

    char   Answer;

    gotoxy( 1, 15 );
    printf("     ============================================================\n" );
    printf("        IS THE DATA IN INCHES OR MILLIMETRES?\n" );
    printf("     ============================================================\n" );
    printf("        MILLIMETERS    (M):\n");
    printf("        INCHES         (I):\n");

    do
    {
      gotoxy( 10, 20 );
      printf("CHOOSE (M) (I): ? ");
      Answer = toupper( getch() );
      putch( Answer ) ;

      switch( Answer )
      {
          /*===========================================================*/
          case 'M':
          /*===========================================================*/
              System = METRIC;
              gotoxy( 10, 14 );
              printf( "MEASUREMENT SYSTEM IS : MILLIMETERS" );
              Continue = FALSE;
          break;

          /*===========================================================*/
          case 'I':
          /*===========================================================*/
              System = IMPERIAL;
              gotoxy( 10, 14 );
              printf( "MEASUREMENT SYSTEM IS : INCHES" );
              Continue = FALSE;
          break;


          /*===========================================================*/
          /* REJECT ALL OTHER CHARACTERS                               */
          /*===========================================================*/
              default:
```

Appendix 11

```c
    /*==========================================================*/
        beep();
        Continue = TRUE;
        break;
    }

   }while( Continue == TRUE );

    /*==========================================================*/
    /* CLEAN USED PART OF THE SCREEN                            */
    /*==========================================================*/
    gotoxy( 1, 15 );
    printf( "                                                \n" );
    printf( "                                                \n" );
    printf( "                                                \n" );
    printf( "                                                \n" );
    printf( "                                                \n" );
    printf( "                                                \n" );

    return( System );
}


/************************************************************************/
void make_gcode_for_surface
    (
        char *CutterLocationFile,
        char *GCodeFile
    )
/************************************************************************/
{
    int     Procedure;

    FILE    *fCutterLocationFile,
            *fGCodeFile;

    float   ZClear,
            XStart,
            YStart,
            ZStart,
            FeedRate;

    int     SpindleSpeed;

    int     System;

    int     CodeLine;

    /*==========================================================*/
    /* OPEN THE CUTTER LOCATION FOR INPUT AND THE G-CODE FILE   */
    /* FOR OUTPUT                                               */
    /*==========================================================*/
    fCutterLocationFile = open_input_file(  CutterLocationFile );
    fGCodeFile          = open_output_file( GCodeFile );


    /*==========================================================*/
    /* READ THE SET UP DATA                                     */
```

MAKEGCDC - Translate a Cutter Location file into G-Code

```c
/*================================================================*/
gotoxy( 10, 7 );
ZClear = (float) get_one_double_value
                ( "ENTER CLEARANCE OVER ZERO:  " );

gotoxy( 10, 8 );
XStart = (float) get_one_double_value
                ( "ENTER X START LOCATION VALUE:  " );

gotoxy( 10, 9 );
YStart = (float) get_one_double_value
                ( "ENTER Y START LOCATION VALUE:  " );

gotoxy( 10, 10 );
ZStart = (float) get_one_double_value
                ( "ENTER Z START LOCATION VALUE:  " );

gotoxy( 10, 11 );
FeedRate = (float) get_one_double_value
                ( "ENTER FEED RATE:              " );

gotoxy( 10, 12 );
SpindleSpeed = (int) get_one_double_value
                ( "SPINDLE SPEED:                " );

/*================================================================*/
/* GET THE CUTTING PROCEDURE                                      */
/*================================================================*/
Procedure = get_machining_procedure( );


/*================================================================*/
/* GET THE MEASUREMENT SYSTEM                                     */
/*================================================================*/
System = get_measurement_system( );

/*================================================================*/
/* WRITE THE HEADER INTO G-CODE FILE                              */
/*================================================================*/
CodeLine = header( fGCodeFile, XStart, YStart, ZStart,
                   SpindleSpeed, System );

switch( Procedure )
{
    /*============================================================*/
    case FORWARD:
    /*============================================================*/
        CodeLine = make_gcode_forward( fCutterLocationFile, fGCodeFile,
                                ZClear, FeedRate, CodeLine );
        break;

    /*============================================================*/
    case BACKWARD:
    /*============================================================*/
        CodeLine = make_gcode_backward( fCutterLocationFile,
                   fGCodeFile, ZClear, FeedRate, CodeLine );
        break;
```

```c
        /*==============================================================*/
        case ALTERNATING:
        /*==============================================================*/
            CodeLine = make_gcode_alternating( fCutterLocationFile,
                    fGCodeFile, ZClear, FeedRate, CodeLine );
        break;
    }

    /*==================================================================*/
    /* RETURN THE CUTTER TO THE START POSITION                          */
    /*==================================================================*/
    ender( fGCodeFile, CodeLine, XStart, YStart, ZStart, ZClear );

}

/**********************************************************************/
int make_gcode_forward
    (
        FILE    *fCutterLocationFile,
        FILE    *fGCodeFile,
        float   ZClear,
        float   FeedRate,
        int     CodeLine
    )
/**********************************************************************/
{
    int NLines,
        Line,
        NVertices;

    float *X,
          *Y,
          *Z;

    /*==================================================================*/
    /* ALLOCATE MEMORY FOR THE VERTICES                                 */
    /*==================================================================*/
    X = (float *) malloc( (unsigned) ( sizeof(float) * MAX_VERTICES ) );
    Y = (float *) malloc( (unsigned) ( sizeof(float) * MAX_VERTICES ) );
    Z = (float *) malloc( (unsigned) ( sizeof(float) * MAX_VERTICES ) );
    if( (X==NULL) || (Y==NULL) ||(Z==NULL) )
    {
        printf( "ERROR: OUT OF MEMORY" );
        exit( 1 );
    }

    /*==================================================================*/
    /* READ THE NUMBER OF LINES                                         */
    /*==================================================================*/
    fscanf( fCutterLocationFile, "%d", &NLines );

    /*==================================================================*/
    /* PREPARE SCREEN FOR PROCESS INFO                                  */
    /*==================================================================*/
    gotoxy( 10, 22 );
    printf( "THERE ARE %d LINES TO PROCESS" );
    gotoxy( 10, 23 );
    printf( "PROCESSING LINE" );
```

MAKEGCDC - Translate a Cutter Location file into G-Code

```c
   /*================================================================*/
   /* LOOP THROUGH ALL LINES                                         */
   /*================================================================*/
   for( Line=0 ; Line<NLines ; Line++ )
   {
      /*=============================================================*/
      /* UPDATE LINE IN PROCESS                                      */
      /*=============================================================*/
      gotoxy( 25, 23 );
      printf( "%4d", Line+1 );

      /*=============================================================*/
      /* READ ONE LINE OF CUTTER LOCATION VERTICES                   */
      /*=============================================================*/
      read_one_line( X, Y, Z, &NVertices, fCutterLocationFile );

      /*=============================================================*/
      /* MAKE THE G-CODE FOR MACHINING THE LINE FORWARD              */
      /*=============================================================*/
      CodeLine = make_gcode_forward_one_line( fGCodeFile, X, Y, Z,
                  NVertices, ZClear, CodeLine, FeedRate, G00_START );

   }

   return( CodeLine );

}
/******************************************************************/
int  make_gcode_backward
   (
      FILE    *fCutterLocationFile,
      FILE    *fGCodeFile,
      float   ZClear,
      float   FeedRate,
      int     CodeLine
   )
/******************************************************************/
{
   int NLines,
       Line,
       NVertices;

   float *X,
         *Y,
         *Z;

   /*================================================================*/
   /* ALLOCATE MEMORY FOR THE VERTICES                               */
   /*================================================================*/
   X = (float *) malloc( (unsigned) ( sizeof(float) * MAX_VERTICES ) );
   Y = (float *) malloc( (unsigned) ( sizeof(float) * MAX_VERTICES ) );
   Z = (float *) malloc( (unsigned) ( sizeof(float) * MAX_VERTICES ) );
   if( (X==NULL) || (Y==NULL) ||(Z==NULL) )
   {
      printf( "ERROR: OUT OF MEMORY" );
      exit( 1 );
```

```c
      }

      /*======================================================================*/
      /* READ THE NUMBER OF LINES                                             */
      /*======================================================================*/
      fscanf( fCutterLocationFile, "%d", &NLines );

      /*======================================================================*/
      /* PREPARE SCREEN FOR PROCESS INFO                                      */
      /*======================================================================*/
      gotoxy( 10, 20 );
      printf( "THERE ARE %d LINES TO PROCESS" );
      gotoxy( 10, 21 );
      printf( "PROCESSING LINE" );

      /*======================================================================*/
      /* LOOP THROUGH ALL LINES                                               */
      /*======================================================================*/
      for( Line=0 ; Line<NLines ; Line++ )
      {
         /*===================================================================*/
         /* UPDATE LINE IN PROCESS                                            */
         /*===================================================================*/
         gotoxy( 25, 21 );
         printf( "%4d", Line+1 );

         /*===================================================================*/
         /* READ ONE LINE OF CUTTER LOCATION VERTICES                         */
         /*===================================================================*/
         read_one_line( X, Y, Z, &NVertices, fCutterLocationFile );

         /*===================================================================*/
         /* MAKE THE G-CODE FOR MACHINING THE LINE BACKWARD                   */
         /*===================================================================*/
         CodeLine = make_gcode_backward_one_line( fGCodeFile, X, Y, Z,
                    NVertices, ZClear, CodeLine, FeedRate, G00_START );

      }

      return( CodeLine );

}
/***************************************************************************/
int   make_gcode_alternating
         (
            FILE    *fCutterLocationFile,
            FILE    *fGCodeFile,
            float   ZClear,
            float   FeedRate,
            int     CodeLine
         )
/***************************************************************************/
{
   int NLines,
       Line,
       NVertices;
```

MAKEGCDC - Translate a Cutter Location file into G-Code

```
    float *X,
          *Y,
          *Z;

    /*===========================================================*/
    /* ALLOCATE MEMORY FOR THE VERTICES                          */
    /*===========================================================*/
    X = (float *) malloc( (unsigned) ( sizeof(float) * MAX_VERTICES ) );
    Y = (float *) malloc( (unsigned) ( sizeof(float) * MAX_VERTICES ) );
    Z = (float *) malloc( (unsigned) ( sizeof(float) * MAX_VERTICES ) );
    if( (X==NULL) || (Y==NULL) ||(Z==NULL) )
    {
       printf( "ERROR: OUT OF MEMORY" );
       exit( 1 );
    }

    /*===========================================================*/
    /* READ THE NUMBER OF LINES                                  */
    /*===========================================================*/
    fscanf( fCutterLocationFile, "%d", &NLines );

    /*===========================================================*/
    /* PREPARE SCREEN FOR PROCESS INFO                           */
    /*===========================================================*/
    gotoxy( 10, 20 );
    printf( "THERE ARE %d LINES TO PROCESS" );
    gotoxy( 10, 21 );
    printf( "PROCESSING LINE" );

    /*===========================================================*/
    /* MAKE FIRST LINE WITH G00 START                            */
    /*===========================================================*/
    Line=0;

    /*===========================================================*/
    /* UPDATE LINE IN PROCESS                                    */
    /*===========================================================*/
    gotoxy( 25, 21 );
    printf( "%4d", Line+1 );

    /*===========================================================*/
    /* READ ONE LINE OF CUTTER LOCATION VERTICES                 */
    /*===========================================================*/
    read_one_line( X, Y, Z, &NVertices, fCutterLocationFile );

    /*===========================================================*/
    /* MAKE THE G-CODE FOR MACHINING THE LINE FORWARD            */
    /*===========================================================*/
    CodeLine = make_gcode_forward_one_line( fGCodeFile, X, Y, Z,
                   NVertices,ZClear, CodeLine, FeedRate,G00_START );
    Line++;

    /*===========================================================*/
    /* LOOP THROUGH ALL LINES                                    */
    /*===========================================================*/
    while( Line < NLines )
    {
       /*===========================================================*/
```

Appendix 11

```c
        /* UPDATE LINE IN PROCESS                                        */
        /*==============================================================*/
        gotoxy( 25, 21 );
        printf( "%4d", Line+1 );

        /*==============================================================*/
        /* READ ONE LINE OF CUTTER LOCATION VERTICES                    */
        /*==============================================================*/
        read_one_line( X, Y, Z, &NVertices, fCutterLocationFile );

        /*==============================================================*/
        /* MAKE THE G-CODE FOR MACHINING THE LINE BACKWARD              */
        /*==============================================================*/
        CodeLine = make_gcode_backward_one_line( fGCodeFile, X, Y, Z,
                    NVertices, ZClear, CodeLine, FeedRate, G01_START );
        Line++;

        if( Line < NLines )
        {
            /*==============================================================*/
            /* UPDATE LINE IN PROCESS                                        */
            /*==============================================================*/
            gotoxy( 25, 21 );
            printf( "%4d", Line+1 );

            /*==============================================================*/
            /* READ ONE LINE OF CUTTER LOCATION VERTICES                    */
            /*==============================================================*/
            read_one_line( X, Y, Z, &NVertices, fCutterLocationFile );

            /*==============================================================*/
            /* MAKE THE G-CODE FOR MACHINING THE LINE FORWARD               */
            /*==============================================================*/
            make_gcode_forward_one_line( fGCodeFile, X, Y, Z, NVertices,
                            ZClear, CodeLine, FeedRate, G01_START );
            Line++;
        }

    }

    return( CodeLine );

}
/************************************************************************/
int header
    (
        FILE    *fGCodeFile,
        float   XStart,
        float   YStart,
        float   ZStart,
        int     SpindleSpeed,
        int     MetricImperial
    )
/************************************************************************/
{
```

```c
    /*===================================================================*/
    /* ONE EMPTY LINE                                                    */
    /*===================================================================*/
    fprintf(fGCodeFile, "%\n" );

    /*===================================================================*/
    /* STANDARD FANUC START SEQUENCE                                     */
    /*===================================================================*/
    fprintf(fGCodeFile, "N1 O23;\n" );

    /*===================================================================*/
    /* PROGRAM START COORDINATES                                         */
    /*===================================================================*/
    fprintf(fGCodeFile, "N2 G92 X%5.3f Y%5.3f Z%5.3f;\n",
                        XStart, YStart, ZStart );


    if( MetricImperial == METRIC )
    {
        /*===============================================================*/
        /* THE DATA IS IN METRIC UNITS                                   */
        /*===============================================================*/
        fprintf(fGCodeFile, "N3 G21;\n" );
    }
    else
    {
        /*===============================================================*/
        /* THE DATA IS IN IMPERIAL UNITS                                 */
        /*===============================================================*/
        fprintf(fGCodeFile, "N3 G20;\n" );
    }

    /*===================================================================*/
    /* THE DATA COMING IS IN ABSOLUTE FORMAT                             */
    /*===================================================================*/
    fprintf(fGCodeFile, "N4 G90;\n" );

    /*===================================================================*/
    /* SET SPINDLE SPEED AND SWITCH SPINDLE ON                           */
    /*===================================================================*/
    fprintf(fGCodeFile, "N5 S%4d M03;\n", SpindleSpeed );

    return( 6 );
}
/*********************************************************************/
int ender
    (
        FILE    *fGCodeFile,
        int     CodeLine,
        float   XStart,
        float   YStart,
        float   ZStart,
        float   ZClear
    )
/*********************************************************************/
{
```

Appendix 11

```c
    /*====================================================================*/
    /* LIFT CUTTER TO CLEARANCE HEIGHT                                    */
    /*====================================================================*/
    fprintf(fGCodeFile, "N%d G00 Z%6.4f;\n",         CodeLine++, ZClear );

    /*====================================================================*/
    /* MOVE CUTTER TO START COORDINATES X & Y                             */
    /*====================================================================*/
    fprintf(fGCodeFile, "N%d G00 X%6.4f Y%6.4f;\n", CodeLine++,
            XStart, YStart  );

    /*====================================================================*/
    /* MOVE CUTTER TO START COORDINATE X                                  */
    /*====================================================================*/
    fprintf(fGCodeFile, "N%d G00 Z%6.4f;\n",         CodeLine++, ZStart );

    /*====================================================================*/
    /* SPINDLE OFF                                                        */
    /*====================================================================*/
    fprintf(fGCodeFile, "N%d M05;\n",                CodeLine++ );

    /*====================================================================*/
    /* END OF G-CODE                                                      */
    /*====================================================================*/
    fprintf(fGCodeFile, "N%d M30;\n",                CodeLine );

    fprintf(fGCodeFile, "%\n");
}
/**********************************************************************/
int gcode_begin_line_g00
    (
    FILE    *fGCodeFile,
    float   X,
    float   Y,
    float   Z,
    float   ZClear,
    float   FeedRate,
    int     CodeLine
    )
/**********************************************************************/
{
    /*====================================================================*/
    /* GO FAST TO ZCLEAR                                                  */
    /*====================================================================*/
    fprintf(fGCodeFile, "N%d G00 Z%6.4f;\n",         CodeLine++, ZClear );

    /*====================================================================*/
    /* GO FAST TO DIRECTLY OVER NEXT VERTEX                               */
    /*====================================================================*/
    fprintf(fGCodeFile, "N%d G00 X%6.4f Y%6.4f;\n", CodeLine++, X, Y );

    /*====================================================================*/
    /* GO SLOW DOWN TO NEXT VERTEX                                        */
    /*====================================================================*/
    fprintf(fGCodeFile, "N%d G01 Z%6.4f F%3.1f;\n", CodeLine++, Z,
            FeedRate );
```

MAKEGCDC - Translate a Cutter Location file into G-Code

```
      return( CodeLine );
}
/*******************************************************************/
int  make_gcode_forward_one_line
     (
         FILE   *fGCodeFile,
         float  *X,
         float  *Y,
         float  *Z,
         int    NVrtcs,
         float  ZClear,
         int    CodeLine,
         float  FeedRate,
         int    StartCondition
     )
/*******************************************************************/
{
   register int Vrtx;

   /*===============================================================*/
   /* MANY CONTROLLERS ONLY ACCEPT LINE NUBER UP 9999               */
   /*===============================================================*/
   if( CodeLine > 9000 )
   {
      CodeLine = 1;
   }

   /*===============================================================*/
   /* START WITH G00 IF FASTSTART IS TRUE                           */
   /*===============================================================*/
   Vrtx = 0;

   if( StartCondition == G00_START )
   {
      /*===========================================================*/
      /* START LINE WITH G00 TO FIRST VERTEX                       */
      /*===========================================================*/
      CodeLine = gcode_begin_line_g00( fGCodeFile,
                  X[Vrtx], Y[Vrtx], Z[Vrtx], ZClear,
                  FeedRate, CodeLine );
   }
   else
   {
      /*===========================================================*/
      /* GO SLOW TO FIRST VERTEX                                   */
      /*===========================================================*/
      fprintf( fGCodeFile, "N%d G01 X%6.4f Y%6.4f Z%6.4f;\n",
                         CodeLine++, X[Vrtx], Y[Vrtx],Z[Vrtx] );
   }

   Vrtx++;

   /*===============================================================*/
   /* LOOP THROUGH ALL VERTICES AND WRITE G01'S                     */
   /*===============================================================*/
   for( ; Vrtx<NVrtcs ; Vrtx++ )
   {
```

```c
      /*==========================================================*/
      /* GO SLOW TO NEXT VERTEX                                   */
      /*==========================================================*/
      fprintf(fGCodeFile, "N%d G01 X%6.4f Y%6.4f Z%6.4f;\n",
               CodeLine++, X[Vrtx],    Y[Vrtx],    Z[Vrtx] );
   }


   return( CodeLine );
}
/**********************************************************************/
int   make_gcode_backward_one_line
      (
         FILE   *fGCodeFile,
         float  *X,
         float  *Y,
         float  *Z,
         int    NVrtcs,
         float  ZClear,
         int    CodeLine,
         float  FeedRate,
         int    StartCondition
      )
/**********************************************************************/
{
   register int Vrtx;

   /*==========================================================*/
   /* MANY CONTROLLERS ONLY ACCEPT LINE NUBER UP 9999          */
   /*==========================================================*/
   if( CodeLine > 9000 )
   {
      CodeLine = 1;
   }

   /*==========================================================*/
   /* START WITH G00 IF FASTSTART IS TRUE                      */
   /*==========================================================*/
   Vrtx = NVrtcs-1;

   if( StartCondition == G00_START )
   {
      /*==========================================================*/
      /* GO FAST TO LAST VERTEX                                   */
      /*==========================================================*/
      CodeLine = gcode_begin_line_g00( fGCodeFile, X[Vrtx], Y[Vrtx],
                     Z[Vrtx], ZClear,  FeedRate, CodeLine );
   }
   else
   {
      /*==========================================================*/
      /* GO SLOW TO LAST VERTEX                                   */
      /*==========================================================*/
      fprintf( fGCodeFile, "N%d G01 X%6.4f Y%6.4f Z%6.4f;\n",
                     CodeLine++, X[Vrtx], Y[Vrtx],Z[Vrtx] );
   }
```

```
      Vrtx--;

      /*================================================================*/
      /* LOOP THROUGH ALL VERTICES AND WRITE G01'S                      */
      /*================================================================*/
      for(  ; Vrtx>=0 ; Vrtx-- )
      {
         /*=============================================================*/
         /* GO SLOW TO PREVIOUS VERTEX                                  */
         /*=============================================================*/
         fprintf(fGCodeFile, "N%d G01 X%6.4f Y%6.4f Z%6.4f;\n",
                  CodeLine++, X[Vrtx],   Y[Vrtx],    Z[Vrtx] );
      }

      return( CodeLine );
}
/*******************************************************************/
void read_one_line
    (
        float   *X,
        float   *Y,
        float   *Z,
        int     *NVertices,
        FILE    *fin
    )
/*******************************************************************/
{
   register int    Vertex;

   /*===================================================================*/
   /* READ THE NUMBER OF VERTICES ON THIS LINE                          */
   /*===================================================================*/
   fscanf( fin, "%d", NVertices );

   /*===================================================================*/
   /* READ THE VALUES INTO THE ALLOCATED MEMORY                         */
   /*===================================================================*/
   for( Vertex=0  ; Vertex<*NVertices ; Vertex++ )
   {
      fscanf( fin, "%f %f %f", &(X[Vertex]),
                               &(Y[Vertex]),
                               &(Z[Vertex]) );
   }

}
```

Appendix 12
C-Program: INVNORM - Invert the Surface Normal Vector

```c
/**********************************************************************/
/*              I N V N O R M . C                                     */
/**********************************************************************/

#include <stdio.h>
#include <math.h>
#include <stdlib.h>
#include <ctype.h>
#include <conio.h>
#include <dos.h>

#include <library.h>


/**********************************************************************/
/* PROTOYPES OF FUNCTIONS                                             */
/**********************************************************************/

void information
    (
        char **FileNames
    );


void invert_normal_vector
    (
        char    *InputFile,
        char    *OutputFile
    );

/**********************************************************************/
main
    (
        int NumberOfParameters,
        char **ParameterStrings
    )
/**********************************************************************/
{

    /*================================================================*/
    /* CHECK FOR RIGHT NUMBER OF PARAMETERS BEING PASSED              */
    /*================================================================*/
    check_parameters_two( NumberOfParameters, ParameterStrings );

    /*================================================================*/
```

```
    /* DISPLAY INFORMATION ABOUT THE PROGRAM                          */
    /*================================================================*/
    information( ParameterStrings );

    /*================================================================*/
    /* CALCULATE THE OFFSETS FOR ONE LINE                             */
    /*================================================================*/
    invert_normal_vector( ParameterStrings[1], ParameterStrings[2] );

    printf( "\n\nPROGRAM SUCCESSFULLY COMPLETED\n" );
}
/******************************************************************/
void information
    (
        char **FileNames
    )
/******************************************************************/
{
    /*================================================================*/
    /* CLEAR THE SCREEN                                               */
    /*================================================================*/
    clrscr( );

    /*================================================================*/
    /* DISPLAY THE INFORMATION ABOUT THE PROGRAM                      */
    /*================================================================*/
    printf( "         PROGRAM *** I N V N O R M ***\n");
    printf( "         INVERTS THE SURFACE NORMAL VECTOR\n\n");

    printf( " INPUT FILE NAME  = %s\n", FileNames[1] );
    printf( " OUTPUT FILE NAME = %s\n", FileNames[2] );
}
/******************************************************************/
void  invert_normal_vector
    (
        char    *InputFile,
        char    *OutputFile
    )
/******************************************************************/
{
    register int    Line,
                    Vertex;

    FILE    *fIn,
            *fOut;

    double  NX,
            NY,
            NZ;

    int     NLines,
            NVertices;

    /*================================================================*/
```

Appendix 12

```c
/* OPEN THE INPUT AND THE OUTPUT FILE                             */
/*===============================================================*/
fIn  = open_input_file(  InputFile );
fOut = open_output_file( OutputFile );


/*===============================================================*/
/* READ AND WRITE THE NUMBER OF LINES IN THIS FILE               */
/*===============================================================*/
fscanf( fIn,  "%d",   &NLines );
fprintf( fOut, "%d\n",  NLines );


/*===============================================================*/
/* DISPLAY HOW MANY LINES TO PROCESS                             */
/*===============================================================*/
gotoxy( 10, 10 );
printf("%d LINES TO PROCESS\n", NLines);
gotoxy( 10, 11 );
printf( "PROCESSING LINE # " );


/*===============================================================*/
/* LOOP THROUGH EVERY LINE                                       */
/*===============================================================*/
for( Line=0 ; Line<NLines ; Line++ )
{
   /*===========================================================*/
   /* UPDATE THE SCREEN - SHOW WHICH LINE NUMBER IN PROCESS     */
   /*===========================================================*/
   gotoxy( 27, 11 );
   printf("%4d", Line+1);


   /*===========================================================*/
   /* READ AND WRITE THE NUMBER OF VERTICES IN THIS LINE        */
   /*===========================================================*/
   fscanf( fIn, "%d",   &NVertices );
   fprintf( fOut,     "%d\n",  NVertices );


   /*===========================================================*/
   /* LOOP THROUGH ALL VERTICES ON THIS LINE                    */
   /*===========================================================*/
   for( Vertex=0 ; Vertex<NVertices ; Vertex++ )
   {
      /*========================================================*/
      /* READ THE NORMAL VECTOR                                 */
      /*========================================================*/
      fscanf( fIn, "%lf %lf %lf", &NX, &NY, &NZ );

      /*========================================================*/
      /* WRITE THE NORMAL VECTOR INVERTED TO THE OUTPUT FILE    */
      /*========================================================*/
      fprintf( fOut, "%10.6lf %10.6lf %10.6lf\n", -NX, -NY, -NZ );
   }
}
}
```

Appendix 13
C-Program: TRANSPOS - Transpose the Surface Matrix

```c
/*******************************************************************/
/*                T R A N S P O S . C                              */
/*******************************************************************/

#include <stdio.h>
#include <math.h>
#include <stdlib.h>
#include <ctype.h>
#include <conio.h>
#include <dos.h>
#include <alloc.h>

/*=================================================================*/
/* INCLUDE FILE FOR OUR OWN LIBRARY FUNCTIONS                      */
/*=================================================================*/
#include <library.h>


/*******************************************************************/
/* PROTYPES OF FUNCTIONS                                           */
/*******************************************************************/
void information
    (
        char **FileNames
    );

void  transpose_surface
    (
        char    *SurfaceFile,
        char    *TransposedFile
    );

void write_file_transposed
    (
        float **X,
        float **Y,
        float **Z,
        int    NLines,
        int    NVrtcs,
        char   *FileName
    );


/*******************************************************************/
```

Appendix 13

```c
main
   (
       int NumberOfParameters,
       char **ParameterStrings
   )
/************************************************************************/
{

   /*====================================================================*/
   /* CHECK FOR RIGHT NUMER OF PARAMETERS BEING PASSED                   */
   /*====================================================================*/
   check_parameters_two( NumberOfParameters, ParameterStrings );

   /*====================================================================*/
   /* DISPLAY INFORMATION ABOUT THE PROGRAM                              */
   /*====================================================================*/
   information( ParameterStrings );

   /*====================================================================*/
   /* CALCULATE THE OFFSETS FOR ONE LINE                                 */
   /*====================================================================*/
   transpose_surface( ParameterStrings[1], ParameterStrings[2] );

   printf( "\n\nPROGRAM SUCCESSFULLY COMPLETED\n" );

}
/************************************************************************/
void information
     (
         char **FileNames
     )
/************************************************************************/
{
   /*====================================================================*/
   /* CLEAR THE SCREEN                                                   */
   /*====================================================================*/
   clrscr( );

   /*====================================================================*/
   /* DISPLAY THE INFORMATION ABOUT THE PROGRAM                          */
   /*====================================================================*/
   printf( "         PROGRAM  ***  T R A N S P O S  ***\n");
   printf( "   TRANSPOSES THE SURFACE OF THE INPUT FILE \n\n");

   printf( "   SURFACE FILE NAME      = %s\n", FileNames[1] );
   printf( "   TRANSPOSED FILE NAME   = %s\n", FileNames[2] );
}
/************************************************************************/
void transpose_surface
       (
           char    *SurfaceFile,
           char    *TransFile
       )
/************************************************************************/
```

TRANSPOSE - Transpose the Surface Matrix

```
{
    float   **XSrfc,
            **YSrfc,
            **ZSrfc;
    int     NLinesSrfc,
            NVrtcsSrfc;


    /*================================================================*/
    /* READ THE SURFACE FILE                                          */
    /*================================================================*/
    read_file_allocate( &XSrfc, &YSrfc, &ZSrfc, &NLinesSrfc, &NVrtcsSrfc,
                        SurfaceFile );

    /*================================================================*/
    /* WRITE THE SURFACE TRANSPOSED TO A NEW FILE                     */
    /*================================================================*/
    write_file_transposed( XSrfc, YSrfc, ZSrfc, NLinesSrfc, NVrtcsSrfc,
                           TransFile );

}

/******************************************************************/
void write_file_transposed
    (
        float   **X,
        float   **Y,
        float   **Z,
        int     NLines,
        int     NVrtcs,
        char    *FileName
    )
/******************************************************************/
{
    register int Line,
                 Vrtx;

    FILE *fTrans;

    /*================================================================*/
    /* OPEN TRANSPOSED FILE FOR OUTPUT                                */
    /*================================================================*/
    fTrans = open_output_file( FileName );

    /*================================================================*/
    /* WRITE THE NUMBER OF VERTICES WHICH WILL BE LINES NOW           */
    /*================================================================*/
    fprintf( fTrans, "%d\n", NVrtcs );

    gotoxy( 5, 12 );
    printf( "WRITING %s: %d LINES TO WRITE", FileName, NVrtcs );
    gotoxy( 5, 13 );
    printf( "WRITING LINE: " );

    /*================================================================*/
    /* LOOP THROUGH EVERY VERTEX                                      */
    /*================================================================*/
```

```
    for( Vrtx=0 ; Vrtx<NVrtcs ; Vrtx++ )
    {
       /*===============================================================*/
       /* UPDATE THE SCREEN - SHOW WHICH VERTEX NUMBER IN PROCESS      */
       /*===============================================================*/
       gotoxy( 20, 13 );
       printf("%4d", Vrtx+1);

       /*===============================================================*/
       /*    WRITE THE NUMBER OF VERTICES IN THIS LINE                 */
       /*===============================================================*/
       fprintf( fTrans, "%d\n",  NLines );

       /*===============================================================*/
       /* LOOP THROUGH ALL LINES ON THIS LINE                          */
       /*===============================================================*/
       for( Line=0 ; Line<NLines ; Line++)
       {

          /*===============================================================*/
          /* WRITE VERTICES TRANPOSED TO THE FILE                         */
          /*===============================================================*/
          fprintf( fTrans, "%10.6f %10.6f %10.6f\n",
                           X[Line][Vrtx],
                                Y[Line][Vrtx],
                                     Z[Line][Vrtx] );
       }

    }

}
```

Appendix 14
C-Program: RSKIP - Reduce Surface Density by Skipping Lines and Vertices

```
/********************************************************************/
/*                S K I P . C                                       */
/********************************************************************/

#include <stdio.h>
#include <math.h>
#include <stdlib.h>
#include <ctype.h>
#include <conio.h>
#include <dos.h>

#include <library.h>

#define SKIP      TRUE
#define NO_SKIP   FALSE


/********************************************************************/
/* PROTOYPES OF FUNCTIONS                                           */
/********************************************************************/
void information
    (
        char **FileNames
    );


void  skip
    (
        char    *InputFile,
        char    *OutputFile,
        int     SkipNLines,
        int     SkipNVrtcs
    );

int how_many_new_values
    (
        int NOldValues,
        int NSkip
    );

int skip_this_one
    (
        int LastNoSkip,
        int Now,
```

```
            int NSkip,
            int NValues
        );

    void lines_and_vertices_to_skip
        (
            int *SkipNLines,
            int *SkipNVrtcs
        );
/***********************************************************************/
main
    (
        int NumberOfParameters,
        char **ParameterStrings
    )
/***********************************************************************/
{
    int SkipNLines,
        SkipNVrtcs;

    /*===================================================================*/
    /* CHECK FOR RIGHT NUMBER OF PARAMETERS BEING PASSED                 */
    /*===================================================================*/
    check_parameters_two( NumberOfParameters, ParameterStrings );

    /*===================================================================*/
    /* DISPLAY INFORMATION ABOUT THE PROGRAM                             */
    /*===================================================================*/
    information( ParameterStrings );

    /*===================================================================*/
    /* GET FROM USER HOW MANY LINES AND VERTICES TO SKIP                 */
    /*===================================================================*/
    lines_and_vertices_to_skip( &SkipNLines, &SkipNVrtcs );

    /*===================================================================*/
    /* CALCULATE THE OFFSETS FOR ONE LINE                                */
    /*===================================================================*/
    skip( ParameterStrings[1], ParameterStrings[2], SkipNLines,
          SkipNVrtcs );

    printf( "\n\nPROGRAM SUCCESSFULLY COMPLETED\n" );

}

/***********************************************************************/
void information
    (
        char **FileNames
    )
/***********************************************************************/
{
    /*===================================================================*/
    /* CLEAR THE SCREEN                                                  */
    /*===================================================================*/
    clrscr( );
```

RSKIP - Reduce Surface Density by Skipping Lines and Vertices

```c
    /*==============================================================*/
    /* DISPLAY THE INFORMATION ABOUT THE PROGRAM                    */
    /*==============================================================*/
    printf( "          PROGRAM *** S K I P ***\n");
    printf( "SKIPS THE SPECIFIED NUMBER OF LINES AND VERTICES \n\n");

    printf( " INPUT FILE NAME = %s\n", FileNames[1] );
    printf( " OUTPUT FILE NAME = %s\n", FileNames[2] );
}
/****************************************************************/
void lines_and_vertices_to_skip
     (
        int *SkipNLines,
        int *SkipNVrtcs
     )
/****************************************************************/
{
   char Answer = 'Y';

   do
   {
      /*==============================================================*/
      /* CLEAR OLD COMMENTS OFF THE SCREEN                            */
      /*==============================================================*/
      gotoxy( 10, 10 );
      printf( "                                                  " );

      /*==============================================================*/
      /* ASK HOW MANY LINES TO SKIP                                   */
      /*==============================================================*/
      gotoxy( 10, 10 );
      printf( "HOW MANY LINES TO SKIP? " );
      scanf( "%d", SkipNLines );

      /*==============================================================*/
      /* ASK HOW MANY VERTICES TO SKIP                                */
      /*==============================================================*/
      gotoxy( 10, 10 );
      printf( "HOW MANY VERTICES TO SKIP? " );
      scanf( "%d", SkipNVrtcs );

      /*==============================================================*/
      /* DISPLAY ENTERED VALUES                                       */
      /*==============================================================*/
      gotoxy( 10, 10 );
      printf( "SKIPPING %3d LINES AND %d VERTICES", *SkipNLines,
                                                    *SkipNVrtcs );

      /*==============================================================*/
      /* MAKE SURE THESE VALUES ARE OK                                */
      /*==============================================================*/
      gotoxy( 10, 11 );
      printf( "DATA OK?" );
      Answer = toupper( getch() );
      gotoxy( 10, 11 );
      printf( "                                   " );
```

```
   }while( Answer != 'Y' );
}
/**********************************************************************/
void  skip
      (
         char   *InputFile,
         char   *OutputFile,
         int    SkipNLines,
         int    SkipNVrtcs
      )
/**********************************************************************/
{
   register int    Line,
                   Vrtx;

   FILE    *fIn,
           *fOut;

   double  X,
           Y,
           Z;

   int     NOldLines,
           NOldVrtcs;

   int     NNewLines,
           NNewVrtcs;

   int     LastLineWritten = -9999,
           LastVrtxWritten = -9999;

   /*===================================================================*/
   /* OPEN THE INPUT AND THE OUTPUT FILE                                */
   /*===================================================================*/
   fIn  = open_input_file(  InputFile );
   fOut = open_output_file( OutputFile );

   /*===================================================================*/
   /* READ THE NUMBER OF LINES IN THIS FILE                             */
   /*===================================================================*/
   fscanf(  fIn,   "%d",    &NOldLines );

   /*===================================================================*/
   /* CALCULATE HOW MANY LINES WILL BY IN THIS FILE                     */
   /*===================================================================*/
   NNewLines = how_many_new_values( NOldLines, SkipNLines );
   fprintf( fOut, "%d\n",  NNewLines );

   /*===================================================================*/
   /* DISPLAY HOW MANY LINES TO PROCESS                                 */
   /*===================================================================*/
   gotoxy( 10, 13 );
   printf("%d LINES TO PROCESS\n", NOldLines);
   gotoxy( 10, 14 );
   printf( "PROCESSING LINE # " );
```

RSKIP - Reduce Surface Density by Skipping Lines and Vertices

```c
/*==================================================================*/
/* LOOP THROUGH EVERY LINE                                          */
/*==================================================================*/
for( Line=0 ; Line<NOldLines ; Line++ )
{
    /*==============================================================*/
    /* UPDATE THE SCREEN - SHOW WHICH LINE NUMBER IN PROCESS        */
    /*==============================================================*/
    gotoxy( 27, 14 );
    printf("%4d", Line+1);

    /*==============================================================*/
    /* READ THE NUMBER OF VERTICES IN THIS LINE                     */
    /*==============================================================*/
    fscanf( fIn, "%d",   &NOldVrtcs );

    /*==============================================================*/
    /* CHECK IF THIS LINE IS TO SKIP, THEN ONLY READ DON'T WRITE    */
    /*==============================================================*/
    if( skip_this_one( LastLineWritten, Line, SkipNLines, NOldLines )
        == SKIP )
    {
        /*==========================================================*/
        /* LOOP THROUGH ALL VERTICES ON THIS LINE                   */
        /*==========================================================*/
        for( Vrtx=0 ; Vrtx<NOldVrtcs ; Vrtx++ )
        {
            /*======================================================*/
            /* READ THE NORMAL VECTOR                               */
            /*======================================================*/
            fscanf( fIn, "%lf %lf %lf", &X, &Y, &Z );
        }
    }
    /*==============================================================*/
    /* THIS LINE IS NOT TO BE SKIPPED, SO READ AND WRITE THE        */
    /* VERTICES                                                     */
    /*==============================================================*/
    else
    {
        /*==========================================================*/
        /* REMEMBER THAT THIS LINE WAS THE LAST WRITTEN TO THE OUTPUT*/
        /* FILE                                                     */
        /*==========================================================*/
        LastLineWritten = Line;

        /*==========================================================*/
        /* CALCULATE HOW MANY VERTICES WILL BY IN THIS LINE         */
        /*==========================================================*/
        NNewVrtcs = how_many_new_values( NOldVrtcs, SkipNVrtcs );
        fprintf( fOut,      "%d\n",  NNewVrtcs );

        /*==========================================================*/
        /* LOOP THROUGH ALL VERTICES ON THIS LINE                   */
        /*==========================================================*/
        for( Vrtx=0 ; Vrtx<NOldVrtcs ; Vrtx++ )
        {
```

```
            /*============================================================*/
            /* READ THIS VERTEX                                           */
            /*============================================================*/
            fscanf( fIn, "%lf %lf %lf", &X, &Y, &Z );

            /*============================================================*/
            /*IF THIS VERTEX IS NOT TO BE SKIPPED WRITE IT TO OUTPUT      */
            /* FILE                                                       */
            /*============================================================*/
            if( skip_this_one( LastVrtxWritten, Vrtx, SkipNVrtcs,
                NOldVrtcs )
                == NO_SKIP )
            {
                /*============================================================*/
                /* REMEMBER THAT THIS VERTEX WAS THE LAST ONE WRITTEN         */
                /* OUT                                                        */
                /*============================================================*/
                LastVrtxWritten = Vrtx;

                /*============================================================*/
                /* WRITE THIS VERTEX TO THE FILE                              */
                /*============================================================*/
                fprintf( fOut, "%10.6lf %10.6lf %10.6lf\n", X, Y, Z );
            }

        }

        /*============================================================*/
        /* MAKE SURE IN THE NEXT LINE THE FIRST VERTEX WILL BE        */
        /* WRITTEN                                                    */
        /*============================================================*/
        LastVrtxWritten = -9999;

    }

  }

}
/*******************************************************************/
int how_many_new_values
    (
        int NOldValues,
        int NSkip
    )
/*******************************************************************/
{
   register int n;

   int NNewValues;

   int LastNoSkip = -9999;

   NNewValues = 0;

   /*============================================================*/
   /* LOOP THROUGH ALL VALUES                                    */
   /*============================================================*/
```

RSKIP - Reduce Surface Density by Skipping Lines and Vertices

```
   for( n=0 ; n<NOldValues ; n++ )
   {
      /*================================================================*/
      /* IF NOT SKIPPED INCREASE COUNTER AND REMENBER WHICH ONE WAS    */
      /* LAST                                                          */
      /*================================================================*/
      if( skip_this_one( LastNoSkip, n, NSkip, NOldValues ) != SKIP )
      {
         NNewValues++;
         LastNoSkip = n;
      }
   }

   return( NNewValues );

}



/********************************************************************/
int skip_this_one
    (
        int LastNoSkip,
        int Now,
        int NSkip,
        int NValues
    )
/********************************************************************/
{
   /*================================================================*/
   /* CHECK IF STILL TO SKIP                                         */
   /*================================================================*/
   if( Now > (LastNoSkip+NSkip)   )
   {
      return( NO_SKIP );
   }
   /*================================================================*/
   /* MAKE SURE THAT THE LAST VALUE AT THE SURFACE BORDER IS ALWAYS  */
   /* USED                                                           */
   /*================================================================*/
   else if( Now == (NValues-1) )
   {
      return( NO_SKIP );
   }
   else
   {
      return( SKIP );
   }

}
```

Appendix 15
C-Program: RADDEND - Add End Vertices to Extend the Surface

```c
/**********************************************************************/
/*     R A D D E N D . C                                              */
/**********************************************************************/

#include <stdio.h>
#include <conio.h>

#include "library.h"

/**********************************************************************/
/* PROTOTYPES OF FUNCTIONS                                            */
/**********************************************************************/
void  information
      (
          char  **FileNames
      );

void  add_ends_to_surface
      (
          char    *SurfaceFile,
          char    *ExtendedFile,
          double  DistanceStart,
          double  DistanceEnd
      );

void  make_new_vertex
      (
          float   XStart,
          float   XEnd,
          float   YStart,
          float   YEnd,
          float   ZStart,
          float   ZEnd,
          float   *ExtX,
          float   *ExtY,
          float   *ExtZ,
          float   NewXLoc
      );
```

RADDEND - Add End Vertices to Extend the Surface

```c
/*****************************************************************/
main
    (
        int     NParameters,
        char    **ParameterStrings
    )
/*****************************************************************/
{
    double  DistanceStart,
            DistanceEnd;

    /*===============================================================*/
    /* CHECK IF RIGHT NUMBER OF PARAMETERS HAVE BEEN PASSED          */
    /*===============================================================*/
    check_parameters_two( NParameters, ParameterStrings );

    /*===============================================================*/
    /* DISPLAY INFORMATION ABOUT THE PROGRAM                         */
    /*===============================================================*/
    information( ParameterStrings );

    /*===============================================================*/
    /* GET THE DISTANCES FOR THE SURFACE EXTENSIONS                  */
    /*===============================================================*/
    DistanceStart = get_one_double_value( "HOW MUCH TO ADD AT START OF
                                           LINES:?" );
    DistanceEnd   = get_one_double_value( "HOW MUCH TO ADD AT END OF
                                           LINES:?" );

    /*===============================================================*/
    /* EXTEND THE SURFACE AT THE END OF THE LINES BY THE GIVEN VALUES */
    /*===============================================================*/
    add_ends_to_surface( ParameterStrings[1], ParameterStrings[2],
                         DistanceStart, DistanceEnd );

    gotoxy( 10, 22 );
    printf( "PROGRAM SUCCESSFUL\n" );
}

/*****************************************************************/
void information
    (
        char **FileNames
    )
/*****************************************************************/
{
    /*===============================================================*/
    /* CLEAR THE SCREEN                                              */
    /*===============================================================*/
    clrscr( );

    /*===============================================================*/
    /* DISPLAY THE INFORMATION ABOUT THE PROGRAM                     */
    /*===============================================================*/
    printf( "          PROGRAM *** R A D D E N D ***\n");
    printf( " EXTENDS A SURFACE BY ADDING VERTICES AT THE END OF
```

```
                        LINES\n\n");

    printf( "  UN EXTENDED SURFACE FILE = %s\n", FileNames[1] );
    printf( "     EXTENDED SURFACE FILE = %s\n", FileNames[2] );
}
/**********************************************************************/
void  add_ends_to_surface
      (
          char    *SurfaceFile,
          char    *ExtendedFile,
          double  DistanceStart,
          double  DistanceEnd
      )
/**********************************************************************/
{
    /*--------------------------------------------------------------*/
    /* VARIABLES FOR SURFACE FILE                                   */
    /*--------------------------------------------------------------*/
    FILE     *fSurfaceFile;
    double   X[2],
             Y[2],
             Z[2];
    int      NLines,
             NVrtcs;

    /*--------------------------------------------------------------*/
    /* VARIABLES FOR EXTENDED SURFACE FILE                          */
    /*--------------------------------------------------------------*/
    FILE     *fExtendedFile;
    float    ExtX,
             ExtY,
             ExtZ;

    /*--------------------------------------------------------------*/
    /* OTHER VARIABLES                                              */
    /*--------------------------------------------------------------*/
    register int      Line,
                      Vertex;

    /*==============================================================*/
    /* OPEN THE INPUT AND THE OUTPUT FILE                           */
    /*==============================================================*/
    fSurfaceFile     = open_input_file( SurfaceFile );
    fExtendedFile    = open_output_file( ExtendedFile );

    /*==============================================================*/
    /* READ AND WRITE THE NUMBER OF LINES IN THIS FILE              */
    /*==============================================================*/
    fscanf( fSurfaceFile, "%d", &NLines );
    fprintf( fExtendedFile, "%d\n", NLines );

    /*==============================================================*/
    /* DISPLAY HOW MANY LINES TO PROCESS                            */
    /*==============================================================*/
    gotoxy( 10, 17 );
    printf("%d LINES TO PROCESS\n", NLines );
    gotoxy( 10, 18 );
```

RADDEND - Add End Vertices to Extend the Surface

```c
    printf( "PROCESSING LINE # " );

    /*================================================================*/
    /* LOOP THROUGH EVERY LINE                                        */
    /*================================================================*/
    for( Line=0 ; Line<NLines ; Line++ )
    {
        /*============================================================*/
        /* UPDATE THE SCREEN - SHOW WHICH LINE NUMBER IN PROCESS      */
        /*============================================================*/
        gotoxy( 27, 18 );
        printf("%4d", Line+1);

        /*============================================================*/
        /*    READ AND WRITE THE NUMBER OF VERTICES IN THIS LINE      */
        /*============================================================*/
        fscanf( fSurfaceFile, "%d",    &NVrtcs );
        fprintf( fExtendedFile, "%d\n",  NVrtcs+2 );

        /*============================================================*/
        /* READ THE FIRST TWO VERTICES                                */
        /*============================================================*/
        fscanf( fSurfaceFile, "%lf %lf %lf", &(X[0]), &(Y[0]), &(Z[0]) );
        fscanf( fSurfaceFile, "%lf %lf %lf", &(X[1]), &(Y[1]), &(Z[1]) );

        /*============================================================*/
        /* CALCULATE LOCATION OF VERTEX OF EXTENDED SURFACE AT THE START*/
        /*============================================================*/
        make_new_vertex( X[0], X[1], Y[0], Y[1], Z[0], Z[1],
                         &ExtX, &ExtY, &ExtZ,
                         X[0] + DistanceStart );

        /*============================================================*/
        /* WRITE THE NEW VERTEX AND THE FIRST TWO VERTCIES TO THE OUTPUT*/
        /* FILE                                                       */
        /*============================================================*/
        fprintf( fExtendedFile, " %9.6f %9.6f %9.6f\n", ExtX, ExtY, ExtZ);
        fprintf( fExtendedFile, " %9.6f %9.6f %9.6f\n", X[0], Y[0], Z[0]);
        fprintf( fExtendedFile, " %9.6f %9.6f %9.6f\n", X[1], Y[1], Z[1]);

        /*============================================================*/
        /* LOOP THROUGH ALL VERTICES ON THIS LINE                     */
        /*============================================================*/
        for( Vertex=2 ; Vertex<(NVrtcs-1) ; Vertex++)
        {
            /*========================================================*/
            /* READ THE SURFACE VERTEX                                */
            /*========================================================*/
            fscanf( fSurfaceFile,"%lf %lf %lf", &(X[0]), &(Y[0]), &(Z[0]));

            /*========================================================*/
            /* WRITE OFFSETS FOR THIS VERTEX TO OUTPUT FILE           */
            /*========================================================*/
            fprintf( fExtendedFile," %9.6f %9.6f %9.6f\n",X[0], Y[0],Z[0]);
        }

        /*============================================================*/
        /* READ THE LAST VERTEX AND WRITE TO THE OUTPUT FILE          */
```

Appendix 15

```c
      /*==============================================================*/
      fscanf( fSurfaceFile, "%lf %lf %lf", &(X[1]), &(Y[1]), &(Z[1]) );
      fprintf( fExtendedFile, " %9.6f %9.6f %9.6f\n", X[1], Y[1], Z[1]);

      /*==============================================================*/
      /* CALCULATE LOCATION OF VERTEX OF EXTENDED SURFACE AT THE END  */
      /*==============================================================*/
      make_new_vertex( X[0], X[1],
                       Y[0], Y[1],
                       Z[0], Z[1],
                       &ExtX, &ExtY, &ExtZ,
                       X[1] + DistanceEnd );

      fprintf( fExtendedFile, " %9.6f %9.6f %9.6f\n", ExtX, ExtY, ExtZ);

   }

}
/**********************************************************************/
void   make_new_vertex
       (
            float    XStart,
            float    XEnd,
            float    YStart,
            float    YEnd,
            float    ZStart,
            float    ZEnd,
            float    *ExtX,
            float    *ExtY,
            float    *ExtZ,
            float    NewXLoc
       )
/**********************************************************************/
{
   double MY,
          MZ;

   double BY,
          BZ;

   /*==============================================================*/
   /* CALCULATE SLOPE FOR STRAIGHT LINE EQUATION                   */
   /*==============================================================*/
   MY = (YEnd-YStart) / (XEnd-XStart);
   MZ = (ZEnd-ZStart) / (XEnd-XStart);

   /*==============================================================*/
   /* CALCULATE THE INTERSECTION WITH THE Y AND Z AXIS             */
   /*==============================================================*/
   BY = YStart - (XStart * MY);
   BZ = ZStart - (XStart * MZ);

   /*==============================================================*/
   /*   CALCULATE THE COORDINATES OF THE NEW VERTEX                */
   /*==============================================================*/
   *ExtX = NewXLoc;
   *ExtY = BY + *ExtX * MY;
   *ExtZ = BZ + *ExtX * MZ;
```

Appendix 16
C-Program: SIMULATE.C - Simulate Tool Path

```
/********************/
/*  DEFINE CONTANTS  */
/********************/
#if __STDC__
#define _Cdecl
#else
#define _Cdecl  cdecl
#endif

#define    MOUSE       0x33
#define    UNION       union

#define    UP          1072
#define    DOWN        1080
#define    LEFT        1075
#define    RIGHT       1077
#define    RETURN      13
#define    PLUS        43
#define    MINUS       45
#define    ESC         27
#define    SPACE       32
#define    BACKSPACE   8
#define    TRUE        1
#define    FALSE       0
#define    FULL        0
#define    TOP         1
#define    ISO         2
#define    SIDE        3
#define    MONITOR     4
#define    GCODE       5
#define    MAXLINELENGTH 80
#define    BLK_SIZE    513
#define    XMAXR       99999.999       /* define valid ranges */
#define    XMINR      -99999.999
#define    YMAXR       99999.999
#define    YMINR      -99999.999
#define    ZMAXR       99999.999
#define    ZMINR      -99999.999
#define    TMAXR       9999
#define    TMINR       0
#define    SMAXR       30000
#define    SMINR       0
#define    GMAXR       99
#define    GMINR       0
```

Appendix 16

```c
#define   MMAXR       99
#define   MMINR       0
#define   FMAXR       15000.0
#define   FMINR       0
#define   RMAXR       99999.999
#define   RMINR       -99999.999
#define   IMAXR       99999.999
#define   IMINR       -99999.999
#define   JMAXR       99999.999
#define   JMINR       -99999.999
#define   KMAXR       99999.999
#define   KMINR       -99999.999
#define   LATHE       0
#define   MILL        1
#define   max_pnts    1000
#define   move        0
#define   cut         1
#define   PI          3.141592654
#define   CW          2
#define   CCW         3

/*********************/
/*    SUB-PROGRAMS   */
/*********************/

void     system_sub(int v);
void     machine_sub(int v);
void     views_sub(int v);
void     colors_sub(int v);
void     speed_sub(int v);

/*********************/
/* GRAPHICS ROUTINES */
/*********************/

void     demo( void);
void     init_graphics( void);
void     title_screen( void);
void     set_screen( void);
void     iso_view( void);
void     top_view( void);
void     side_view( void);
void     all_view( void);
void     d_gcode(char newcode[40]);
void     d_line(int x0,int x1,int y0,int y1);
void     d_path( void);
void     monitor(float x,float y,float z);
void     scale_path(float factor);
void     initial_pnt(int xi, int yi, int zi);
void     line_pts(int xn, int yn, int zn);
void     normalize(float *x,float *y,float *z,float sxmax,float
                   symax,float szmax);
void     getxyz(float *x,float *y,float sxmin,float symin,float
                   sxmax,float symax,int zflag);
void     set_view(int view);


/**************************/
```

SIMULATE.C - Simulate Tool Path

```
/*      GENERAL ROUTINES     */
/************************/

void        doscommand( void);
void        fileread( void);
void        firstpass( void);
void        secondpass( void);
void        first( void);
void        process(char blk[]);
void        int_error( void);

void        doxyz(float x, float y, float z, int draw );
void        circ(float Xs,float Ys,float Zs,float Xe,float Ye,
                 float Ze,float R,float Xc,float Yc,float Zc,int dir,
                 int draw);
void        error_num(int err_num, int par1, float par2);
void        second( void);
void        process2(char blk[]);
void        hot_key( void);
int         get_key( void );


/*********************/
/*   PARSE    ROUTINES */
/*********************/

int         parse(char file[12]);
int         getlinechar(char ln[MAXLINELENGTH],int *j);
void        getnextparm(char ln[MAXLINELENGTH],int *j,char *s,int *pc);
void        getnext(char ln[MAXLINELENGTH],int *j,char *s,int *pc);
void        echoline(char line[MAXLINELENGTH],int *writeorig);
int         isgcode(char code[6]);
int         is_empty_string( char * s );

/*********************/
/*   MENU ROUTINES   */
/*********************/

int         getkey( void);
void        set_menu(char *s[15][15]);
void        set_bar(char *s[15][15], int h, int v);
void        use_menu(char *s[15][15], int *ph, int *pv, int *status);
void        show_string(char *s[15][15], int h, int v);
int         input_str(char *s, char *m, char *p, int h, int v, int option);
int         input_int(int  *i, char *m, char *p, int h, int v, int option);


/***********/
/* MOUSE.C */
/***********/

void        _Cdecl flagreset(int *mouseStatusPtr, int *numberOfButtonsPtr);
void        _Cdecl showcursor( void );
void        _Cdecl hidecursor( void );
void        _Cdecl getposbut(int *buttonStatus, int *horizontal,
                             int *vertical);
void        _Cdecl setcursorpos(int horizontal, int vertical);
void        _Cdecl sethorizontallimits(int minPos, int maxPos);
```

```c
void    _Cdecl setverticallimits(int minPos, int maxPos);
```

```
/**********************************************************************/
/*                                                                    */
/*                                                                    */
/*                       PROGRAM   CAM.C                              */
/*    This program is written in TURBO C  and uses the Borland        */
/*    graphics library (GRAPHICS.LIB).The appropriete fonts and       */
/*    screen driver have to be added to graphics.lib to link.         */
/*                                                                    */
/*    The program can be compiled and linked by typing in the         */
/*    following  command in DOS.                                      */
/*                                                                    */
/*    TCC -ml [-f87] CAM1 CAM2 CAM3 GRAPHICS.LIB                      */
/*                                                                    */
/*    The following files must be present in the graphics library    */
/*    to successfully run the program. Refer to the TurboC manuals.   */
/*                                                                    */
/*    TRIP.CHR        - character font file                           */
/*    SANS.CHR        - character font file                           */
/*    EGAVGA.BGI      - graphics driver                               */
/*                                                                    */
/*                                                                    */
/*    Note: This program will run on a computer with an VGA or EGA   */
/*          graphics card.                                            */
/*                                                                    */
/*                                                                    */
/**********************************************************************/


/********************/
/*    INCLUDES      */
/********************/

#include <stdio.h>
#include <conio.h>
#include <ctype.h>
#include <io.h>
#include <fcntl.h>
#include <mem.h>
#include <dos.h>
#include <bios.h>
#include <stdlib.h>
#include <string.h>
#include <alloc.h>
#include <math.h>
#include <graphics.h>
#include "cam.h"

/**********************************************************************/
/*                    M O U S E     M O D U L E                       */
/*                                                                    */
/*                                                                    */
/**********************************************************************/
/**********************************************************************/
```

SIMULATE.C - Simulate Tool Path

```c
void    flagreset(int *mouseStatusPtr, int *numberOfButtonsPtr)
{
        UNION REGS regs;
        regs.x.ax = 0;
        int86(MOUSE, &regs, &regs);
        *mouseStatusPtr = regs.x.ax;
        *numberOfButtonsPtr = regs.x.bx;
}


/*******************************************************************/

void    showcursor( void )
{
        UNION REGS regs;
        regs.x.ax = 1;
        int86(MOUSE, &regs, &regs);
}


/*******************************************************************/

void    hidecursor( void )
{
        UNION REGS regs;
        regs.x.ax = 2;
        int86(MOUSE, &regs, &regs);
}


/*******************************************************************/

void    getposbut(int *buttonStatus, int *horizontal, int *vertical)
{
        UNION REGS regs;
        regs.x.ax = 3;
        int86(MOUSE, &regs, &regs);
        *buttonStatus = regs.x.bx;
        *horizontal = regs.x.cx;
        *vertical = regs.x.dx;
}


/*******************************************************************/

void    setcursorpos(int horizontal, int vertical)
{
        UNION REGS regs;
        regs.x.ax = 4;
        regs.x.cx = horizontal;
        regs.x.dx = vertical;
        int86(MOUSE, &regs, &regs);
}


/*******************************************************************/
```

Appendix 16

```c
void    sethorizontallimits(int minPos, int maxPos)
{
        UNION REGS regs;
        regs.x.ax = 7;
        regs.x.cx = minPos;
        regs.x.dx = maxPos;
        int86(MOUSE, &regs, &regs);
}
/**********************************************************************/

void    setverticallimits(int minPos, int maxPos)
{
        UNION REGS regs;
        regs.x.ax = 8;
        regs.x.cx = minPos;
        regs.x.dx = maxPos;
        int86(MOUSE, &regs, &regs);
}


                        /*********************/
                        /* GLOBAL VARIABLES */
                        /*********************/
/**********************************************************************/
/*                                                                    */
/*                  VARIABLES USED FOR DISPLAY                        */
/*                                                                    */
/**********************************************************************/
int     foreground, background;         /* System colors              */
int     mouse_flag, loop_flag;          /* System flags               */
int     scale_flag;                     /* Set when in zoomed views   */
int     machine;                        /* Machine flag (LATHE or MILL */
int     num_pnts;                       /* Index into data base       */

int     move_color;                     /* Rapid traverse color       */
int     cut_color;                      /* Linear interpolation color */
int     xasp,yasp;                      /* Aspect ratio parameters    */
int     speed;                          /* Speed of processing g-code */

float   xyratio;                        /* Aspect ratio       */

char    single_step;                    /* For single step mode       */
char    dgcode[6][40];                  /* Array of current g-codes   */
char    blank[40];                      /* Used to re-display g-codes */
                                        /* after a screen clear.      */

struct pnt                              /*         DATA   BASE        */
{
    int   tool;                         /* Cut or rapid traverse      */
    float x,y,z,xz,xiso,yiso;           /* Points used to draw path   */
    float xwc,ywc,zwc;                  /* World coordinates of path  */
};

struct pnt point[max_pnts];


/**********************************************************************/
```

SIMULATE.C - Simulate Tool Path

```c
/*                                                                      */
/*                  VARIABLES USED IN PARSE ROUTINES                    */
/*                                                                      */
/**********************************************************************/

FILE       *instream;                 /* Original g-code               */
FILE       *outstream;                /* Parsed g-code                 */
FILE       *errorfile;                /* Error file                    */

char       gcode[6];                  /* Current g-code being read     */
char       last_gcode[6];             /* Last g-code in file           */
char       X[15],Y[15],Z[15];         /* G-code parameters             */
char       I[15],J[15],K[15],R[15];


/**********************************************************************/
/*                                                                      */
/*                  VARIABLES USED FOR ERROR ANALYSIS                   */
/*                                                                      */
/**********************************************************************/

char       spindle[15],               /* Spindle speed number          */
           feedrate[15],              /* Feedrate string               */
           toolnum[15],               /* Tool number string            */
           gcodefile[12],             /* G-code file name              */
           gcodest[40];               /* G-code string                 */

float      xmax,ymax,zmax,xmin,ymin,zmin;/* Extreme X,Y,Zs of workpiece */
float      xprev,yprev,zprev;         /* Previous X,Y,Z values         */

int        error_flag,                /* Error in Gcode file           */
           spin_flag,                 /* Spindle speed specified?      */
           tool_flag,                 /* Has tool been specified?      */
           feed_flag,                 /* Feedrate specified?*/
           strt_flag,                 /* When was spindle started?     */
           g_flag,                    /* Gcode other than G00 found    */
           stop_flag,                 /* When was spindle stopped?     */
           pend_flag,                 /* End of program encountered    */
           g92_flag,                  /* G92 encountered       */
           gz_flag,                   /* G code (0,1,2,3) before 92    */

           g17_flag,                  /* G17 xy-plane selection        */
           g18_flag,                  /* G18 xz-plane selection        */
           g19_flag,                  /* G19 yz-plane selection        */

           g20_flag,                  /* Flags whether inches or mm    */
           g90_flag,                  /* Absolute or incremental       */
           initial_flag,              /* Initial point flag */

           counter,                   /* Tracks the order that the     */
                                      /* above conditions occured      */

           err_array[2][10];          /* Errors found in G-code        */


/**********************************************************************/
/* err_array is a two dimensional array that stores information        */
/* pertaining to errors that may exist in the Gcode file.              */
```

Appendix 16

```c
/*                                                                      */
/*                                                                      */
/*  Row 0 - stores a flag = 1 when condition is true (Code located)     */
/*  Row 1 - stores a counter that tracks order in which conditions      */
/*          have been located.                                          */
/*                                                                      */
/*  Elements-Row 0                                                      */
/*  [0]: Spindle Speed specified                                        */
/*  [1]: Tool specified                                                 */
/*  [2]: Feedrate specified                                             */
/*  [3]: Spindle started                                                */
/*  [4]: !=G00 encountered                                              */
/*  [5]: Spindle stop                                                   */
/*  [6]: End of program                                                 */
/*  [7]: G92 found                                                      */
/*  [8]: G00 encountered                                                */
/************************************************************************/




/************************************************************************/
/************************************************************************/
/*                    M A I N    P R O G R A M                 */
/************************************************************************/
/************************************************************************/

void main()
{

  char     *s[15][15];
  int      h, v, i, a;
  int      status, numofbut;

/*************************/
/* INITAILIZE MENU PROMPTS */
/*************************/

  i = 0;
  s[i][0]   = "SYSTEM";
  s[i][1]   = "DOS COMMAND ";
  s[i][2]   = "SIMULATE    ";
  s[i][3]   = "            ";
  s[i][4]   = "EXIT PROGRAM";
  s[i][5]   = "";

  i = 1;
  s[i][0]   = "MACHINE";
  s[i][1]   = "LATHE";
  s[i][2]   = "MILL ";
  s[i][3]   = "";

  i = 2;
  s[i][0]   = "VIEWS";
  s[i][1]   = "TOP      ";
  s[i][2]   = "ISOMETRIC";
  s[i][3]   = "SIDE     ";
```

SIMULATE.C - Simulate Tool Path

```c
   s[i][4]     = "ALL VIEWS";
   s[i][5]     = "";

   i = 3;
   s[i][0]     = "COLORS";
   s[i][1]     = "CUT      ";
   s[i][2]     = "TRAVERSE";
   s[i][3]     = "";

   i = 4;
   s[i][0]     = "SPEED";
   s[i][1]     = "SET SPEED    ";
   s[i][2]     = "SINGLE STEP ";
   s[i][3]     = "";

   i = 5;
   s[i][0]     = "";


/*******************************/
/* SET SYSTEM FLAGS & VARIABLES */
/*******************************/

   mouse_flag   = FALSE;
   loop_flag    = TRUE;
   scale_flag   = FALSE;
   machine      = MILL;
   num_pnts     = 0;
   move_color   = WHITE;
   cut_color    = RED;
   speed        = 0;
   single_step  = 'N';
   blank[0]     = 32;
   blank[1]     = 0;

   flagreset(&status,&numofbut);
   if (status == -1)
   {
      mouse_flag = TRUE;
      sethorizontallimits(2, 317);
      setverticallimits(131, 341);
   }

   for (i=0 ; i<=5 ; i++)
         strcpy(dgcode[i]," ");


   init_graphics();
   getaspectratio(&xasp,&yasp);
   xyratio = 1.29;
   background = BLUE;
   foreground = WHITE;
   set_menu(s);
   title_screen();
   set_screen();
```

```
/*********************/
/* MAIN CONTROL LOOP */
/*********************/

   h          = 0;
   v          = 0;

   do
   {
      a = get_key();
      if (a == RETURN)
          a = ESC;
      switch(a)
      {
          case ESC:
              use_menu(s,&h,&v,&a);
              if (a == 1)
              {
                  switch(h)
                  {
                     case 0:
                         system_sub(v);
                         if (v == 1)
                             set_menu(s);
                         break;
                     case 1:
                         machine_sub(v);
                         break;
                     case 2:
                         views_sub(v);
                         break;
                     case 3:
                         colors_sub(v);
                         break;
                     case 4:
                         speed_sub(v);
                         break;
                  }
              }
              break;

      }
   }
   while (loop_flag == TRUE);

   closegraph();
}


/**********************************************************************/
/*                                                                    */
/*                                                                    */
/*                    MAIN   MENU   FUNCTIONS                         */
/*                                                                    */
/*                                                                    */
/**********************************************************************/
```

SIMULATE.C - Simulate Tool Path

```c
/*******************************************************************/
/*                                                                 */
/*   FUNCTION system_sub                                           */
/*                                                                 */
/*   Allows user to enter DOS shell, simulate g-code, or exit program. */
/*   Called by main program.                                       */
/*                                                                 */
/*******************************************************************/
void system_sub(int choice)
{
   switch(choice)
   {
      case 1:
         if (scale_flag == FALSE) /* Can't execute when zoomed  */
            doscommand();         /* Enter a DOS command*/
         break;
      case 2:
         if (scale_flag == FALSE)
         {
            fileread();           /* Read g-code file name      */
            firstpass();          /* Start simulation   */
         }
         break;
      case 3:
         break;
      case 4:
         loop_flag = FALSE;       /* Results in program exit    */
         break;
   }
   return;
}


/*******************************************************************/
/*                                                                 */
/*   FUNCTION machine_sub                                          */
/*                                                                 */
/* Flags which machine is being used.  MACHINE can be LATHE/MILL (0/1)*/
/*                                                                 */
/*******************************************************************/
void machine_sub(int choice)
{
   switch(choice)
   {
      case 1:
         if (scale_flag == FALSE)
         {
            machine = LATHE;
            setcolor(BLUE);
            setfillstyle(1,BLUE);
            bar3d(495,189,550,199,0,1);
```

```
                        setcolor(WHITE);
                        outtextxy(495,189, "LATHE");
                    }
                    break;
            case 2:
                    if (scale_flag == FALSE)
                    {
                        machine = MILL;
                        setcolor(BLUE);
                        setfillstyle(1,BLUE);
                        bar3d(495,189,550,199,0,1);
                        setcolor(WHITE);
                        outtextxy(495,189, "MILL");
                    }
                    break;
        }
        return;
}


/**********************************************************************/
/*                                                                    */
/*   FUNCTION views_sub                                               */
/*                                                                    */
/*   Brings desired view to full screen.                              */
/*   Called by main program.                                          */
/*                                                                    */
/**********************************************************************/

void views_sub(int choice)
{

    setviewport(1,21,638,348,1);
    clearviewport();

    switch(choice)
    {
            case 1:
                    top_view();
                    set_view(FULL);
                    break;
            case 2:
                    iso_view();
                    set_view(FULL);
                    break;
            case 3:
                    side_view();
                    set_view(FULL);
                    break;
            case 4:
                    set_screen();
                    if (scale_flag == TRUE)
                    {
                        set_view(TOP);
                        outtextxy(110,75,"PLEASE WAIT...");
                        scale_path(FALSE);
                        setcolor(BLACK);
                        bar3d(110,75,270,85,0,1);
```

```
                      scale_flag = FALSE;
                   }
                   d_path();               /* Re-draw cutter path      */
                   d_gcode(blank);         /* Display last 6 g-codes   */
                   break;
      }
      return;
   }


/****************************************************************/
/*                                                              */
/*   FUNCTION colors_sub                                        */
/*                                                              */
/*   Allows user to change colors. Display prompt in a diff. area */
/*   of   screen depending on whether the display is zoomed or not. */
/*                                                              */
/*   Called by main program.                                    */
/*                                                              */
/****************************************************************/
void colors_sub(int choice)
{
   setfillstyle(1,BLACK);

   switch(choice)
   {
         case 1:

                if (scale_flag == TRUE)
                {
                input_int(&cut_color,
                       "ENTER COLOR FOR CUTTING: ","99",400,30,4);
                setcolor(BLACK);
                bar3d(390,25,638,47,0,1);
                }

                else
                {
                set_view(GCODE);
                setfillstyle(1,BLUE);
                setcolor(BLUE);
                bar3d(0,0,317,73,0,1);
                input_int(&cut_color,
                  "ENTER COLOR FOR CUTTING: ","99",50,30,4);
                setfillstyle(1,BLUE);
                setcolor(BLUE);
                bar3d(0,0,317,73,0,1);
                d_gcode(blank);
                }

                break;
         case 2:
                if (scale_flag == TRUE)
                {
                input_int(&move_color,
                    "ENTER COLOR FOR TRAVERSING: ","99",390,30,4);
```

```
                    setcolor(BLACK);
                    bar3d(380,25,638,47,0,1);
                    }

                    else
                    {
                    set_view(GCODE);
                    setfillstyle(1,BLUE);
                    setcolor(BLUE);
                    bar3d(0,0,317,73,0,1);
                    input_int(&move_color,"ENTER COLOR FOR TRAVERSING:
","99",40,30,4);
                    setfillstyle(1,BLUE);
                    setcolor(BLUE);
                    bar3d(0,0,317,73,0,1);
                    d_gcode(blank);
                    }

                    break;
        }
        setcolor(WHITE);
        set_view(FULL);
        return;
}



/**********************************************************************/
/*                                                                    */
/*   FUNCTION speed_sub                                               */
/*                                                                    */
/*   Allows user to set speed of g-code processing.                   */
/*   Called by main program.                                          */
/*                                                                    */
/**********************************************************************/

void speed_sub(int choice)
{
    if (scale_flag == FALSE)              /* Can't enter in zoom view   */
    {
        set_view(GCODE);
        setfillstyle(1,BLUE);
        setcolor(BLUE);
        bar3d(0,0,317,73,0,1);

    switch(choice)
        {
            case 1:
                input_int(&speed,"ENTER SPEED (FAST(0)-SLOW(999)):
","999",14,30,4);
                break;
            case 2:
                do
                {
                    input_str(&single_step,"SINGLE STEP (Y/N)? ","A", 65,
30,4);
                    if (single_step >= 'a' &&
                        single_step <= 'z') single_step =
```

```
                    single_step - 'a' + 'A';
            }
            while ( (single_step != 'Y') && (single_step != 'N') );
            break;
    }

    setfillstyle(1,BLUE);
    setcolor(BLUE);
    bar3d(0,0,317,73,0,1);
    d_gcode(blank);
    setcolor(WHITE);
    set_view(FULL);
    }

    return;
}



/***********************************************************************/
/*                                                                     */
/*  FUNCTION init_graphics                                             */
/*                                                                     */
/*  Sets up Turbo C graphics system. Load EGAVGA driver file.          */
/*  Note: This is set up for VGA or EGA graphics cards only.           */
/*                                                                     */
/*  Called by main program                                             */
/*                                                                     */
/***********************************************************************/
void init_graphics()
{
    int   gdriver = EGA,
          gmode   = EGAHI;

    if( registerbgidriver( EGAVGA_driver)  < 0 )
        {
            printf( "UNABLE TO FIND GRAPHIC DRIVER\n" );
            exit( 1 );
        }

    initgraph(&gdriver,&gmode,"");


    if( registerbgifont( triplex_font ) < 0 )
    {   printf( "UNABLE TO FIND TRIPLEX FONT" );
        exit( 1 );
    }

    if( registerbgifont( sansserif_font ) < 0 )
    {   printf( "UNABLE TO FIND TRIPLEX FONT" );
        exit( 1 );
    }

    return;
```

```c
}


/************************************************************************/
/*                                                                      */
/*   FUNCTION title_screen()                                            */
/*                                                                      */
/*   Displays title screen.                                             */
/*   Called by main program.                                            */
/*                                                                      */
/************************************************************************/

void title_screen()
{
  struct textsettingstype oldset;

  bar3d( 0, 20,639,349,0,1);
  gettextsettings(&oldset);
  setcolor(YELLOW);
  settextjustify(LEFT_TEXT, TOP_TEXT);
  settextstyle(TRIPLEX_FONT, HORIZ_DIR, 6);
  outtextxy( 65,  80, "G-CODE SIMULATION");
  settextstyle(SANS_SERIF_FONT, HORIZ_DIR, 3);
  settextjustify(oldset.horiz, oldset.vert);
  settextstyle(oldset.font, oldset.direction, oldset.charsize);

  setcolor(WHITE);
  bar3d( 210, 308, 407, 324,0,1);
  outtextxy(213, 313, " PRESS ANY KEY TO BEGIN ");
  get_key();

  return;
}




/************************************************************************/
/*                                                                      */
/*   FUNCTION set_screen()                                              */
/*                                                                      */
/*   Sets different windows on screen.                                  */
/* Called by main program, views_sub(), doscommand(), and firstpass().*/
/*                                                                      */
/************************************************************************/

void set_screen()
{

  setviewport(1,21,638,348,1);
  clearviewport();
  set_view(FULL);
  setcolor(WHITE);
  setfillstyle(0,foreground);
  bar3d( 0, 20,639,349,0,1);
  line(320,21,320,349);
```

```c
        line(1,185,639,185);
        setfillstyle(1,BLUE);
        bar3d(320,185,639,349,0,1);
        setfillstyle(1,BLACK);
        outtextxy(425,189, "MONITOR");
        if (machine == 0)
             outtextxy(495,189, "LATHE");
        else
             outtextxy(495,189, "MILL");
        outtextxy(330,210, "Tool: ");
        outtextxy(346,225, "FR:    ");
        outtextxy(346,240, "SS:    ");
        outtextxy(511,210, "X:     ");
        outtextxy(511,225, "Y:     ");
        outtextxy(511,240, "Z:     ");
        line(321,255,639,255);
        outtextxy(460,259, "G-CODE");
        outtextxy(145, 24, "TOP");
        outtextxy(450, 24, "ISOMETRIC");
        outtextxy(140, 189, "SIDE");

        return;
}


/***********************************************************************/
/*                                                                     */
/*  FUNCTION iso_view()                                                */
/*                                                                     */
/*  Brings isometric view to full screen.                              */
/*  Called by views_sub()                                              */
/*                                                                     */
/***********************************************************************/
void iso_view()
{
   int    i;

   outtextxy(240,15,"ISOMETRIC VIEW");
   outtextxy(240,16,"_____");

   setviewport(10,60,630,340,1);

   if (scale_flag == FALSE)
   {
        outtextxy(240,150,"PLEASE WAIT...");
        scale_path(TRUE);
        setcolor(BLACK);
        bar3d(240,150,380,160,0,1);
        scale_flag = TRUE;
   }

   for (i=1 ; i<num_pnts ; i++)
   {
        if (point[i].tool == move)
             setcolor(move_color);
```

384   **Appendix 16**

```
            else
                  setcolor(cut_color);

            d_line(point[i-1].xiso,point[i].xiso,
                   point[i-1].yiso,point[i].yiso);
     }

     return;
}


/**********************************************************************/
/*                                                                    */
/*   FUNCTION top_view()                                              */
/*                                                                    */
/*   Brings top view to full screen.                                  */
/*   Called by views_sub()                                            */
/*                                                                    */
/**********************************************************************/

void top_view()
{
   int    i;

   outtextxy(265,15,"TOP VIEW");
   outtextxy(265,16,"_____");

   setviewport(10,60,630,340,1);

   if (scale_flag == FALSE)
   {
        outtextxy(240,150,"PLEASE WAIT...");
        scale_path(TRUE);
        setcolor(BLACK);
        bar3d(240,150,380,160,0,1);
        scale_flag = TRUE;
   }

   for (i=1 ; i<num_pnts ; i++)
   {
        if (point[i].tool == move)
             setcolor(move_color);
        else
             setcolor(cut_color);

        d_line(point[i-1].x,point[i].x,point[i-1].y,point[i].y);
   }

   return;
}



/**********************************************************************/
/*                                                                    */
/*   FUNCTION side_view()                                             */
/*                                                                    */
```

SIMULATE.C - Simulate Tool Path 385

```c
/*  Brings side view to full screen.                                  */
/*  Called by views_sub()                                             */
/*                                                                    */
/**********************************************************************/
void side_view()
{
  int     i;

  outtextxy(263,15,"SIDE VIEW");
  outtextxy(263,16,"_____");

  setviewport(10,60,630,340,1);

  if (scale_flag == FALSE)
  {
       outtextxy(240,150,"PLEASE WAIT...");
       scale_path(TRUE);
       setcolor(BLACK);
       bar3d(240,150,380,160,0,1);
       scale_flag = TRUE;
  }

  for (i=1 ; i<num_pnts ; i++)
  {
       if (point[i].tool == move)
            setcolor(move_color);
       else
            setcolor(cut_color);

       d_line(point[i-1].xz,point[i].xz,point[i-1].z,point[i].z);
  }

  return;
}




/**********************************************************************/
/*                                                                    */
/*  FUNCTION set_view                                                 */
/*                                                                    */
/*  Activates viewport based on the value of view:                    */
/*                                                                    */
/*  i = 0    viewport is set to full screen.                          */
/*  i = 1    viewport is set to TOP viewport.                         */
/*  i = 2    viewport is set to ISOMETRIC viewport.                   */
/*  i = 3    viewport is set to SIDE viewport.                        */
/*  i = 4    viewport is set to MONITOR viewport.                     */
/*  i = 5    viewport is set to G-CODE  viewport.                     */
/*                                                                    */
/**********************************************************************/
void set_view(int view)
{
  switch(view)
  {
```

```
        case 0:
             setviewport(0,0,639,349,1);
             break;
        case 1:
             setviewport(8,40,313,180,1);
             break;
        case 2:
             setviewport(321,34,638,184,1);
             break;
        case 3:
             setviewport(8,204,313,344,1);
             break;
        case 4:
             setviewport(321,198,638,254,1);
             break;
        case 5:
             setviewport(321,275,638,348,1);
             break;
   }
   return;
}


/************************************************************************/
/*                                                                      */
/*   FUNCTION d_gcode()                                                 */
/*                                                                      */
/*   If called with a blank string, re-displays last six g-codes.       */
/*   If called with a g-code, scrolls the display of g-codes.           */
/*                                                                      */
/*   Called by views_sub()                                              */
/*                                                                      */
/************************************************************************/

void d_gcode(char newcode[40])
{

   int i;

   set_view(GCODE);
   setfillstyle(1,BLUE);

   if (newcode[0] == 32)
        for (i=0 ; i<=60 ; i+=12)
        {
             setcolor(BLUE);
             bar3d(3,i,317,i+9,0,1);
             setcolor(WHITE);
             outtextxy(3,i,dgcode[(i/12)]);
        }

   else

   if (strcmp(dgcode[5],newcode))
   {
        set_view(GCODE);
        setfillstyle(1,BLUE);
```

```c
            for (i=0 ; i<=4 ; i++)
                strcpy(dgcode[i],dgcode[i+1]);

            strcpy(dgcode[5],newcode);

            for (i=0 ; i<=60 ; i+=12)
            {
                setcolor(BLUE);
                bar3d(3,i,317,i+9,0,1);
                setcolor(WHITE);
                outtextxy(3,i,dgcode[(i/12)]);
            }
    }
    set_view(FULL);
    return;
}


/*********************************************************************/
/*                                                                   */
/*  FUNCTION d_line()                                                */
/*                                                                   */
/*  Draws a line between the two points sent to it.                  */
/*  Called by iso_view(),top_view(),side_view(),d_path(),line_pts(). */
/*                                                                   */
/*********************************************************************/
void d_line(int x0,int x1,int y0,int y1)
{
  moveto(x0,y0);
  lineto(x1,y1);

  return;
}




/*********************************************************************/
/*                                                                   */
/*  FUNCTION d_path()                                                */
/*                                                                   */
/*  Draws the cutter path in all views.                              */
/*  Called by views_sub(), and doscommand().                         */
/*                                                                   */
/*********************************************************************/
void d_path()
{
  int    i;

  for (i=1 ; i<num_pnts ; i++)
  {
        if (point[i].tool == move)
            setcolor(move_color);
        else
```

388   **Appendix 16**

```
                   setcolor(cut_color);

         set_view(TOP);
         d_line(point[i-1].x,point[i].x,point[i-1].y,point[i].y);

         set_view(SIDE);
         d_line(point[i-1].xz,point[i].xz,point[i-1].z,point[i].z);

         set_view(ISO);
         d_line(point[i-1].xiso,point[i].xiso,
                point[i-1].yiso,point[i].yiso);
  }

  set_view(FULL);
  return;
}



/**********************************************************************/
/*                                                                    */
/*    FUNCTION monitor()                                              */
/*                                                                    */
/*    Displays tool, spindle speed, feed rate, x, y, and z.           */
/*    Called by process2()                                            */
/*                                                                    */
/**********************************************************************/

void monitor(float x,float y,float z)
{
  char    buf[20];

  set_view(MONITOR);
  setfillstyle(1,BLUE);
  setcolor(BLUE);

  bar3d(57,12,187,22,0,1);
  setcolor(WHITE);
  outtextxy(57,12,toolnum);
  setcolor(BLUE);

  bar3d(57,27,187,37,0,1);
  setcolor(WHITE);
  outtextxy(57,27,feedrate);
  setcolor(BLUE);

  bar3d(57,42,187,52,0,1);
  setcolor(WHITE);
  outtextxy(57,42,spindle);
  setcolor(BLUE);

  bar3d(214,12,315,22,0,1);
  setcolor(WHITE);
  outtextxy(214,12,gcvt(x,4,buf));
  setcolor(BLUE);

  bar3d(214,27,315,37,0,1);
```

```
    setcolor(WHITE);
    outtextxy(214,27,gcvt(y,4,buf));
    setcolor(BLUE);

    bar3d(214,42,315,.52,0,1);
    setcolor(WHITE);
    outtextxy(214,42,gcvt(z,4,buf));
    setcolor(BLUE);

    setcolor(WHITE);
    return;
}
#include <stdio.h>
#include <conio.h>
#include <ctype.h>
#include <io.h>
#include <fcntl.h>
#include <mem.h>
#include <dos.h>
#include <bios.h>
#include <stdlib.h>
#include <string.h>
#include <alloc.h>
#include <math.h>
#include <graphics.h>
#include "cam.h"

/********************************************************************/
/*                      EXTERNAL VARIABLES                          */
/********************************************************************/
extern int      foreground, background;
extern int      mouse_flag, loop_flag;
extern int      scale_flag;
extern int      machine;
extern int      num_pnts;
extern int      move_color;
extern int      cut_color;
extern int      xasp,yasp;
extern int      speed;

extern float    xyratio;

extern char     single_step;
extern char     dgcode[6][40];
extern char     blank[40];


extern struct pnt
{
    int     tool;
    float   x,y,z,xz,xiso,yiso;
    float   xwc,ywc,zwc;
};

extern struct pnt point[max_pnts];
```

Appendix 16

```
extern FILE      *instream;
extern FILE      *outstream;
extern FILE      *errorfile;

extern char      gcode[6];
extern char      last_gcode[6];
extern char      X[15],Y[15],Z[15];
extern char      I[15],J[15],K[15],R[15];


extern char      spindle[15],
         feedrate[15],
         toolnum[15],
         gcodefile[12],
         gcodest[40];

extern float     xmax,ymax,zmax,xmin,ymin,zmin;
extern float     xprev,yprev,zprev;

extern int       error_flag,
         spin_flag,
         tool_flag,
         feed_flag,
         strt_flag,
         g_flag,
         stop_flag,
         pend_flag,
         g92_flag,
         gz_flag,

         g17_flag,
         g18_flag,
         g19_flag,

         g20_flag,
         g90_flag,
         initial_flag,

         counter,


         err_array[2][10];




/**********************************************************************/
/*                                                                    */
/*   FUNCTION scale_path()                                            */
/*                                                                    */
/*   Scales the cutter path up or down depending on the parameter     */
/*   sent to it.                                                      */
/*   Called by views_sub(), iso_view(), top_view(), side_view().      */
/*                                                                    */
/**********************************************************************/
```

```c
void scale_path(float scale)
{
   int      i;
   float    xztemp;

   for (i=0 ; i<num_pnts ; i++)
   {
        point[i].x = point[i].xwc;
        point[i].y = point[i].ywc;
        point[i].z = point[i].zwc;
   }

   if (scale == TRUE)
        for (i=0 ; i<num_pnts ; i++)
        {
             xztemp = point[i].x;
             if (machine==MILL)
                  getxyz(&point[i].x,&point[i].y,0,0,620,280,TOP);
             getxyz(&xztemp,&point[i].z,0,0,620,280,SIDE);
             point[i].xz = xztemp;

             point[i].xiso = ((0.7071*point[i].x) -
                             (0.7071*(point[i].z)));
             point[i].yiso=((-0.4082*point[i].x)+(0.8156*point[i].y)-
                             (0.4082*(point[i].z)));

             point[i].xiso = point[i].xiso*0.7 + 225;
             point[i].yiso = point[i].yiso*0.7 + 190;

        }
   else
        for (i=0 ; i<num_pnts ; i++)
        {
             xztemp = point[i].x;
             if (machine==MILL)
                  getxyz(&point[i].x,&point[i].y,0,0,305,140,TOP);
             getxyz(&xztemp,&point[i].z,0,0,305,140,SIDE);
             point[i].xz = xztemp;

             point[i].xiso = ((0.7071*point[i].x) -
                             (0.7071*(point[i].z)));
             point[i].yiso=((-0.4082*point[i].x)+(0.8156*point[i].y)-
                             (0.4082*(point[i].z)));

             point[i].xiso = point[i].xiso*0.7 + 100;
             point[i].yiso = point[i].yiso*0.7 +  90;

        }

   return;
}


/*******************************************************************/
/*                                                                 */
```

```
/*                                                                    */
/*   FUNCTION line_pts()                                               */
/*                                                                    */
/*   Puts point in data base and draws a line from the previous point */
/*   Called by process2(), and doxyz().                                */
/*                                                                    */
/**********************************************************************/

void line_pts(int xn, int yn, int zn)
{
  int    i;

  i = num_pnts;

  point[i].x = xn;
  point[i].y = yn;
  point[i].z = zn;
  point[i].xiso = ((0.7071*point[i].x) - (0.7071*(point[i].z)));
  point[i].yiso=((-0.4082*point[i].x)+(0.8156*point[i].y)-
                 (0.4082*(point[i].z)));

  point[i].xiso = point[i].xiso*0.7 + 100;
  point[i].yiso = point[i].yiso*0.7 +  90;

  if (point[i].tool == move)
       setcolor(move_color);
  else
       setcolor(cut_color);

  set_view(TOP);
  d_line(point[i-1].x,point[i].x,point[i-1].y,point[i].y);

  set_view(SIDE);
  d_line(point[i-1].xz,point[i].xz,point[i-1].z,point[i].z);

  set_view(ISO);
  d_line(point[i-1].xiso,point[i].xiso,point[i-1].yiso,point[i].yiso);

  num_pnts = num_pnts + 1;

  set_view(FULL);

  return;
}




/**********************************************************************/
/*                                                                    */
/*   FUNCTION initial_pnt()                                            */
/*                                                                    */
/*   Stores very first point in data base.                             */
/*   Called by process2()                                              */
/*                                                                    */
/**********************************************************************/

void initial_pnt(int xi, int yi, int zi)
{
  int    i;
```

```c
    i = num_pnts;

    point[i].x = xi;
    point[i].y = yi;
    point[i].z = zi;

    point[i].xiso = ((0.7071*point[i].x) - (0.7071*(point[i].z)));
    point[i].yiso=((-0.4082*point[i].x)+(0.8156*point[i].y)-
                    (0.4082*(point[i].z)));

    point[i].xiso = point[i].xiso*0.7 + 100;
    point[i].yiso = point[i].yiso*0.7 +  90;

    num_pnts = num_pnts + 1;

    return;
}




/**********************************************************************/
/*                                                                    */
/*  FUNCTION normalize()                                              */
/*                                                                    */
/*  Puts world coordinates into data base and calls getxyz() to       */
/*  obtain pixel coordinates that maximize the viewports.             */
/*                                                                    */
/*  Called by process2(), and doxyz().                                */
/*                                                                    */
/**********************************************************************/
void normalize(float *x,float *y,float *z,float sxmax,float symax,
               float szmax)
{
    float   xztemp = *x;

    point[num_pnts].xwc = *x;
    point[num_pnts].ywc = *y;
    point[num_pnts].zwc = *z;

    if (machine==MILL)
        getxyz(x,y,0,0 ,sxmax,symax,TOP);
    getxyz(&xztemp,z,0,0 ,sxmax,szmax,SIDE);

    point[num_pnts].xz = xztemp;

    return;
}



/**********************************************************************/
/*                                                                    */
/*  FUNCTION getxyz()                                                 */
/*                                                                    */
```

```c
/*  Automatically scales pixel coordinates to fill viewports without  */
/*  changing the object's proportions.                                */
/*                                                                    */
/*  Called by scale_path(), and normalize().                          */
/*                                                                    */
/**********************************************************************/

void getxyz(float *x,float *y,float sxmin,float symin,float sxmax,float
symax,int zflag)
{
   float    obratio,scratio,xfactor,yfactor;
   float    scalefact;
   float    xcorrect = 0;
   float    ycorrect = 0;
   float    ymaxmin,xmaxmin;

   if (zflag == TOP)
        ymaxmin = ymax - ymin;
   else
        ymaxmin = zmax - zmin;

   xmaxmin = xmax - xmin;

   obratio = fabs(xmaxmin / ymaxmin);
   scratio = fabs(((sxmax - sxmin) / (symax - symin)) / 1.29);
   xfactor = fabs((sxmax - sxmin) / xmaxmin);
   yfactor = fabs(((symax - symin)*1.29) / ymaxmin);

   if (obratio >= scratio)
   {
        scalefact = xfactor;
        ycorrect  = ( (1.29 * (symax-symin)) - (scalefact * ymaxmin) ) /
2;
   }
   else
   {
        scalefact = yfactor;
        xcorrect  = ( (sxmax-sxmin) - (scalefact * xmaxmin) ) / 2;
   }

   *x = ((*x - xmin)*scalefact) + xcorrect;

   if (zflag == TOP)
        *y = symax - (((((*y - ymin)*scalefact) + ycorrect)/1.29);
   else
        *y = symax - (((((*y - zmin)*scalefact) + ycorrect)/1.29);


   return;
}




/**********************************************************************/
/*                                                                    */
/*  FUNCTION doscommand()                                             */
/*                                                                    */
/*  This function allows the user execute to DOS commands from        */
```

```
/*   within the CAM program.                                          */
/*                                                                    */
/**********************************************************************/
void doscommand()
{
    char dos[63];                       /* User's DOS request*/

    /* Define input viewport and find out what user's dos request is.*/

    setviewport(10,75,639,105,1);
    strcpy(dos, "");
    input_str(dos,"DOS Command: ",
"XXXXXXXXXXXXXXXXXXXXXXXXXXXXXXXXXXXXXXXXXXXXXXXXXXXXXXXXXXXXX",
        10,10,5);
    dos[62] = 0;

    set_view(FULL);                     /* Clear viewport and screen */
    system("CLS");
    clearviewport();

    printf("%s\n",dos);                 /* Execute DOS command       */
    system(dos);
    printf("Hit any key to continue\n");
    get_key();
    system("CLS");

    set_screen();

    if (num_pnts > 0)
    {
        d_path();
        d_gcode(blank);
    }

    return;
}                                       /* End of function doscommand*/


/**********************************************************************/
/*                                                                    */
/*   FUNCTION fileread()                                              */
/*                                                                    */
/*   This function reads in the name of the gcode file to be          */
/*   simulated.                                                       */
/*                                                                    */
/**********************************************************************/
void fileread()
{
    set_view(GCODE);
    setfillstyle(1,BLUE);
    setcolor(BLUE);
    bar3d(0,0,317,73,0,1);
    input_str(gcodefile,
        "ENTER G-CODE FILE NAME: ","XXXXXXXXXXXX",11,30,4);
```

Appendix 16

```c
    setfillstyle(1,BLUE);
    setcolor(BLUE);
    bar3d(0,0,317,73,0,1);

    return;
}



/**********************************************************************/
/*                                                                    */
/*   FUNCTION firstpass()                                             */
/*                                                                    */
/*   This function checks the gcode file for syntax and finds         */
/*   max's and mins of the object to be machined.                     */
/*                                                                    */
/**********************************************************************/

void firstpass()
{
    int    i;
    int    cnt1,cnt2;                    /* Initializing variables    */

    char   dumm[2];  dumm[0] = 32;  dumm[1] = 0;

    set_screen();

    num_pnts = 0;                    /* reset num_pnts to 0, new object */

    for (i=0 ; i<=5 ; i++)           /* Blank out gcode string area*/
        strcpy(dgcode[i]," ");

    errorfile = fopen("gcode.err","w");

    printf("\n");
    printf("\n");

    if (parse(gcodefile) == 0)              /* parse raw gcode file */
        return;

    xprev=0;        yprev=0;        zprev=0;          /* old gcode point */
    xmax=-3.0E38; ymax=-3.0E38; zmax=-3.0E38;
    xmin= 3.0E38; ymin= 3.0E38; zmin= 3.0E38;

    error_flag=FALSE;  spin_flag=FALSE;  tool_flag=FALSE;
    feed_flag=FALSE;
    strt_flag=FALSE;   stop_flag=FALSE;  pend_flag=FALSE;  g_flag=FALSE;
    g92_flag=FALSE;    gz_flag=FALSE;
    g17_flag=TRUE;     g18_flag=FALSE;   g19_flag=FALSE;

    if (machine==LATHE)
    {
        g17_flag=FALSE;
        g18_flag=TRUE;
    }
```

SIMULATE.C - Simulate Tool Path

```
    counter=0;

    for (cnt1=0;cnt1<2;cnt1++)
       for (cnt2=0;cnt2<10;cnt2++)
          err_array[cnt1][cnt2]=FALSE;

    g20_flag=FALSE;                     /* Using mm, True if inches */
    g90_flag=TRUE;                      /* Absolute, False if incremental */

    first();                            /* process max/min/errors */

    printf("\n");
    int_error();            /* Interpret err_array from first pass */
    setcolor(BLACK);
    bar3d(110,75,270,85,0,1);

    fclose(errorfile);

    if (error_flag==FALSE)
    {
       g20_flag=FALSE;                  /* Using mm, True if inches */
       g90_flag=TRUE;                   /* Absolute, False if incremental */

       g17_flag=TRUE;  g18_flag=FALSE;  g19_flag=FALSE;

       if (machine==LATHE)
       {
          g17_flag=FALSE;
          g18_flag=TRUE;
       }

       secondpass();                    /* Draw tool path */
    }
    else
    {
       set_view(GCODE);
       setfillstyle(1,BLUE);
       setcolor(BLUE);
       bar3d(0,0,317,73,0,1);
       input_str(dumm," G-CODE File contains errors","X",41,18,2);
       input_str(dumm," Errors placed in GCODE.ERR","X",45,45,2);
       get_key();
       setfillstyle(1,BLUE);
       setcolor(BLUE);
       bar3d(0,0,317,73,0,1);
       set_view(FULL);
    }

    return;
}                                       /* End of function firstpass */


/***************************************************************/
/*                                                             */
/* FUNCTION secondpass()                                       */
```

```
/*                                                                    */
/*   This function process the parsed gcode file and draws the        */
/*   tool path on the screen.                                         */
/*                                                                    */
/**********************************************************************/

void secondpass()
{
   xprev=0;        yprev=0;        zprev=0;        /* old gcode point */

   g20_flag=FALSE;                    /* Using mm, True if inches */
   g90_flag=TRUE;                     /* Absolute, False if incremental */
   initial_flag=TRUE;                 /* First point? True until found */

   strcpy(spindle,"");     strcpy(feedrate,"");
   strcpy(toolnum,"");     strcpy(gcodest,"");

   second();                          /* Draw tool path of object */

   return;
}                                     /* End of function secondpass */


/**********************************************************************/
/*                                                                    */
/* Function first()                                                   */
/*                                                                    */
/* Opens the parsed gcode file and reads in one block.  A block       */
/* may contain up to 352 characters and is delimited by a             */
/* linefeed <10>. FIRST then calls PROCESS which sifts through        */
/* the gcode to determine Xmax, Ymax, Zmax, Xmin, Ymin, Zmin          */
/* of the part.  It also checks for gcode errors. (no feedrate,       */
/* spindle not turned on, etc).                                       */
/*                                                                    */
/* Note : A Block can have no more than 19 TOKENS.  Tokens            */
/*        alternate Char, Num, Char, Num, ..... until an *            */
/*        is encountered followed by SNum, String.                    */
/*                                                                    */
/*        Char token is character, Num token is up to 15 chars,       */
/*        SNum is up to 5 characters. String token is up to 200       */
/*        characters (but will be truncated to 39 characters).        */
/*                                                                    */
/*        Therefore a Block is a maximum of 352 characters            */
/*                                                                    */
/**********************************************************************/

void first()
{
   char block[BLK_SIZE];    /* One block of gcode commands*/
   FILE *gcode;

/******************************/
/*   Open GCODE Parsed File   */
/******************************/

   if((gcode=fopen("gcode.dat","r"))==NULL)
```

SIMULATE.C - Simulate Tool Path

```c
    {
       printf("Cannot Open GCODE.DAT\n");
       exit(1);
    }
/*****************************/
/*        Process Data       */
/*****************************/
    strcpy(block,"");
    while(fgets(block,BLK_SIZE,gcode)!=NULL)   /* read in data*/
    {
       block[strlen(block)]='\0';          /* terminate block*/
       process(block);
       strcpy(block,"");
    }                         /* End while block!=EOF*/

    fclose(gcode);      /* Close GCODE Parsed File*/
    return;

} /* End of function first*/



/*********************************************************************/
/*                                                                   */
/* Function process(char blk[])                                      */
/*                                                                   */
/* PROCESS is passed a block of gcode and sifts through it to        */
/* determine errors and min/max values.                              */
/*                                                                   */
/*********************************************************************/

void process(char blk[])
{
    char    let_tok[2],  /* one letter token*/
            gc[200],     /* gcode string*/
            gctrun[40];  /* truncated gcode string*/

    int     fin_flag,    /* flags when block has been processed*/
            indx,        /* index for a loop*/

            x_flag,      /* new x coordinate ?*/
            y_flag,      /* new y coordinate ?*/
            z_flag,      /* new z coordinate ?*/
            i_flag,      /* new i coordinate ?*/
            j_flag,      /* new j coordinate ?*/
            k_flag,      /* new k coordinate ?*/
            rad_flag,    /* new r coordinate ?*/

            tnum,        /* Tool Number*/
            mnum,        /* M code Number*/
            gnum,        /* G code Number*/
            snum;        /* Spindle Speed*/

    float   x,y,z,       /* Coordinates*/
            feed,        /* Feed rate*/
            rad,         /* Radius*/
```

Appendix 16

```c
        i,j,k;          /* Arc information*/

   char  *TokPnt;

   char  CheckString[512];

/*********************************/
/*    Initialize variables       */
/*********************************/

   x=0.0;   y=0.0;   z=0.0;   feed=0.0;   rad=0.0;   i=0.0;   j=0.0;   k=0.0;
   tnum=0;  mnum=0;  gnum=0;  snum=0;

   fin_flag=FALSE;     x_flag=FALSE;      y_flag=FALSE;      z_flag=FALSE;
   i_flag=FALSE;       j_flag=FALSE;      k_flag=FALSE;      rad_flag=FALSE;
   strcpy(gc,"");
   strcpy(gctrun,"");


   strcpy( CheckString, blk );
   if( (TokPnt = strtok(blk," ") ) == NULL )
   {
      printf( "ERROR: EMPTY BLOCK IN GCODE FILE\n" );

      exit( 1 );
   }

   strcpy(let_tok,TokPnt );           /* Get first letter token*/
   counter++;                         /* increment counter*/

   while( fin_flag==FALSE )
   {
      switch( let_tok[0] )            /* Branch on character and do*/
      {                               /* appropriate processing    */
         case 'G':
             gnum=atoi(strtok(NULL," "));

             if ((gnum==50)&&(machine==LATHE))
                gnum=92;

             if ((gnum<GMINR)||(gnum>GMAXR))    /* check ranges*/
             {
                error_flag=TRUE;
                error_num(2,gnum,x);
             }

             if (((gnum==1)||(gnum==2)||
               (gnum==3))&&(g_flag==FALSE)) /* !=G00 encountered*/
             {
                g_flag=TRUE;
                err_array[0][4]=TRUE;
                err_array[1][4]=counter;
             }

             if (((gnum==0)||(gnum==1)||(gnum==2)||
                 (gnum==3))&&(gz_flag==FALSE))
             {
                gz_flag=TRUE;
```

SIMULATE.C - Simulate Tool Path

```c
            err_array[0][8]=TRUE;
            err_array[1][8]=counter;
        }

        if ((gnum == 92)&&(g92_flag==FALSE))
        {
            g92_flag=TRUE;
            err_array[0][7]=TRUE;
            err_array[1][7]=counter;
        }


        if (gnum==20)
            g20_flag=TRUE;          /* inches selected*/

        if (gnum==21)
            g20_flag=FALSE;         /* mm selected*/

        if (gnum==90)
            g90_flag=TRUE;          /* absoulute mode*/

        if (gnum==91)
            g90_flag=FALSE;         /* incremental mode*/

        if (machine==MILL)
        {
            if (gnum==17)
            {
                g17_flag=TRUE;
                g18_flag=FALSE;
                g19_flag=FALSE;
            }

            if (gnum==18)
            {
                g17_flag=FALSE;
                g18_flag=TRUE;
                g19_flag=FALSE;
            }

            if (gnum==19)
            {
                g17_flag=FALSE;
                g18_flag=FALSE;
                g19_flag=TRUE;
            }
        }
        break;

case 'X':
    if( (TokPnt = strtok(NULL," ") )== NULL )
    {
      printf( "ERROR: PREMATURE END OF CODE LINE\n" );
      printf( "%s", CheckString );
      exit( 1 );
    }
```

Appendix 16

```c
            x=atof(TokPnt);

            if ((x<XMINR)||(x>XMAXR)) /* check ranges*/
            {
                error_flag=TRUE;
                error_num(3,gnum,x);
                break;
            }

            if (machine==LATHE)
                x=x/2;

            x_flag=TRUE;
            break;

    case 'Y':
            if( (TokPnt = strtok(NULL," ") ) == NULL )
            {
              printf( "ERROR: PREMATURE END OF CODE LINE\n" );
              printf( "%s", CheckString );
              exit( 1 );
            }
            y=atof( TokPnt );

            if ((y<YMINR)||(y>YMAXR)) /* check ranges*/
            {
                error_flag=TRUE;
                error_num(4,gnum,y);
                break;
            }

            y_flag=TRUE;
            break;

    case 'Z':
            if( (TokPnt = strtok(NULL," ") ) == NULL )
            {
              printf( "ERROR: PREMATURE END OF CODE LINE\n" );
              printf( "%s", CheckString );
              exit( 1 );
            }

            z=atof( TokPnt );

            if ((z<ZMINR)||(z>ZMAXR)) /* check ranges*/
            {
                error_flag=TRUE;
                error_num(5,gnum,z);
                break;
            }

            z_flag=TRUE;
            break;

    case 'S':
```

SIMULATE.C - Simulate Tool Path

```c
        if( (TokPnt = strtok(NULL," ") ) == NULL )
        {
          printf( "ERROR: PREMATURE END OF CODE LINE\n" );
          printf( "%s", CheckString );
          exit( 1 );
        }

        snum=atoi( TokPnt );

        if ((snum<SMINR)||(snum>SMAXR))    /* check ranges*/
        {
            error_flag=TRUE;
            error_num(6,snum,x);
        }

        if (spin_flag==FALSE)     /* spindle speed set*/
        {
            spin_flag=TRUE;
            err_array[0][0]=TRUE;
            err_array[1][0]=counter;
        }
        break;

  case 'I':
        if( (TokPnt = strtok(NULL," ") ) == NULL )
        {
          printf( "ERROR: PREMATURE END OF CODE LINE\n" );
          printf( "%s", CheckString );
          exit( 1 );
        }

        i=atof( TokPnt );

        if ((i<IMINR)||(i>IMAXR)) /* check ranges*/
        {
            error_flag=TRUE;
            error_num(7,gnum,i);
            break;
        }

        i_flag=TRUE;
        break;

  case 'J':
        if( (TokPnt = strtok(NULL," ") ) == NULL )
        {
          printf( "ERROR: PREMATURE END OF CODE LINE\n" );
          printf( "%s", CheckString );
          exit( 1 );
        }
        j=atof( TokPnt );

        if ((j<JMINR)||(j>JMAXR)) /* check ranges*/
        {
            error_flag=TRUE;
            error_num(8,gnum,j);
```

```
                break;
            }

            j_flag=TRUE;
            break;

    case 'K':
            if( (TokPnt = strtok(NULL," ") ) == NULL )
            {
              printf( "ERROR: PREMATURE END OF CODE LINE\n" );
              printf( "%s", CheckString );
              exit( 1 );
            }

            k=atof( TokPnt );

            if ((k<KMINR)||(k>KMAXR)) /* check ranges*/
            {
                error_flag=TRUE;
                error_num(9,gnum,k);
                break;
            }

            k_flag=TRUE;
            break;


    case 'R':
            if( (TokPnt = strtok(NULL," ") ) == NULL )
            {
              printf( "ERROR: PREMATURE END OF CODE LINE\n" );
              printf( "%s", CheckString );
              exit( 1 );
            }

            rad=atof( TokPnt );

            if ((rad<RMINR)||(rad>RMAXR)) /* check ranges*/
            {
                error_flag=TRUE;
                error_num(10,gnum,rad);
                break;
            }

            rad_flag=TRUE;
            break;


    case 'F':
            if( (TokPnt = strtok(NULL," ") ) == NULL )
            {
              printf( "ERROR: PREMATURE END OF CODE LINE\n" );
              printf( "%s", CheckString );
              exit( 1 );
            }
            feed=atof( TokPnt );
```

```
         if ((feed<FMINR)||(feed>FMAXR))    /* check ranges*/
         {
            error_flag=TRUE;
            error_num(11,gnum,feed);
         }

         if (feed_flag==FALSE)     /* feedrate set*/
         {
            feed_flag=TRUE;
            err_array[0][2]=TRUE;
            err_array[1][2]=counter;
         }
         break;

      case 'T':
         if( (TokPnt = strtok(NULL," ") ) == NULL )
         {
           printf( "ERROR: PREMATURE END OF CODE LINE\n" );
           printf( "%s", CheckString );
           exit( 1 );
         }

         tnum=atoi( TokPnt );

         if ((tnum<TMINR)||(tnum>TMAXR))    /* check ranges*/
         {
            error_flag=TRUE;
            error_num(12,tnum,x);
         }

         if (tool_flag==FALSE)     /* tool selected*/
         {
            tool_flag=TRUE;
            err_array[0][1]=TRUE;
            err_array[1][1]=counter;
         }
         break;

      case 'M':
         if( (TokPnt = strtok(NULL," ") ) == NULL )
         {
           printf( "ERROR: PREMATURE END OF CODE LINE\n" );
           printf( "%s", CheckString );
           exit( 1 );
         }

         mnum=atoi( TokPnt );

         if ((mnum<MMINR)||(mnum>MMAXR))    /* check ranges*/
         {
            error_flag=TRUE;
            error_num(13,mnum,x);
         }

         if (((mnum==3)||(mnum==4))&&(strt_flag==FALSE))
         {                         /* spindle started*/
```

```c
                    strt_flag=TRUE;
                    err_array[0][3]=TRUE;
                    err_array[1][3]=counter;
                }

                if ((mnum==5)&&(stop_flag==FALSE)) /* spindle stop*/
                {
                    stop_flag=TRUE;
                    err_array[0][5]=TRUE;
                    err_array[1][5]=counter;
                }

                if ((mnum==30)&&(pend_flag==FALSE)) /* End of program*/
                {
                    pend_flag=TRUE;
                    err_array[0][6]=TRUE;
                    err_array[1][6]=counter;
                }
                break;


        case '*':
                if( (TokPnt = strtok(NULL,"\n") ) == NULL )
                {
                  printf( "ERROR: PREMATURE END OF CODE LINE\n" );
                  printf( "%s", CheckString );
                  exit( 1 );
                }
                strcpy(gc,TokPnt );    /* read in gcode string*/

                for (indx=0; indx<=38; indx++)
                {
                    gctrun[indx]=gc[indx];
                }
                gctrun[39]='\0';

                break;

        default:
                break;

    }  /* End of Switch (let_tok)*/

    if( (TokPnt = strtok(NULL," ") ) == NULL )
    {
        fin_flag=TRUE;
    }
    else
    {
        strcpy( let_tok, TokPnt );
    }

  }  /* End of While (fin_flag == FALSE)*/

/*******************************************************************/
/* Determine minimums and maximums                                 */
/* Take into consideration incremental or absolute and mm or inches*/
/*******************************************************************/
```

SIMULATE.C - Simulate Tool Path

```
if ((machine==LATHE)&&(y!=0.0))   /* Y must equal 0*/
{
   error_flag=TRUE;
   error_num(27,gnum,y);
}

if ((gnum==0)||(gnum==1)||(gnum==2)||(gnum==3)||(gnum==92))
{
   if (g20_flag==TRUE)
   {
      x*=25.4;
      y*=25.4;
      z*=25.4;
   }

   if (x_flag==TRUE)
   {
      if (g90_flag==FALSE)        /* incremental mode*/
         x+=xprev;
   }
   else       /* x_flag==FALSE*/
      x=xprev;

   if (y_flag==TRUE)
   {
      if (g90_flag==FALSE)        /* incremental mode*/
         y+=yprev;
   }
   else       /* y_flag==FALSE*/
      y=yprev;

   if (z_flag==TRUE)
   {
      if (g90_flag==FALSE)        /* incremental mode*/
         z+=zprev;
   }
   else       /* z_flag==FALSE*/
      z=zprev;

   if (machine==LATHE)
      x=x/2;

   if ((gnum!=2)&&(gnum!=3))      /* Not circle adjust x,y,z*/
   {
      if (x>xmax)                 /* determine xmax,xmin of part*/
         xmax=x;

      if (x<xmin)
         xmin=x;

      if (y>ymax)                 /* determine ymax,ymin of part*/
         ymax=y;

      if (y<ymin)
         ymin=y;

      if (z>zmax)                 /* determine zmax,zmin of part*/
```

```
                    zmax=z;

            if (z<zmin)
                zmin=z;
        }
        else       /* have a circle, adjust x,y,z*/
        {
            if (((i_flag==TRUE)||(j_flag==TRUE))||(k_flag==TRUE))
                circ(xprev,yprev,zprev,x,y,z,0.0,x+i,y+j,z+k,gnum,0);

            if (rad_flag==TRUE)
                circ(xprev,yprev,zprev,x,y,z,rad,0.0,0.0,0.0,gnum,0);
        }

     xprev=x;   yprev=y;   zprev=z;
                       /* Save new values of x,y,z as previous*/

     }/* End of if gnum=0|1|2|3*/
     return;
} /* End of Function Process*/




/**********************************************************************/
/*                                                                    */
/* Function int_error()                                               */
/*                                                                    */
/* Interpret the err_array and display error messages.                */
/* Uses global variables: counter, error_flag, err_array              */
/*                                                                    */
/**********************************************************************/

void int_error()
{
    int dum1;
    float dum2;

    dum1=0;  dum2=0.0;


    if (err_array[0][0]==FALSE)
    {
        error_flag=TRUE;
        error_num(14,dum1,dum2);
    }

    if (err_array[0][1]==FALSE)
    {
        error_flag=TRUE;
        error_num(15,dum1,dum2);
    }

    if (err_array[0][2]==FALSE)
    {
        error_flag=TRUE;
        error_num(16,dum1,dum2);
    }
```

```c
if (err_array[0][3]==FALSE)
{
   error_flag=TRUE;
   error_num(17,dum1,dum2);
}
else
{
   if (err_array[1][0]>err_array[1][3])
   {
      error_flag=TRUE;
      error_num(18,dum1,dum2);
   }

   if (err_array[1][1]>err_array[1][3])
   {
      error_flag=TRUE;
      error_num(19,dum1,dum2);
   }

   if (err_array[1][3]>err_array[1][4])
   {
      error_flag=TRUE;
      error_num(21,dum1,dum2);
   }

   if (err_array[1][3]>err_array[1][6])
   {
      error_flag=TRUE;
      error_num(22,dum1,dum2);
   }
}
if (err_array[0][4]==FALSE)
{
   error_flag=TRUE;
   error_num(23,dum1,dum2);
}

if (err_array[0][5]==FALSE)
{
   error_flag=TRUE;
   error_num(24,dum1,dum2);
}

if (err_array[0][6]==FALSE)
{
   error_flag=TRUE;
   error_num(25,dum1,dum2);
}
else
{
   if ((err_array[1][6]!=(counter-1))&&(err_array[1][6]!=
            (counter-2)))
   {
      error_flag=TRUE;
      error_num(26,dum1,dum2);
   }
```

Appendix 16

```c
    }

    if (err_array[0][7]==FALSE)
    {
        error_flag=TRUE;
        error_num(20,dum1,dum2);
    }
    else
        if (err_array[1][8]<=err_array[1][7])
        {
            error_flag=TRUE;
            error_num(28,dum1,dum2);
        }

    return;
}

#include <stdio.h>
#include <conio.h>
#include <ctype.h>
#include <io.h>
#include <fcntl.h>
#include <mem.h>
#include <dos.h>
#include <bios.h>
#include <stdlib.h>
#include <string.h>
#include <alloc.h>
#include <math.h>
#include <graphics.h>
#include "cam.h"

/*********************************************************************/
/*                    EXTERNAL VARIABLES                             */
/*********************************************************************/

extern int      foreground, background;
extern int      mouse_flag, loop_flag;
extern int      scale_flag;
extern int      machine;
extern int      num_pnts;
extern int      move_color;
extern int      cut_color;
extern int      xasp,yasp;
extern int      speed;

extern float    xyratio;

extern char     single_step;
extern char     dgcode[6][40];
extern char     blank[40];


extern struct pnt
{
    int   tool;
    float x,y,z,xz,xiso,yiso;
    float xwc,ywc,zwc;
```

SIMULATE.C - Simulate Tool Path

```c
};

extern struct pnt point[max_pnts];



extern FILE     *instream;
extern FILE     *outstream;
extern FILE     *errorfile;

extern char     gcode[6];
extern char     last_gcode[6];
extern char     X[15],Y[15],Z[15];
extern char     I[15],J[15],K[15],R[15];


extern char     spindle[15],
        feedrate[15],
        toolnum[15],
        gcodefile[12],
        gcodest[40];

extern float    xmax,ymax,zmax,xmin,ymin,zmin;
extern float    xprev,yprev,zprev;

extern int      error_flag,
        spin_flag,
        tool_flag,
        feed_flag,
        strt_flag,
        g_flag,
        stop_flag,
        pend_flag,
        g92_flag,
        gz_flag,

        g17_flag,
        g18_flag,
        g19_flag,

        g20_flag,
        g90_flag,
        initial_flag,

        counter,


        err_array[2][10];


/************************************************************/
/*                                                          */
/* Function error_num(int err_num, float par1)-             */
/*                                                          */
/* This function receives an err_num, parameter 1 and       */
/* parameter 2 and prints out an error message on the       */
```

```c
/* screen.                                                            */
/*                                                                    */
/**********************************************************************/

void error_num(int err_num, int par1, float par2)
{
    switch(err_num)                 /* Branch on error number */
    {
        case 2:
            fprintf(errorfile,"G code out of range %d\n",par1);
            break;

        case 3:
            fprintf(errorfile,"X Coordinate out of range %f\n",par2);
            break;

        case 4:
            fprintf(errorfile,"Y Coordinate out of range %f\n",par2);
            break;

        case 5:
            fprintf(errorfile,"Z Coordinate out of range %f\n",par2);
            break;

        case 6:
            fprintf(errorfile,"S code out of range %d\n",par1);
            break;

        case 7:
            fprintf(errorfile,"I Coordinate out of range %f\n",par2);
            break;

        case 8:
            fprintf(errorfile,"J Coordinate out of range %f\n",par2);
            break;

        case 9:
            fprintf(errorfile,"K Coordinate out of range %f\n",par2);
            break;

        case 10:
            fprintf(errorfile,"Radius out of range %f\n",par2);
            break;

        case 11:
            fprintf(errorfile,"Feedrate out of range %f\n",par2);
            break;

        case 12:
            fprintf(errorfile,"T code out of range %d\n",par1);
            break;

        case 13:
            fprintf(errorfile,"M code out of range %d\n",par1);
            break;

        case 14:
            fprintf(errorfile,"Spindle speed has not been set\n");
```

```
            break;

    case 15:
        fprintf(errorfile,"Tool has not been selected\n");
        break;

    case 16:
        fprintf(errorfile,"Feedrate has not been specified\n");
        break;

    case 17:
        fprintf(errorfile,"Spindle has not been started\n");
        break;

    case 18:
        fprintf(errorfile,
            "Spindle started before spindle speed was set\n");
        break;

    case 19:
        fprintf(errorfile,"Spindle started before tool selected\n");
        break;

    case 20:
        fprintf(errorfile,"G92/50 Absolute Zero Point not found\n");
        break;

    case 21:
        fprintf(errorfile,
"G code other than G00 encountered before spindle was started\n");
        break;

    case 22:
        fprintf(errorfile,
 "End of program encountered before spindle was started\n");
        break;

    case 23:
        fprintf(errorfile,
   "No G codes other than G00 encountered, cannot process\n");
        break;

    case 24:
        fprintf(errorfile,"Spindle has not been stopped\n");
        break;

    case 25:
        fprintf(errorfile,"End of program not encountered\n");
        break;

    case 26:
        fprintf(errorfile,"Tape Rewind ,%, not found\n");
        break;

    case 27:
        fprintf(errorfile,
            "Program set up for Lathe, Y is %f must be 0\n",par2);
        break;
```

```
            case 28:
                fprintf(errorfile,
                    "G code 0/1/2/3 encountered before G92/50\n");
                break;

    } /* End of switch(err_num) */
    return;
} /* End of function error_num */




/***********************************************************************/
/*                                                                     */
/* Function second()                                                   */
/*                                                                     */
/* Opens the error checked parsed gcode file and reads in one          */
/* block.  SECOND then calls PROCESS2 which reads the gcode            */
/* and takes the appropriate action: cut, move, and display            */
/* data.                                                               */
/*                                                                     */
/***********************************************************************/

void second()
{
    char block[BLK_SIZE];  /* One block of gcode commands */
    FILE *gcode;

/*******************************/
/*   Open GCODE Parsed File   */
/*******************************/

    if((gcode=fopen("gcode.dat","r"))==NULL)
    {
        printf("Cannot Open File\n");
        exit(1);
    }

/*******************************/
/*        Process Data         */
/*******************************/

    strcpy(block,"");
    while(fgets(block,BLK_SIZE,gcode)!=NULL)   /* read in data */
    {
        block[strlen(block)]=NULL;          /* end block */
        process2(block);
        strcpy(block,"");
    }                       /* End while block!=EOF */

    fclose(gcode);          /* Close GCODE Parsed File */
    return;

} /* End of function first */
```

SIMULATE.C - Simulate Tool Path

```c
/*********************************************************************/
/*                                                                   */
/* Function process2(char blk[BLK_SIZE])                             */
/*                                                                   */
/* PROCESS is passed a block of gcode and displays the tool          */
/* paths and information.                                            */
/*                                                                   */
/*********************************************************************/

void process2(char blk[BLK_SIZE])
{
    char    let_tok[1],  /* one letter token */
            gc[200],     /* gcode string */
            dumm[2];     /* dummy character string used in input_str */

    int     fin_flag,    /* flags when block has been processed */
            indx,        /* index for a loop */

            gen_flag,    /* G 0,1,2,3,92 encountered */
            x_flag,      /* new x coordinate ? */
            y_flag,      /* new y coordinate ? */
            z_flag,      /* new z coordinate ? */
            i_flag,      /* new i coordinate ? */
            j_flag,      /* new j coordinate ? */
            k_flag,      /* new k coordinate ? */
            rad_flag,    /* new r coordinate ? */

            mnum,        /* M code Number */
            gnum;        /* G code Number */

    float   x,y,z,       /* Coordinates */
            rad,         /* Radius */
            i,j,k,       /* Arc information */
            tempx,tempy,tempz;   /* temporary x,y,z variables */

    char    *TokPnt;

/*********************************/
/*      Initialize variables     */
/*********************************/

    x=0.0;   y=0.0;   z=0.0;   rad=0.0;   i=0.0;   j=0.0;   k=0.0;
    mnum=0;  gnum=0;
    tempx=0.0;  tempy=0.0;  tempz=0.0;  dumm[0]=32;  dumm[1]=0;

    fin_flag=FALSE;   x_flag=FALSE;   y_flag=FALSE;   z_flag=FALSE;
    i_flag=FALSE;     j_flag=FALSE;   k_flag=FALSE;   rad_flag=FALSE;
    gen_flag=FALSE;
    strcpy(gc,"");

    if( (TokPnt=strtok(blk," ")) == NULL )
    {
        printf( "ERROR: EMPTY GCODE LINE\n" );
        exit( 1 );
    }

    strcpy( let_tok,TokPnt );         /* Get first letter token */
```

Appendix 16

```c
while (fin_flag==FALSE)
{
   switch(let_tok[0])              /* Branch on character and do */
   {                               /* appropriate processing     */
      case 'G':
         if( (TokPnt=strtok(NULL," ")) == NULL )
         {
            printf( "ERROR: PREMATURE END OF GCODE LINE\n" );
            exit( 1 );
         }
         gnum=atoi( TokPnt );

         if ((gnum==50)&&(machine==LATHE))
            gnum=92;

         if ((gnum==0)||(gnum==1)||
            (gnum==2)||(gnum==3)||(gnum==92))
            gen_flag=TRUE;

         if (gnum==20)
            g20_flag=TRUE;          /* inches selected */

         if (gnum==21)
            g20_flag=FALSE;         /* mm selected */

         if (gnum==90)
            g90_flag=TRUE;          /* absoulute mode */

         if (gnum==91)
            g90_flag=FALSE;         /* incremental mode */

         if (machine==MILL)
         {
            if (gnum==17)
            {
               g17_flag=TRUE;
               g18_flag=FALSE;
               g19_flag=FALSE;
            }

            if (gnum==18)
            {
               g17_flag=FALSE;
               g18_flag=TRUE;
               g19_flag=FALSE;
            }

            if (gnum==19)
            {
               g17_flag=FALSE;
               g18_flag=FALSE;
               g19_flag=TRUE;
            }
         }
         break;

      case 'X':
```

```c
            if( (TokPnt=strtok(NULL," ")) == NULL )
            {
               printf( "ERROR: PREMATURE END OF GCODE LINE\n" );
               exit( 1 );
            }
            x=atof( TokPnt );

            if (machine==LATHE)
               x=x/2;

            x_flag=TRUE;
            break;

      case 'Y':
            if( (TokPnt=strtok(NULL," ")) == NULL )
            {
               printf( "ERROR: PREMATURE END OF GCODE LINE\n" );
               exit( 1 );
            }
            y=atof( TokPnt );
            y_flag=TRUE;
            break;

      case 'Z':
            if( (TokPnt=strtok(NULL," ")) == NULL )
            {
               printf( "ERROR: PREMATURE END OF GCODE LINE\n" );
               exit( 1 );
            }
            z=atof( TokPnt );
            z_flag=TRUE;
            break;

      case 'S':
            if( (TokPnt=strtok(NULL," ")) == NULL )
            {
               printf( "ERROR: PREMATURE END OF GCODE LINE\n" );
               exit( 1 );
            }
            strcpy(spindle,TokPnt );
            break;

      case 'I':
            if( (TokPnt=strtok(NULL," ")) == NULL )
            {
               printf( "ERROR: PREMATURE END OF GCODE LINE\n" );
               exit( 1 );
            }
            i=atof( TokPnt );
            i_flag=TRUE;
            break;

      case 'J':
```

```c
            if( (TokPnt=strtok(NULL," ")) == NULL )
            {
               printf( "ERROR: PREMATURE END OF GCODE LINE\n" );
               exit( 1 );
            }
            j=atof( TokPnt );
            j_flag=TRUE;
            break;

      case 'K':
            if( (TokPnt=strtok(NULL," ")) == NULL )
            {
               printf( "ERROR: PREMATURE END OF GCODE LINE\n" );
               exit( 1 );
            }
            k=atof( TokPnt );
            k_flag=TRUE;
            break;

      case 'R':
            if( (TokPnt=strtok(NULL," ")) == NULL )
            {
               printf( "ERROR: PREMATURE END OF GCODE LINE\n" );
               exit( 1 );
            }
            rad=atof( TokPnt );
            rad_flag=TRUE;
            break;

      case 'F':
            if( (TokPnt=strtok(NULL," ")) == NULL )
            {
               printf( "ERROR: PREMATURE END OF GCODE LINE\n" );
               exit( 1 );
            }
            strcpy(feedrate,TokPnt );
            break;

      case 'T':
            if( (TokPnt=strtok(NULL," ")) == NULL )
            {
               printf( "ERROR: PREMATURE END OF GCODE LINE\n" );
               exit( 1 );
            }
            strcpy(toolnum, TokPnt );
            break;

      case 'M':
            if( (TokPnt=strtok(NULL," ")) == NULL )
            {
               printf( "ERROR: PREMATURE END OF GCODE LINE\n" );
               exit( 1 );
            }
```

```
    }

    if (mnum==1)                 /* optional stop - hit key to continue */
    {
        set_view(GCODE);
        setfillstyle(1,BLUE);
        setcolor(BLUE);
        bar3d(0,0,317,73,0,1);
        input_str(dumm,
                "Optional Stop M01, Hit Key to cont.","X",15,30,3);
        get_key();
        setfillstyle(1,BLUE);
        setcolor(BLUE);
        bar3d(0,0,317,73,0,1);
        d_gcode(blank);
    }

    if (single_step == 'Y')
        if (get_key() == ESC)
            single_step = 'N';
        else;
    else
        delay (speed);

    hot_key();

    return;
} /* End of Function Process2 */




/*****************************************************************/
/*                                                               */
/* FUNCTION getlinechar() returns the character at index *j and  */
/* increments *j.                                                */
/*                                                               */
/*****************************************************************/
int getlinechar( ln, j )
char ln[MAXLINELENGTH];
int *j;
{
        int i;

        i = *j;
        (*j)++;
        return( ln[i] );
}




/*****************************************************************/
/*                                                               */
/* FUNCTION getnextparm() reads in characters up to but not including */
/* the  next letter stripping excess blanks and inserting a single    */
/* leading blank and a single trailing blank. The current character of*/
```

```
/* the instream is updated using pc (pointer to character).           */
/*                                                                    */
/**********************************************************************/

void getnextparm( ln, j, s, pc )
char ln[MAXLINELENGTH];/* input string */
int *j;              /* input string index */
char *s;             /* output string */
int *pc;             /* character pointer */
{
        int i;

/* insert leading space */
        s[0] = ' ';
        i = 1;

        *pc = getlinechar( ln, j );
        while ( !isalpha(*pc) && *pc != ';' )
        {
        /* copy nonblank characters to output string and increment i */
                if ( !isspace(*pc) )
                {
                        s[i] = *pc;
                        i++;
                }

        /* get next character */
                *pc = getlinechar( ln, j );

        }
/* insert trailing space */
        s[i] = ' ';
        s[i+1] = '\0';
}




/**********************************************************************/
/*                                                                    */
/* FUNCTION getnext() calls getnextparm repeatedly until a G,F,T,S,   */
/* or M is encountered.                                               */
/*                                                                    */
/**********************************************************************/

void getnext( ln, j, s, pc )
char ln[MAXLINELENGTH];/* input string */
int *j;              /* input string index */
char *s;             /* output string */
int *pc;             /* character pointer */

{
        while( *pc != 'G' &&
                        *pc != 'F' &&
                        *pc != 'T' &&
                        *pc != 'S' &&
                        *pc != 'M' &&
                        *pc != ';' )
```

```
                {
                        getnextparm( ln, j, s, pc );
                }
}




/****************************************************************/
/*                                                              */
/* FUNCTION echoline() prints original gcode line at end of parsed */
/* gcode line in output file.  * character marks beginning of original*/
/* line.'\n' character appended to output line.                 */
/*                                                              */
/****************************************************************/
void echoline( line, writeorig )
char line[MAXLINELENGTH];
int *writeorig;
{

/* set gcode if necessary */
        if (gcode[0] == '\0' &&
                   (X[0] != '\0' || Y[0] != '\0' || Z[0] != '\0'))
                strcpy( gcode, last_gcode );

        fputs( gcode, outstream );
        fputs( X, outstream );
        fputs( Y, outstream );
        fputs( Z, outstream );
        fputs( I, outstream );
        fputs( J, outstream );
        fputs( K, outstream );
        fputs( R, outstream );
        gcode[0] = '\0';
        X[0] = '\0';
        Y[0] = '\0';
        Z[0] = '\0';
        I[0] = '\0';
        J[0] = '\0';
        K[0] = '\0';
        R[0] = '\0';

/* write out original gcode comment line on first output line only   */
        if (*writeorig)
        {
                fprintf( outstream, "\t\t * " );
                fputs( line, outstream );
                *writeorig = FALSE;
        }
}



/****************************************************************/
/*                                                              */
```

Appendix 16

```c
/* FUNCTION isgcode() returns true if the gcode is one of the         */
/* following:  G00,G01,G02,G03,G20,G21,G90,G91.                        */
/*                                                                     */
/***********************************************************************/

int isgcode(char code[6])
{
        int num;

/* parse digits from code and check if in valid list */
        if ( ( (num = atoi( &code[2] )) == 0 ||
                        num == 1 ||
                        num == 2 ||
                        num == 3 ||
                        num == 17 ||
                        num == 18 ||
                        num == 19 ||
                        num == 20 ||
                        num == 21 ||
                        num == 50 ||
                        num == 90 ||
                        num == 91 ||
                        num == 92 )
                return( TRUE );
        else return( FALSE );
}




/***********************************************************************/
/*                                                                     */
/* FUNCTION parse() reads an input gcode file and produces a           */
/* parsed intermediate file for input to the validation routines.      */
/*  G F N T O S M   X Y Z I J K R                                      */
/*                                                                     */
/***********************************************************************/

int parse( char file[12] )
{
        char temp[15];
        char line[MAXLINELENGTH];
        char dumm[2]; /* dummy character string used in input_str */

        int i;
        int c;     /* last character read */
        int first_ln;
        /* newline flag used to determine when to write
           original gcodes */
        int SaveLine = TRUE;

        dumm[0] = 32; dumm[1] = 0;


/* open gcode file and parse file for output */
        if ((instream = fopen( file, "r")) == NULL)
        {
```

SIMULATE.C - Simulate Tool Path

```c
                set_view(GCODE);
                setfillstyle(1,BLUE);
                setcolor(BLUE);
                bar3d(0,0,317,73,0,1);
                input_str(dumm," FILE NOT FOUND","X",91,30,3);
                get_key();
                setfillstyle(1,BLUE);
                setcolor(BLUE);
                bar3d(0,0,317,73,0,1);
                d_gcode(blank);
                return 0;
        }

        set_view(TOP);
        outtextxy(110,75,"PLEASE WAIT...");

        outstream = fopen( "gcode.dat", "w"); /*###*/

        first_ln = TRUE;
        gcode[0] = '\0';
        X[0] = '\0';
        Y[0] = '\0';
        Z[0] = '\0';
        I[0] = '\0';
        J[0] = '\0';
        K[0] = '\0';
        R[0] = '\0';

/* determine first line */
        fgets( line, MAXLINELENGTH, instream );
        i = 0;

        while (!(feof( instream )))
        {
                c = getlinechar( line, &i );

                do
                {
                   switch (c)
                   {
                     case 'O':
                     case 'o':
                     case 'N':
                     case 'n':
                     {
                             temp[0] = c;
                             getnextparm( line, &i, &temp[1], &c );
                             break;
                     }

                     case 'X':
                     case 'x':
                     {
                             X[0] = c;
                             getnextparm( line, &i, &X[1], &c );
                             break;
                     }
                     case 'Y':
```

Appendix 16

```
case 'y':
{
        Y[0] = c;
        getnextparm( line, &i, &Y[1], &c );
        break;
}
case 'Z':
case 'z':
{
        Z[0] = c;
        getnextparm( line, &i, &Z[1], &c );
        break;
}

case 'I':
case 'i':
{
        I[0] = c;
        getnextparm( line, &i, &I[1], &c );
        break;
}
case 'J':
case 'j':
{
        J[0] = c;
        getnextparm( line, &i, &J[1], &c );
        break;
}
case 'K':
case 'k':
{
        K[0] = c;
        getnextparm( line, &i, &K[1], &c );
        break;
}
case 'R':
case 'r':
{
        R[0] = c;
        getnextparm( line, &i, &R[1], &c );
        break;
}

case 'G':
case 'g':
{
/* if G then if valid then store G and code in
     gcode */
/* if gcode not valid print tabs *, read and echo
     up to EOL */

        gcode[0] = c;
        getnextparm( line, &i, &gcode[1], &c );
        if ( isgcode( gcode ) )
        {
                strcpy( last_gcode, gcode );
        }
        else
```

SIMULATE.C - Simulate Tool Path

```c
                           {
                                  gcode[0] = '\0';
                                  getnext( line, &i, &temp[1], &c );
                           }
                           break;
              }
              case 'F':
              case 'f':
              case 'T':
              case 't':
              case 'S':
              case 's':
              case 'M':
              case 'm':
                    temp[0] = c;

/* write out G code + parameters only if necessary */
                    if (gcode[0] != '\0' ||
                        X[0] != '\0' ||
                        Y[0] != '\0' ||
                        Z[0] != '\0')
                          echoline( line, &first_ln );

/* parse and write out F,T,S,M code */
                    getnextparm( line, &i, &temp[1], &c );
                    fputs( temp, outstream );

/* if not first output line start new output line */
                    if (!first_ln)
                          fputc( '\n', outstream );

                    else
                              echoline( line, &first_ln );
              break;

              default:
                    if (c != 'G' && c != 'g' &&
                        c != 'F' && c != 'f' &&
                        c != 'S' && c != 's' &&
                        c != 'T' && c != 't' &&
                        c != 'M' && c != 'm' &&
                        c != 'N' && c != 'n' &&
                        c != 'O' && c != 'o' &&
                        c != 'X' && c != 'x' &&
                        c != 'Y' && c != 'y' &&
                        c != 'Z' && c != 'z' &&
                        c != 'I' && c != 'i' &&
                        c != 'J' && c != 'j' &&
                        c != 'K' && c != 'k' &&
                        c != 'R' && c != 'r' &&
                        c != ' ' && c != '%' &&
                        c != 'H' && c != 'D')
                    {

                          if( is_empty_string( line ) )
                          {
                              fprintf( errorfile,
```

```
                                "FOUND AND SKIPPED EMPTY LINE\n");
                            SaveLine = FALSE;
                        }
                        else
                        {
                            fprintf( errorfile,
                    "ERROR invalid letter specified in G-code\n" );
                            fprintf( errorfile,
                                "ERROR IN>> %s ", line );
                            fprintf( errorfile,
                                "INVALID LETTER >>%c\n", c );
                        }

                        set_view(TOP);
                        outtextxy(10,95,
                            "ERROR-CHECK FILE: GCODE.ERR" );

                    }

                    if( c != '\n' )
                    {
                        getnextparm( line, &i, &temp[1], &c );
                    }
                break;

            } /* end switch(c) */

        } while (c != ';' && c != '%' && c!='\n');

    /* write out string to parse file */
        if( SaveLine == TRUE )
        {
            echoline( line, &first_ln );
        }

        SaveLine = TRUE;

    /* determine new line */
        fgets( line, MAXLINELENGTH, instream );
        i = 0;
        first_ln = TRUE;

    } /* end while not EOF */

    fclose( instream );
    fclose( outstream );

    return 1;
} /* end function parse */



/********************************************************************/
/*                                                                  */
/* FUNCTION doxyz() checks the point passed to see if it exceeds any */
/* of the current global max or min values if draw is FALSE.  If draw */
/* is TRUE it calls line_pnts() to draw the line from the last point  */
```

```
/* that was passed.                                                         */
/*                                                                          */
/****************************************************************************/
void doxyz( float x, float y, float z, int draw )

{
   /* if draw flag then set point[num_pnts].tool = cut and call
      line_pnts(X Y Z) */
   if (draw)
   {
      point[num_pnts].tool = cut;
      normalize(&x,&y,&z,305,140,140);
      line_pts((int)x,(int)y,(int)z);
   }

   /* if !draw flag then check generated points against xmax, xmin,
      ymax, ymin, zmax, zmin */
   else
   {
      if (x < xmin) xmin = x;
      if (x > xmax) xmax = x;
      if (y < ymin) ymin = y;
      if (y > ymax) ymax = y;
      if (z < zmin) zmin = z;
      if (z > zmax) zmax = z;
   }
}


/****************************************************************************/
/*                                                                          */
/* FUNCTION circle() calculates the center point of the specified arc       */
/* if the center point is not specified in Xc, Yc, Zc,  the parameters      */
/* used are the start point, end point (subscripts s, e) and a radius       */
/* R.Xc, Yc, Zc must all be zero if R is specified and vice versa.          */
/* The flags dir and draw are used to determine the direction of            */
/* rotation   and whether or not to draw the arc using calls to             */
/* line_pnts().  If draw is false then the globals xmax, xmin, ymax,        */
/* ymin should be set if any of the generated arc points exceed the         */
/* current limits. The arc points are generated at 5 degree intervals       */
/* on the arc.                                                              */
/*                                                                          */
/****************************************************************************/
void circ(float Xs,float Ys,float Zs,float Xe,float Ye,float Ze,float
R,float Xc,float Yc,float Zc,int dir,int draw)

{
   double Xm,Ym,Zm;      /* midpoint between start and end points */
   double theta,phi,beta;
   double startangle,endangle;
   double X,Y,Z;
   double arcinterval = PI/36.0;

         if (Xs != Xe && Ys != Ye && Zs != Ze)
         {
```

Appendix 16

```
                fprintf(errorfile, "ERROR: non planar arc described" );
        }
/* calculate center point of circle if not specified in Xc, Yc, Zc */

        if (Xc == 0.0 && Yc == 0.0 && Zc == 0.0)
        {
/* find midpoint between start and end points: Xm, Ym, Zm */
                Xm = (Xe - Xs)/2;
                Ym = (Ye - Ys)/2;
                Zm = (Ze - Zs)/2;

        /* calculate angle between radial and midpoint segment */

                if (sqrt( Xm*Xm + Ym*Ym + Zm*Zm )/fabs(R) > 1.0)
                {
                    fprintf(errorfile,
                        "ERROR: illegal radius specified %d \n",R );
                    theta = 0.0;
                }
                else
                    theta =
                      acos( sqrt( Xm*Xm + Ym*Ym + Zm*Zm )/fabs(R) );

        /* find angle of midpoint vector */

                if (g17_flag)
                        beta = atan2( Ym, Xm );
                if (g18_flag)
                        beta = atan2( Xm, Zm );
                if (g19_flag)
                        beta = atan2( Zm, Ym );

        /* decide which center point is correct */

                if (( dir == CW && R > 0.0 ) ||
                    ( dir == CCW && R < 0.0 ))
                        phi = beta - theta;
                if (( dir == CW && R < 0.0 ) ||
                    ( dir == CCW && R > 0.0 ))
                        phi = beta + theta;

        /* calculate coordinates of center point */

                if (g17_flag)
                {
                        Xc = Xs + fabs(R)*cos( phi );
                        Yc = Ys + fabs(R)*sin( phi );
                        Zc = Ze;
                }
                if (g18_flag)
        {
                        Zc = Zs + fabs(R)*cos( phi );
                        Xc = Xs + fabs(R)*sin( phi );
                        Yc = Ye;
        }
                if (g19_flag)
        {
```

SIMULATE.C - Simulate Tool Path

```
                    Yc = Ys + fabs(R)*cos( phi );
                    Zc = Zs + fabs(R)*sin( phi );
                    Xc = Xe;
            }
    }
    else if (R == 0.0)
/* calculate radius */
        R = sqrt( (Xe-Xc)*(Xe-Xc) + (Ye-Yc)*(Ye-Yc) +
                  (Ze-Zc)*(Ze-Zc) );

/* generate points on the arc at arcinterval radian intervals */
/* calculate start and end angle */

    if (g17_flag)
    {
            startangle = atan2( Ys-Yc, Xs-Xc );
            endangle = atan2( Ye-Yc, Xe-Xc );
    }
    if (g18_flag)
    {
            startangle = atan2( Xs-Xc, Zs-Zc );
            endangle = atan2( Xe-Xc, Ze-Zc );
    }
    if (g19_flag)
    {
            startangle = atan2( Zs-Zc, Ys-Yc );
            endangle = atan2( Ze-Zc, Ye-Yc );
    }

/* convert both angles positive and fix arcinterval sign if necessary */
    if (dir == CW) arcinterval *= -1.0;
    if ((startangle < endangle) && (dir == CW)) startangle += 2*PI;
    if ((startangle > endangle) && (dir == CCW)) startangle -= 2*PI;

    for ( theta = startangle;
                fabs(theta - endangle) > fabs(arcinterval);
                theta += arcinterval )
    {
            if (g17_flag)
            {
                    X = Xc + fabs(R)*cos( theta );
                    Y = Yc + fabs(R)*sin( theta );
                    Z = Ze;
            }
            if (g18_flag)
            {
                    Z = Zc + fabs(R)*cos( theta );
                    X = Xc + fabs(R)*sin( theta );
                    Y = Ye;
            }
```

```
                if (g19_flag)
                {
                        Y = Yc + fabs(R)*cos( theta );
                        Z = Zc + fabs(R)*sin( theta );
                        X = Xe;
                }
      doxyz( X, Y, Z, draw );

   }

   doxyz( Xe, Ye, Ze, draw );
}
/**********************************************************************/
/*                                                                    */
/*  FUNCTION get_key()                                                */
/*                                                                    */
/*  Waits until a key is pressed or the mouse is moved and returns:   */
/*  a) ASCII value for regular keys                                   */
/*  b) SCAN code + 1000 for function keys                             */
/*  c) ASCII 13 (RETURN) if mouse button 1 is pressed.                */
/*  d) ASCII 27 (ESC) if mouse button 2 is pressed                    */
/*  e) SCAN code + 1000 of UP, DOWN, LEFT or RIGHT arrow keys if      */
/*     the mouse is moved in the corresponding direction.             */
/*                                                                    */
/**********************************************************************/

int get_key()

{
    int    x,y,button,a,res,key;
    int    xmin,xmax,ymin,ymax,cx,cy;

    if (mouse_flag == TRUE)
    {
        cx    = 160;
        cy    = 237;
        res   = 45;
        xmin  = cx - 1.5*res;
        xmax  = cx + 1.5*res;
        ymin  = cy - res;
        ymax  = cy + res;
        a     = 0;
        setcursorpos(cx,cy);
        do
        {
          getposbut(&button,&x,&y);
          key = bioskey(1);
          if (key == 0)
          {
              if (y > ymax)/* down */
                  a = DOWN;
              if (x > xmax)/* right */
                  a = RIGHT;
              if (y < ymin)/* up */
                  a = UP;
              if (x < xmin)/* left */
                  a = LEFT;
```

```
            if (button == 1)  /* return */
                a = RETURN;
            if (button == 2)  /* esc */
                a = ESC;
         }
         else
         {
           a = getch();
           if (a == 0)
               a = getch() + 1000;
         }
      }
      while (a == 0);

      do
         getposbut(&button,&x,&y);
      while (button != 0);
      setcursorpos(cx,cy);
   }
   else
   {
     a = getch();
     if (a == 0)
         a = getch() + 1000;
   }
   return a;
}


/****************************************************************/
/*                                                              */
/*  FUNCTION input_str                                          */
/*                                                              */
/*  Controls all user I/O to the screen.                        */
/*  Requires the following parameters:                          */
/*  1) *s       pointer to input string.                        */
/*  2) *m       pointer to message string                       */
/*  3) *p       pointer to picture string                       */
/*  4) h        horizontal position of input area.              */
/*  5) v        vertical position of input area.                */
/*  6) option   controls input area                             */
/*                                                              */
/*  puts input area on the screen in the following format:      */
/*                                                              */
/*     <MESSAGE STR> <INPUT STR>                                */
/*                                                              */
/*                                                              */
/*   picture string:       X - any character                    */
/*                         A - only letters                     */
/*    (similiar to         9 - only numbers                     */
/*     dbase III+ )        # - numbers or decimal point         */
/*                         S - numbers or negative sign         */
/*                                                              */
/*   options:              0 - clears box area                  */
/*                         1 - and places message string        */
/*    (hierarchical)       2 - and places inputs string         */
/*                         3 - and draws box                    */
/*                         4 - and allows input string to be changed */
```

```
/*                      5 - and remove input area on RETURN or ESC  */
/*                                                                   */
/*    Returns the value of the last key pressed if option 4 or 5 is  */
/*    used otherwise returns 0.                                      */
/*                                                                   */
/*********************************************************************/

int  input_str(char *s, char *m, char *p, int h, int v, int option)
{
    char *hold;
    int  x,max,xx,yy,a;
    int  loop,change,null_flag;
    int  l,t,r,b;
    int  pic,msg;

    /***********/
    /* INITIALIZE  */
    /***********/

    a     = 0;
    hold  = malloc(200);
    max   = strlen(p);
    pic   = max*8;
    msg   = strlen(m)*8;
    l     = h - 5;
    t     = v - 5;
    r     = h + msg + pic + 5;
    b     = v + 13;

    /************/
    /* CLEAR AREA   */
    /************/

    setfillstyle(0,foreground);
    setcolor(foreground);
    bar(l,t,r,b);

    /*************/
    /* DRAW MESSAGE */
    /*************/

    if (option > 0)
       outtextxy(h,v,m);

    /***********/
    /* DRAW INPUT */
    /***********/

    if (option > 1)
    {
       null_flag = 0;
       for(x=0; x<max; x++)
       {
          if (*(s+x) == NULL)
             null_flag = 1;
          if (null_flag == 1)
             *(hold + (x*2))  = ' ';
          else
```

434   Appendix 16

```
               *(hold + (x*2))   = *(s+x);
            *(hold + (x*2) + 1) = NULL;
            xx = msg + h + (x*8);
            yy = v;
            outtextxy(xx, yy, hold+(x*2));
         }
      }

      /***********/
      /* DRAW BORDER */
      /***********/

      if (option > 2)
         rectangle(l,t,r,b);

      /***********/
      /* INPUT DATA   */
      /***********/

      if (option > 3)
      {
         setfillstyle(1,foreground);
         setcolor(background);

         for(x=0; x<max; x++)
         {
           xx = msg + h + (x*8);
           yy = v;
           bar(xx-1,yy-2,xx+8,yy+10);
           outtextxy(xx, yy, hold+(x*2));
         }

         loop = TRUE;
         x = 0;
         do
         {
            xx = msg + h + (x*8);
            yy = v+2;
            outtextxy(xx, yy, "_");
            a = get_key();
            bar(xx,yy+7,xx+8,yy+8);

            switch(a)
            {
               case RIGHT:
                  x++;
                  if (x > max-1)
                     x = max-1;
                  break;
               case LEFT:
                  x--;
                  if (x < 0)
                     x = 0;
                  break;
               case RETURN:
                  for(x=0; x<max; x++)
                     *(s+x) = *(hold + (x*2));
                  loop     = 0;
```

```
                    break;
            case ESC:
                    loop    = 0;
                    break;
            case BACKSPACE:
                    xx = msg + h + (x*8);
                    yy = v;
                    bar(xx,yy,xx+7,yy+8);
                    *(hold + (x*2))  = SPACE;
                    outtextxy(xx, yy, hold+(x*2));
                    x--;
                    if (x < 0)
                        x = 0;
                    break;
            default:
                    change = 0;
                    if (*(p+x) == 'X')
                    {
                        if (a > 31 && a < 256)
                            change = 1;
                    }
                    if (*(p+x) == 'A')
                    {
                        if (a > 64 && a < 123 || a == 32)
                            change = 1;
                    }
                    if (*(p+x) == '9')
                    {
                        if (a > 47 && a < 58  || a == 32)
                            change = 1;
                    }
                    if (*(p+x) == 'S')
                    {
                        if (a > 47 && a < 58  || a == 32 || a == 45)
                            change = 1;
                    }
                    if (*(p+x) == '#')
                    {
                        if (a > 47 && a < 58  || a == 32 || a == 46)
                            change = 1;
                    }
                    if (change == 1)
                    {
                        xx = msg + h + (x*8);
                        yy = v;
                        bar(xx,yy,xx+7,yy+8);
                        *(hold + (x*2))  = a;
                        outtextxy(xx, yy, hold+(x*2));
                        x++;
                        if (x > max-1)
                            x = max-1;
                    }
        }
    }
    while(loop == TRUE);
}

/***********/
```

SIMULATE.C - Simulate Tool Path

```
   /* CLEAR AREA  */
   /***********/

   if (option > 4)
   {
      setfillstyle(0,foreground);
      setcolor(foreground);
      bar(l,t,r,b);
   }

   /****/
   /* EXIT */
   /****/

   free(hold);
   setfillstyle(0,foreground);
   setcolor(foreground);
   return a;
}


/***************************************************************/
/*                                                             */
/*   FUNCTION input_int                                        */
/*                                                             */
/*   Same as input_str except for integer values.              */
/*   *i is a pointer to an integer value.                      */
/*                                                             */
/***************************************************************/
int input_int(int *i, char *m, char *p, int h, int v, int option)
{
   char *s;
   int   max,a;

   max = strlen(p);
   s   = malloc(100);
   itoa(*i,s,10);
   a = input_str(s,m,p,h,v,option);
   *(s + max)     = ' ';
   *(s + max + 1) = NULL;
   *i = atoi(s);
   free(s);
   return a;
}


/***************************************************************/
int input_real(double *r, char *m, char *p, int h, int v, int option)
{
   char *s;
   int   max,a;

   max = strlen(p);
   s   = malloc(50);
   gcvt(*r,max,s);
   a = input_str(s,m,p,h,v,option);
```

```
    *(s + max)     = ' ';
    *(s + max + 1) = NULL;
    *r = atof(s);
    free(s);
    return a;
}
/**********************************************************************/
/*                                                                    */
/*  FUNCTION horz_tab()                                               */
/*                                                                    */
/*  Returns the number of spaces between the left margin and the      */
/*  top string.                                                       */
/*  NOTE: 1 space = 8 pixels                                          */
/*                                                                    */
/**********************************************************************/

int horz_tab(char *s[15][15],int h)
{
  int c,i;

  c = 3;
  for (i = 0; i < h; i++)
      c = c + strlen(s[i][0])+3;
  return c;
}




/**********************************************************************/
/*                                                                    */
/*  FUNCTION show_string                                              */
/*                                                                    */
/*  Places the string at array index [h][v] on the screen in the      */
/*  position required by the pop_up menu routine.                     */
/*                                                                    */
/**********************************************************************/

void show_string(char *s[15][15],int h, int v)
{
  int x,y,c;

  c = horz_tab(s,h);
  x = c * 8;
  if (v == 0)
      y = 5;
  else
      y = v * 18 + 6;
  outtextxy(x,y,s[h][v]);
  return;
}



/**********************************************************************/
/*                                                                    */
/*  FUNCTION set_menu                                                 */
/*                                                                    */
```

SIMULATE.C - Simulate Tool Path

```
/*  Places horizontal portion of main menu on the screen.       */
/*  Called by main program.                                     */
/*                                                              */
/****************************************************************/
void set_menu(char *s[15][15])
{
  int   i,hmax;

  for (i = 0; strlen(s[i][0]) != 0; i++);
  hmax = i - 1;
  setfillstyle(1,foreground);
  setcolor(background);
  bar3d(0,0,640,18,0,1);
  for (i = 0; i <= hmax; i++)
      show_string(s,i,0);
  setfillstyle(0,foreground);
  setcolor(foreground);
  return;
}


/****************************************************************/
/*                                                              */
/*  FUNCTION set_bar()                                          */
/*                                                              */
/*  Highlights a string. Calling this function a second time with the */
/*  same string will unhilight the string.                      */
/*                                                              */
/****************************************************************/
void set_bar(char *s[15][15],int h, int v)
{
  int       c,w,l,t,r,b;
  void      *buffer;
  unsigned  size;

  c = horz_tab(s,h);
  w = strlen(s[h][v]);
  l = (c - 1) * 8 + 1;
  r = (c + w + 1) * 8 - 1;
  if (v == 0)
      t = 3;
  else
      t = v * 18 + 4;
  b = t + 10;

  size   = imagesize(l,t,r,b);
  buffer = malloc(size);
  getimage(l,t,r,b,buffer);
  putimage(l,t,buffer,4);
  free(buffer);
  return;
}
```

Appendix 16

```c
/**********************************************************************/
/*                                                                    */
/*   FUNCTION use_menu                                                */
/*                                                                    */
/*   Places main menu on graphics screen. Will return menu selection  */
/*   based on values H and V an a STATUS variable.                    */
/*   if STATUS = 0   ESC was pressed.                                 */
/*   if STATUS = 1   RETURN was pressed.                              */
/*   Called by main program.                                          */
/*                                                                    */
/**********************************************************************/
void use_menu(char *s[15][15], int *ph, int *pv, int *status)

{
   void       *buffer;
   unsigned   size;              /* buffer size */
   int        r,t,l,b;           /* right, top, left, bottom */
   int        c,i,a,loop;        /* counters */
   int        h,v,hmax,vmax,htmp;

   h    = *ph;
   v    = *pv;
   loop = 1;

   for (i = 0; strlen(s[i][0]) != 0; i++);
   hmax = i - 1;
   if (h < 0)
       h = 0;
   if (h > hmax)
       h = hmax;

   setfillstyle(1,foreground);
   setcolor(BLACK);

   do
   {
       for (i = 1; strlen(s[h][i]) != 0; i++);
       vmax = i - 1;
       if (v < 0)
           v = 0;
       if (v > vmax)
           v = vmax;

       if ( vmax > 0)
       {
           c = horz_tab(s,h);
           l = (c - 1) * 8;
           t = 18;
           r = (c + strlen(s[h][1]) + 1) * 8;
           b = vmax * 18 + 18;

           size   = imagesize(l,t,r,b);
           buffer = malloc(size);
           getimage(l,t,r,b,buffer);
           bar3d(l,t,r,b,0,1);
           for (i = 1; i <= vmax; i++)
               show_string(s,h,i);
```

SIMULATE.C - Simulate Tool Path

```c
}
htmp = h;
set_bar(s,htmp,0);

do
{
   if (v > 0)
      set_bar(s,h,v);
   a = get_key();
   if (v > 0)
      set_bar(s,h,v);

   switch(a)
   {
      case DOWN:
         v++;
         if (v > vmax)
         {
            if (vmax == 0)
               v = 0;
            else
               v = vmax;
         }
         break;
      case UP:
         v--;
         if (v < 1)
         {
            if (vmax == 0)
               v = 0;
            else
               v = 1;
         }
         break;
      case RIGHT:
         h++;
         if (h > hmax)
            h = hmax;
         v = 0;
         break;
      case LEFT:
         h--;
         if (h < 0)
            h = 0;
         v = 0;
         break;
      case RETURN:
         *status = 1;
         loop    = 0;
         break;
      case ESC:
         *status = 0;
         loop    = 0;
         break;
   }
}
```

```
            while (h == htmp && loop == 1);
            set_bar(s,htmp,0);
            if (vmax > 0)
            {
                putimage(l,t,buffer,0);
                free(buffer);
            }
    }
    while (loop == 1);
    setfillstyle(0,foreground);
    setcolor(foreground);

    *ph     = h;
    *pv     = v;
    return;
}


/**********************************************************************/
/*                                                                    */
/*   FUNCTION hot_key                                                 */
/*                                                                    */
/*   Check to see if a key was pressed. If the space bar is pressed   */
/*   viewport #1 (tool paths) is cleared.                             */
/*   Returns ASCII value of key pressed.                              */
/*                                                                    */
/**********************************************************************/

void hot_key()
{
    int a;

    a = 0;
    if ( bioskey(1) != 0 )
    {
        a = getch();
        if (a == SPACE)
        {
            set_view(TOP);
            clearviewport();
            set_view(SIDE);
            clearviewport();
            set_view(ISO);
            clearviewport();
        }
        if ((a == ESC)&&(single_step=='Y'))
            single_step='N';

        if ((a == ESC)&&(single_step=='N'))
            single_step='Y';
    }
    return;
/**********************************************************************/
/*                                                                    */
/*   FUNCTION is_empty_string                                         */
/*                                                                    */
```

```c
/*    Check if the string is empty                                 */
/*                                                                 */
/*******************************************************************/

int is_empty_string(  char * s )
{
   int    n = 0;

   for( ; ; )
   {
      if( (s[n]=='\n') || (s[n]==EOF)  )
      {
         return( TRUE );
      }
      else if( (s[n]==' ') || (s[n]=='\t') )
      {
         ;
      }
      else
      {
         return( FALSE );
      }

      n++;
   }

}
```

Bibliography

Amstead, B.H., P.F. Ostwald and M.L. Begeman, *Manufacturing processes*, Wiley, 1979.

Baker, M.P., *Computer graphics*, McGraw-Hill, 1983.

Barnhill, R.E. and R.F. Riesenfeld, *Computer-aided geometric design*, Academic Press, 1974.

Bedi, S., and G.W. Vickers, *Postprocessors for numerically controlled machine tools*, Computers in Industry, vol. 9, no. 1, pp. 3-18, Aug./Sept. 1987.

Bedi, S., W. Chernoff and G.W. Vickers, *Computer-aided fairing and direct numerically controlled machining of ship hull hydrodynamic testing models*, Trans. of Can. Soc. for Mech. Engr., vol. 12, no. 1, pp. 43-48, 1988.

Besant, C.B. and C.W.K. Lui, *Computer-aided design and manufacturing*, Ellis Horwood, 1986.

Bézier, P., *Numerical control mathematics and applications*, Wiley, 1970.

Boothroyd, G., *Fundamentals of metal machining and machine tools*, McGraw-Hill, 1975.

Chang, C. and M.A. Melkanoff, *NC machine programming and software design*, Prentice-Hall, 1989.

Cook, N., *Manufacturing analysis*, Addison-Wesley, 1966.

Duncan, J.P. and K.K. Law, *Computer-aided sculpture*, Cambridge University Press, 1988.

Duncan, J.P. and S.G. Mair, *Sculpture surfaces in engineering and medicine*, Cambridge University Press, 1983.

Faux, I.D., and Pratt, M.J., *Computational geometry for design and manufacture*, Ellis Horwood, 1979.

Foley, J.D., and L.A. Van Dam, *Fundamentals of interactive computer graphics*, Addison-Wesley Company, 1982.

Forsyth, D.G., G.W. Vickers and J.P. Duncan, *Replication of anatomical and other irregular surfaces*, Proc. 4th North American Metalworking Research Conf., Battelle's Columbus Laboratories, Columbus, OH, 1976.

Gallagher, C.C. and W.A. Knight, *Group technology production methods in manufacture*, Ellis Horwood, 1986.

Gasson, P.C., *Geometry of spatial forms*, Ellis Horwood, 1983.

Groover, M.P. and E.W. Zimmers, *CAD/CAM computer-aided design and manufacturing*, Prentice-Hall, 1984.

Groover, M.P., *Automation, production systems, and computer-integrated manufacturing*, Prentice-Hall, 1987.

Harrington, S., *Computer graphics: A programming approach*, McGraw-Hill, 1983.

Hearn, D. and P. Baker, *Computer graphics*, Prentice-Hall, 1986.

Hsu, T.R., *Computer-Aided design*, West Educational Publications, 1990.

Koren, Y., *Computer control of manufacturing systems*, McGraw-Hill, 1983.

Mortenson, M.E., *Geometric Modelling*, Wiley, 1985.

Niebel, B.W., A.B. Draper and R.A. Wysk, *Modern manufacturing process engineering*, McGraw-Hill, 1989.

Oetter, R.G., and G.W. Vickers, *Platable hull surfaces continuing non-platable regions*, CSME Mechanical Engineering Forum, Toronto, June 1990.

Onivubiko, C., *Foundations of computer-aided design*, West Publishing, 1989.

Potts, J., *Computer-aided drafting and design using AutoCAD*, Technology Publications, 1989.

Pressman, R.S. and J.E. Williams, *Numerical control and computer manufacturing*, Wiley, 1977.

Preston, E.J., G.W. Crawford and M.E. Coticchia, *CAD/CAM systems*, Marcel Dekker, 1984.

Ránky, P.G., *Computer integrated manufacturing*, Prentice-Hall, 1985.

Rogers, D.F., *Procedural elements for computer graphics*, McGraw-Hill, 1985.

Rooney, J. and P. Steadman, *Principles of computer-aided design*, Prentice-Hall, 1987.

Schey, J.A., *Introduction to manufacturing processes*, McGraw-Hill, 1977.

Skinner, W., *Manufacturing: The formidable competitive weapon*, Wiley & Sons, 1985.

Smith, G.T., *Advanced machining. The handbook of cutting technology*, IFS Publications, 1989.

Sproull, R.F., W.R. Sutherland and M.K. Ulner, *Device independent graphics*, McGraw-Hill, 1985.

Stanton, G.C., *Numerical control programming - Manual and APT Compact II*, Wiley, 1988.

Ulsoy, A.G. and W.R. DeVries, *Microcomputer applications in manufacturing*, Wiley, 1989.

Vickers, G.W. and C. Bradley, *Machine vision to machined objects*, Pacific Conference on Manufacturing, Melbourne, December, 1990.

Vickers, G.W. and K.W. Quan, *Ball mills versus end-mills for curved surface machining*, ASME J. of Engr. for Industry, vol. 111, no. 1, pp. 22-26, 1989.

Vickers, G.W. and S. Bedi, *The generation of smooth curved surfaces from sparse irregular data*, ASME Computers in Mech. Engr. vol. 6, no. 4, pp. 52-58, January/February 1988.

Vickers, G.W. *Computer-aided manufacture of marine propellers*, Computer-Aided Design, vol. 9, no. 4, pp. 267-274, Oct. 1978.

Vickers, G.W., and C.G. Saunders. *A generalized approach to the replication of cylindrical bodies with compound curvature*, Trans. ASME J. of Mechanisms, Transmissions and Automation in Design, 1983.

Vickers, G.W., J.P. Duncan and V. Lee. *Interactive surface adjustment of marine propellers*, Computer-Aided Design, vol. 10, no. 6, pp. 375-379, Nov. 1978.

Vickers, G.W., S. Bedi and R. Haw. *The definition and manufacture of compound curvature surfaces using G-surf*, Computers in Industry, vol. 6, no. 3, pp. 173-183, June, 1985.

Vickers, G.W., S. Bedi, D. Blake, and D. Dark, *Computer-aided lofting, fairing and manufacture in small shipyards*, Int. J. of Adv. Manufacturing Tech., vol. 2, no. 4, pp. 79-90, November, 1987

Vickers, G.W., and C. Bradley, *Do-it-yourself CNC*, Woodworking, vol. 3, no. 6, pp. 20-21, November, 1989.

Wang, H.P., E. Hewgill and G.W. Vickers, *An efficient evaluation algorithm for generating B-spline interpolation curves and surfaces from B-spline approximations*, Communications in Applied Numerical Methods, May, 1990.

Index

A

anatomical shapes 205
Anvil-5000
   analysis 114
   basic geometry 114
   CLFile 123
   data base management 114
   deep side cutting 123
   display control 114
   drafting 114
   drilling 130
   extended geometry 114
   facing 123
   flange cutting 122
   geometry statements 114
   machining module 122
   pocketing 122
   profiling 122
   side cutting 123
   system modals 114
   track pocketing 129
APT
   contouring surfaces 74
   geometry statements 21
   motion statement 21, 71
   part-program 71
   postprocessor 21, 72
   programming 17
APT IV code 132
artificial intelligence 113
ASCII 17, 19, 132
AutoCAD 21, 71, 101, 154, 232
Autolisp 154
automated handling system 131
automatic cutout insertion 231
automated woodshaper 244

B

B-Spline
   approximation 146, 227
   basis function 146,
   control polygon 146,
   control vertices 146
   curvature 146
   curve 153, 227
   delta-T 153
   divided difference 147
   enclosing polygon 146
   initial conditions 146
   interpolation 228
   knot vector 149, 153
   lines 152, 194
   parametric 146
   recursive algorithm 149
   slope 146
   smoothing 146
   surfaces 146, 194
Ball-mill 173
behind tape-reader (BTR) 132
binary code 14, 17
bit 132, 134
boolean approach 118
buffer memory 22, 132

C

C-programming
   library 151
   linking 151
CAD 21, 71, 101, 143, 231
CAD/CAM 21, 101, 132, 138
CADDS4 113

Index

CAM 21, 71, 101
canned cycles 56
case studies 185
centrifuge desalination unit 120
check surface 73
circular 23
circular arc interpolation 177
   algorithm 179
   augmented matrix 181
   error tolerance 181
CLFile 123, 138
cloud data 143
CNC
   start and end sequences 23
   computer assisted 17, 21
   controller 17, 71
   drilling 113
   flame-cutting 113, 223
   lathe 11, 13, 71
   machine tools 71, 144, 178
   machining 113, 118
   manual programming 21, 71
   manufacture 202
   mechanical drive 11
   milling 113, 226
   milling machine 11
   motor drive 11
   programs 15, 21
   punching 113
   simulation program 235
   testing 235
   turning 113
computer-aided design 114
computer-assisted programming 71, 132
computervision 113
control polyhedron 146
conversational programming 71
cross-feed distance 176
curved surface machining
   aeroplanes 143
   cars 143
   cabinet housings 143
   doubly curved 144
   faces 143
   limbs 143
   matrix of data points 144
   ships 143
   sockets 143
   telephones 143
cusp height 175
custom macros 63
cutter
   ball-mill 162
   carbide inserts 177
   effective radius 173
   elliptical profile 173
   end-mill 162
   generalized shaped 162
   inclination 173
   location 71
   location calculation 143
   location file 113, 167
   offsets 169
   path 72, 104
   reference point 166
   representation 163
cutting speed 177
cutting time analysis 113

D

data input 143, 144
developable surface 229
digital transfer 131
digitization 209
dimensions
   absolute/incremental 39
   inch/millimeter 30
direct numerically controlled machining 131
display processor 240
DNC
   link 136
   remote buffer link 132
   transmission 132
drive surface 73
DXF 233

E

end-mill 173
error checking 239, 242

F

F-codes 51
fairing 22, 223

Fanuc controllers 17
FAPT 81
fast traverse 23
finite element 113
flexible manufacturing FMS 131
fringe order 208
fringe pattern 205

G

G-code 21, 72, 101,132, 167, 225, 243
generalised milling cutter 165
generators 229
graphical simulation 17

H

head shapes 216
heart valve 22, 221
hidden line removal 115
home position 23
homogeneous coordinate transformations 157
host computer 135
HPGL 247
hull definition 194
hull machining 196
hydrodynamic testing 194

I

inspection probe 20
integrated computer aided design 223
interactive computer aided graphic program 21
Intergraph 113
intersecting surface 73
ISO 17, 132
isometric view 112, 117

K

knot vector 149

L

laser scanned surface 143

laser scanning 205, 216
lathe 11
limb shape 205, 210
linear interpolation 178, 249
line fairing 227
lofting 22, 194, 223

M

M-code 23, 74
machine
    alternate 167
    backward 167
    code 17
    forward 167
    tool 131
machining centre 11
manual data input (MDI) 17
manual part programming 23
marine propeller
    helicoidal surface 186
    hub 189
    pitch 187
    rake angle 189
    surface definition 185
MEDUSA 113
milling machine 11, 15
motion statement 73

N

non-developable regions 229
normal vector, inverted 170
numerically controlled machine 11, 133

O

optical encoders 14

P

paper tape 131
parallel communications 134
parser 237
part definition 115
part program
    downloading 131
    storage 131

Index

transmission 131
part surface 73
peck cycles 45
pericontourography 211
perspective views 114
Personal Designer 71, 101
photogrammetry 205
plate expansion 230
pocketing 101
postprocessor 21, 72, 132, 138
primitives 118
program
   coded block 23
   identifying number 23
   storage 22
   transmission 17
propeller 22, 185

R

RAPID 138
recirculating ballbearing nut 14
remote buffer 135
retrofitting machine tools 243
rotary encoder 14
rotary table 11
roughing cut 171
router 243
ruled surface 155
RS-232 17, 134

S

S-code 23
segmenting 249
serial communication 134
serial link 131
servo motor 14
shadow moiré, 205
ship hull forms 194
ship hulls 223
ShipCAM 194, 223
shoe lasts 211
simulation program 21
SmartCAM 71, 101
smoothing 143
solids modelling 22, 113
spreadsheet input 226
stepping motor 14, 243

surface
   AutoCAD 154
   Autolisp 154
   B-spline 228
   Bezier 114, 154
   Coon's 114
   curved 167
   definition 145
   developable 225
   edge-defined 154
   extended 172
   fitting 146, 171
   general polygon meshes 154
   generation 178
   low curvature 173
   manipulation 170
   modelling 143, 146
   normal calculations 143
   normal unit vector 159
   normal vector 143, 165
   orientation 143, 157
   polyhedron 160
   reduced 171
   resolution 143, 167
   rotation 157
   roughness 71, 176
   ruled 154, 229
   scaling 157
   tabulated 154
   translation 157
   transposed 171
   vector offset 159
symbolic FAPT 81

T

T-code 23
table of offsets 226
tape reader 17, 131
tool
   changer 11
   length compensation 23
   path 21, 71, 101
   reference point 162
   selection 23
   storage location 23
transformations 160
transmission 17
turbine blades

452 Index

francis 199
lowhead 204
turgo 199
Turbo-C programming 236
turning centres 11

U

undercutting 213
UNITS 138

V

vector cross product 163
vertices 143, 170

W

wire frame 22, 113
woodshaper 244
world coordinates 241
work coordinate system 38